Communications in Computer and Information Science 1686

More information about this series at https://link.springer.com/bookseries/7899

Boris Villazón-Terrazas ·
Fernando Ortiz-Rodriguez · Sanju Tiwari ·
Miguel-Angel Sicilia ·
David Martín-Moncunill (Eds.)

Knowledge Graphs and Semantic Web

4th Iberoamerican Conference and
third Indo-American Conference, KGSWC 2022
Madrid, Spain, November 21–23, 2022
Proceedings

 Springer

Editors
Boris Villazón-Terrazas ⓘ
EY wavespace/UNIR
Madrid, Spain

Fernando Ortiz-Rodriguez ⓘ
Autonomous University of Tamaulipas
Ciudad Victoria, Mexico

Sanju Tiwari ⓘ
Autonomous University of Tamaulipas
Ciudad Victoria, Mexico

Miguel-Angel Sicilia ⓘ
University of Alcalá
Alcalá de Henares, Spain

David Martín-Moncunill ⓘ
Universidad Camilo José Cela
Madrid, Spain

ISSN 1865-0929   ISSN 1865-0937 (electronic)
Communications in Computer and Information Science
ISBN 978-3-031-21421-9   ISBN 978-3-031-21422-6 (eBook)
https://doi.org/10.1007/978-3-031-21422-6

This Springer imprint is published by the registered company Springer Nature Switzerland AG
The registered company address is: Gewerbestrasse 11, 6330 Cham, Switzerland

Preface

This volume contains the main proceedings of the fourth Iberoamerican and the third Indo-American Knowledge Graphs and Semantic Web Conference (KGSWC 2022), held jointly during November 21-23, 2022, at Camilo José Cela University in Madrid, Spain. KGSWC is established as a yearly venue for discussing the latest scientific results and technology innovations related to the Knowledge Graphs and the Semantic Web. At KGSWC, international scientists, industry specialists, and practitioners meet to discuss knowledge representation, natural language processing/text mining, and machine/deep learning research. The conference's goals are (a) to provide a forum for the AI community, bringing together researchers and practitioners in the industry to share ideas about innovative projects, and (b) to increase the adoption of AI technologies in these regions.

KGSWC 2022 followed on from successful past events in 2019, 2020, and 2021. It was also a venue for broadening the focus of the Semantic Web community to span other relevant research areas in which semantics and web technology play an important role and for experimenting with innovative practices and topics that deliver extra value to the community.

The main scientific program of the conference comprised 25 papers: 22 full research papers and three short research paper selected out of 63 reviewed submissions, which corresponds to an acceptance rate of 39.6%. The program was completed with five workshops sessions, a hackathon, and a Winter School where researchers could present their latest results and advances and learn from experts. The program also included five exciting invited keynotes (Maria Esther Vidal, Ora Lassila, Roberto Navigli, Pascal Hitzler, and Paul Groth), with novel Semantic Web topics. An industry session was also organized with two talks on different industry projects; the first talk on "Semantic data models construction in the H2020 PLATOON project" was given by Sarra Ben Abbes and Lynda Temal of, CSAI Lab, ENGIE, France. The second talk entitled "White Paper 3: Ontologies for Building Digital Twins" was given by Pablo Vicente Legazpi of the, Building Digital Twin Association, Belgium.

The General and Program Committee chairs would like to thank the many people involved in making KGSWC 2022 a success. First, our thanks go to the five co-chairs of the main event and the 50 plus reviewers for ensuring a rigorous XX blind review process, that led to an excellent scientific program, with an average of three reviews per article.

Further, we thank the kind support of all the people at Camilo José Cela University. We are thankful for the editorial team at Springer. We also thank Juan Pablo Martínez, Rubén Barrera, and Gerardo Haces, who administered the website and helped to make strong publicity. We finally thank our sponsors and our community for their vital support of this edition of KGSWC.

The editors would like to close the preface with warm thanks to our supporting keynotes, the Program Committee for rigorous commitment in carrying out reviews, and, last but not least, our enthusiastic authors who made this event truly international.

November 2022   Boris Villazón-Terrazas
                                                                          Fernando Ortiz-Rodríguez
                                                                                     Sanju Tiwari
                                                                              Miguel Angel Sicilia
                                                                              David Martin-Moncunill

Organization

Chairs

Boris Villazón-Terrazas	EY wavespace/UNIR, Spain
Fernando Ortiz-Rodríguez	Universidad Autónoma de Tamaulipas, Mexico
Sanju Tiwari	Universidad Autónoma de Tamaulipas, Mexico

Local Chairs

Miguel Angel Sicilia	University of Alcalá, Spain
David Martin-Moncunill	Universidad Camilo José Cela, Spain

Workshop

Sven Groppe	University of Lübeck, Germany
Shishir Shandilya	VIT Bhopal University, India
Shikha Mehta	JIIT Noida, India

Winter School

Sanju Tiwari	Universidad Autónoma de Tamaulipas, Mexico
Fernando Ortiz-Rodríguez	Universidad Autónoma de Tamaulipas, Mexico
Boris Villazón-Terrazas	EY wavespace/UNIR, Spain

Publicity

Yusniel Hidalgo Delgado	Universidad de las Ciencias Informáticas, Cuba

Program Committee Chairs

Sanju Tiwari	Universidad Autónoma de Tamaulipas, Mexico
Fernando Ortiz-Rodriguez	Universidad Autónoma de Tamaulipas, Mexico
Boris Villazón-Terrazas	EY wavespace/UNIR, Spain

Program Committee

Ahlem Rhayem	Universidad Politécnica de Madrid, Spain
Alba Fernandez-Izquierdo	BASF Digital Solutions, Spain
Alberto Fernández	Universidad Rey Juan Carlos, Spain

Alex Mircoli	Università Politécnica delle Marche, Italy
Alejandro Rodríguez	Universidad Politécnica de Madrid, Spain
Amit Sheth	University of South Carolina, USA
Antonella Carbonaro	University of Bologna, Italy
Adolfo Anton-Bravo	MPVD.es, Spain
Amed Abel Leiva Mederos	Universidad Central de las Villas, Cuba
Ana B. Ríos-Alvarado	Universidad Autónoma de Tamaulipas, Mexico
Anastasija Ņikiforova	University of Latvia, Latvia
Boris Villazón-Terrazas	EY wavespace/UNIR, Spain
Boyan Brodaric	Geological Survey of Canada, Canada
Carlos Bobed	University of Zaragoza, Spain
Cogan Shimizu	Kansas State University, USA
Diego Collarana	Enterprise Information System (EIS), Germany
Diego Rubén Rodríguez Regadera	Universidad Camilo José Cela, Spain
David Martín-Moncunill	Universidad Camilo José Cela, Spain
Dimitris Kontokostas	Diffbot, Greece
Edgar Tello Leal	Universidad Autonóma de Tamaulipas, Mexico
Edgard Marx	Leipzig University of Applied Sciences (HTWK), Germany
Emmanouel Garoufallou	International Hellenic University, Greece
Erick Antezana	Norwegian University of Science and Technology, Norway
Eric Pardede	La Trobe University, Australia
Fatima N. Al-Aswadi	Universiti Sains Malaysia, Malaysia, and Hodeidah University, Yemen
Fatima Zahra Amara	University of Khenchela, Algeria
Fernando Ortiz-Rodríguez	Universidad Autónoma de Tamaulipas, Mexico
Gerardo Haces-Atondo	Universidad Autónoma de Tamaulipas, Mexico
Gustavo De Assis Costa	Federal Institute of Education, Goiás, Brazil
Hong Yung Yip	University of South Carolina, USA
Janneth Alexandra Chicaiza Espinosa	Universidad Tecnica Particular de Loja, Ecuador
Jennifer D'Souza	TIB - Leibniz Information Centre for Science and Technology, Germany
Jose L. Martínez-Rodríguez	Universidad Autónoma de Tamaulipas, Mexico
Jose Melchor Medina-Quintero	Universidad Autónoma de Tamaulipas, Mexico
Julio Emilio Sandubete Galán	Universidad Camilo José Cela, Spain
Lino González García	Universidad Camilo José Cela, Spain
Lu Zhou	Kansas State University, USA
Luis Ramos	TIB Hannover, Germany
Manas Gaur	University of Maryland, Baltimore, USA
Manuel Enrique	Universidad de las Ciencias Informáticas, Cuba

Mayank Kejriwal	University of Southern California, USA
Maria-Esther Vidal	TIB - Leibniz Information Centre for Science and Technology, Germany
Md Kamruzzaman Sarker	Kansas State University, USA
Miguel-Angel Sicilia	University of Alcala, Spain
Nandana Mihindukulasooriya	IBM Research, USA
Neha Keshan	Rensselaer Polytechnic Institute, USA
Oliver Karras	TIB - Leibniz Information Centre for Science and Technology, Germany
Ora Lassila	Amazon Neptune, USA
Panos Alexopoulos	Textkernel B.V., The Netherlands
Pascal Hitzler	Kansas State University, USA
Patience Usoro Usip	University of Uyo, Nigeria
Roberto Navigli	Sapienza University of Rome, Italy
Rosa M. Rodríguez	University of Jaén, Spain
Ronald Ojino	University of Dar es Salaam, Tanzania
Ronak Panchal	Cognizant, India
Russa Biswas	FIZ Karlsruhe, Germany
Sarra Ben Abbes	ENGIE, France
Sanju Tiwari	Universidad Autonoma de Tamaulipas, Mexico
Serge Sonfack	INP Toulouse, France
Soren Auer	TIB, L3S, Germany
Sven Groppe	University of Lubeck, Germany
Takanori Ugai	Fujitsu Laboratories Ltd. Japan
Tassilo Pellegrini	University of Applied Sciences St. Pölten, Austria
Tommaso Soru	University of Leipzig, Germany
Yusniel Hidalgo Delgado	Universidad de las Ciencias Informáticas, Cuba

Contents

DBkWik++- Multi Source Matching of Knowledge Graphs

Sven Hertling$^{(\boxtimes)}$ ⓘ and Heiko Paulheim ⓘ

Data and Web Science Group, University of Mannheim, Mannheim, Germany
{sven,heiko}@informatik.uni-mannheim.de

Abstract. Large knowledge graphs like DBpedia and YAGO are always based on the same source, i.e., Wikipedia. But there are more wikis that contain information about long-tail entities such as wiki hosting platforms like Fandom. In this paper, we present the approach and analysis of DBkWik++, a fused Knowledge Graph from thousands of wikis. A modified version of the DBpedia framework is applied to each wiki which results in many isolated Knowledge Graphs. With an incremental merge based approach, we reuse one-to-one matching systems to solve the multi source KG matching task. Based on this alignment we create a consolidated knowledge graph with more than 15 million instances.

Keywords: Knowledge graph matching · Incremental merge · Fusion

1 Introduction

There are many knowledge graphs (KGs) available in the Linked Open Data Cloud[1], some have a special focus like life science or governmental data. In the course of time, a few hubs evolved which have a high link degree to other datasets – two of them are DBpedia and YAGO. Both cover general knowledge which is extracted from Wikipedia. Thus many applications use these datasets. One drawback is that they all originate from the same source and have thus nearly the same concepts. For many applications like recommender systems [1], information about not so well known entities (also called *long tail entities*) is required to find similar concepts. Additional sources for such entities can be found in other wikis than Wikipedia.

One example is the wiki farm *Fandom*[2]. Everyone can create wikis about any topic. Due to the restricted scope in each of these wikis, also pages about not so well known entities are created. As an example William Riker (the fictional character in the Star Trek universe) appears in Wikipedia because this character is famous enough to be added (see also the notability criterion for people[3]). For other characters, like his mother Betty Riker, this notability is not given, so it only appears in special wikis like *memory alpha*[4] (a Star Trek wiki in Fandom).

---

[1] https://lod-cloud.net.
[2] https://www.fandom.com.
[3] https://en.wikipedia.org/wiki/Wikipedia:Notability_(people).
[4] https://memory-alpha.fandom.com/wiki/Betty_Riker.

© The Author(s), under exclusive license to Springer Nature Switzerland AG 2022
B. Villazón-Terrazas et al. (Eds.): KGSWC 2022, CCIS 1686, pp. 1–15, 2022.
https://doi.org/10.1007/978-3-031-21422-6_1

Table 1. Comparison of public Knowledge graphs based on [10] sorted by the number of instances.

Knowledge graph	# Instances	# Assertions	# Classes	# Relations	Source
Voldemort	55,861	693,428	621	294	Web
Cyc	122,441	2,229,266	116,821	148	Experts
DBpedia	5,044,223	854,294,312	760	1,355	Wikipedia
NELL	5,120,688	60,594,443	1,187	440	Web
YAGO	6,349,359	479,392,870	819,292	77	Wikipedia
CaLiGraph	7,315,918	517,099,124	755,963	271	Wikipedia
BabelNet	7,735,436	178,982,397	6,044,564	22	multiple
DBkWik	11,163,719	91,526,001	12,029	128,566	Fandom
DBkWik++	15,346,033	106,347,347	15,642	215,273	Fandom
Wikidata	52,252,549	732,420,508	2,356,259	6,236	Community

The idea in this work is to use these wikis and apply a modified version of the DBpedia extraction framework to create knowledge graphs out of them containing information about long tail entities. Each resulting KG is isolated and can share same instances, properties, and classes that need to be matched. For this multi source knowledge graph matching task we reuse a one-to-one matcher and apply it multiple times in an incremental merge based setup to create an alignment over all KGs.

After fusing all KGs together based on the generated alignment, we end up with DBkWik++, a fused knowledge graph of more than 40,000 wikis from Fandom. The contributions of this paper are threefold:

- presentation of the overall approach to generate the KGs
- matching of 40,000 KGs on schema and instance level by reusing a one-to-one matcher
- analysis of the resulting knowledge graph

The rest of this paper is structured as follows. Section 2 describes related work such as different general purpose knowledge graphs and matching techniques. Afterward, Sect. 3 shows details about the retrieval of wikis, application of the DBpedia extraction framework, and the incremental merge of the KGs. After the fusion and provenance information of the KG, the resulting alignment is profiled in Sect. 4 and the KG in Sect. 5. We conclude with an outlook and future work.

2   Related Work

This section is divided into two parts: 1) description of other cross-domain knowledge graphs and 2) multi source matching approaches to combine isolated KGs into one large KG.

Table 1 shows the public cross-domain KGs together with the number of instances, assertions, classes, and relations. In addition, the main source of the content is provided in the last column. The KGs are sorted by the number of instances starting with the smallest.

VoldemortKG [34] uses data extracted from webpages (common crawl) via structured annotations using approaches like RDFa, Microdata, and Microformats. The resulting set of KGs (one for each webpage) is then merged together and linked to DBpedia by using Wikipedia links occurring on those webpages. The overall graph is relatively small and contains only 55,861 instances.

Cyc [18] was generated by a small number of experts. They focus on common sense knowledge and create more assertions than instances. The scalability is quite limited because of the manual generation. The total cost was estimated as 120 Million USD. The numbers in the table refer to the openly available subset OpenCyc.

DBpedia [2] instead uses another approach that scales much better. The main source is Wikipedia where a lot of entries contain information in so-called infoboxes (MediaWiki templates). These templates contain attribute value pairs where the values are shown on the webpage. When processing these pages without resolving the templates, those key-value pairs can be extracted and transformed to triples where the wiki page is the subject, the template key is the property, and the template value is the corresponding literal or resource (in case it is a URL). This scales much better and opens the door for other data-driven approaches.

NELL [22] (Never-Ending Language Learning) is an approach to extract information from free text appearing on web pages. Based on some initial facts, textual patterns for these relations are extracted and applied to unseen text to extract more subjects and objects for that relation. The resulting facts are again used to derive new patterns. With a human-in-the-loop approach, the authors try to increase the quality by removing incorrect triples.

YAGO [7] uses the same source as DBpedia (namely Wikipedia) but creates the class hierarchy based on the categories defined in Wikipedia instead of manually creating the hierarchy like DBpedia.

CaLiGraph [11] also uses the category tree but converts the information in category names into formal axioms e.g. "List of people from New York City" where each instance in this category should have the triple <instance, bornIn, New York City>.

Babelnet [24] integrates a lot of sources like Wikipedia and WordNet [21] to collect synonyms and translations in many languages.

DBkWik [12] is generated from Fandom wikis with the DBpedia extraction framework. Thus it has the same structure as DBpedia but includes more long-tail entities, especially from the entertainment domain. It uses information from 12,840 wikis.

Wikidata [36] is a community-driven approach like Wikipedia but allows to add factual information in form of triples instead of free text. Furthermore, it includes and fuses other large-scale datasets such as national libraries' bibliographies.

Many of the above mentioned approaches need some kind of data integration to build up the final KG. Thus matching algorithms are compared in the following by dividing into entity vs KG matching and two sources vs multi source matching.

In the case of entity matching, the schema is fixed and usually already aligned. If the schema alignment is missing, it can be manually created due to the reduced size of attributes. In KG matching, on the other hand, the concepts are of different classes (e.g. persons, places, events, etc) and can be described using many properties (e.g. name, hair color, height) which in addition can also connect entities (e.g. location property which links a person to a location). The schema (including classes and properties) is usually not aligned and cannot be created manually due to its large size.

Entity and KG matching exists in two flavors: one to one and multi source matching. In the former case, two inputs are given whereas in multi source matching a set of data sources needs to be matched. Even though the one to one matching can be seen as a special case of multi source matching, many matchers are only able to align two sources at the same time.

There are many one to one entity matching tools available like rule based systems [32] and deep neural network based ones (DeepMatcher [23]). In addition, configurable frameworks like MOMA [33] are discussed in [16]. SILK [35] matches data sources formatted in RDF but focuses on entities rather than the schema. It uses human generated rules which can be further improved by supervised methods. For multi source entity matching there exists matching systems like FAMER [30] and ALMSER [28] which in addition uses active learning to reduce the size of the initial alignment.

For one to one KG matching, the ontology alignment evaluation initiative (OAEI [27]) allows submitting ontology and KG matching systems which are evaluated by the track/dataset organizers. Starting in 2018, the OAEI contains also a KG track which requires instance as well as schema matches [13]. All systems participating in this track are able to produce a one to one KG alignment like the top performing systems AML [8] and ALOD2Vec [26]. For multi source KG matching [14] reuses one to one matching systems to solve the multi source task. [29] focus on matching entities in multiple KGs using hierarchical agglomerative clustering (HAC).

3   DBkWik++

The overall workflow for creating DBkWik++ is shown in Fig. 1. First, all wiki dumps need to be downloaded from Fandom. Afterward, a modified version of the DBpedia extraction framework is applied to extract information in RDF. The result is many isolated KGs which are linked and fused to create a consolidated KG.

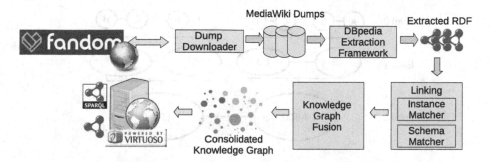

Fig. 1. Overview of the approach to create the knowledge graph.

3.1 Acquisition of Wiki Dumps

The wiki hosting platform Fandom allows wiki creators to provide dumps of their wikis once it is manually triggered[5]. These files can be downloaded easily. Unfortunately, not many creators do it and thus the coverage of wikis to download is rather low. It can also be very outdated depending on the time the dump is initiated. Therefore, we reached out to Fandom to ask for another way of getting these dumps, but did not succeed. As an alternative solutions, all dumps were downloaded by the wikiteam[6] software which archives MediaWikis. This resulted in 307,466 dumps (208,552 of them in English).

3.2 Creating KGs With DBpedia Extraction Framework

The wiki dumps generated in the previous steps consist of MediaWiki markup and need to be transformed to a KG. For this step, a modified version of the DBpedia extraction framework is used. As the name suggests, it is used to create the DBpedia [2] knowledge graph. The transformation of a wiki to a KG works as follows: each wiki page corresponds to a resource and each infobox[7] becomes a class. Due to the fact that templates like infoboxes do not form a hierarchy, the classes do not have a hierarchy as well. In DBpedia, this hierarchy is created manually in the corresponding mappings wiki[8]. Similarly, properties are homogenized in case the attribute key of the template is different e.g. the template properties `birthdate`, `birth_date`, and `dateofbirth` are mapped to the ontological property `birthDate`. All extractors in the DBpedia extraction framework which use knowledge of the mappings wiki are excluded because no mapping exists for Fandom and a manual creation would not scale to hundreds of thousands of wikis. Instead, the extractors which do not make use of the mapping are included e.g. types and properties are created from infobox template

---

[5] https://community.fandom.com/wiki/Help:Database_download.

[6] https://github.com/WikiTeam/wikiteam.

[7] a template in MediaWiki which usually contains the text `infobox` to visualize important information at the top right corner of a page.

[8] http://mappings.dbpedia.org.

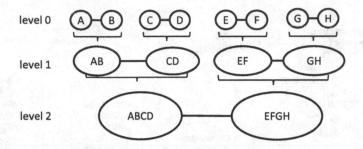

Fig. 2. Incremental merge strategy for wikis A to H.

names and infobox keys. In the usual DBpedia extraction, the abstracts of a wiki are extracted by setting up a MediaWiki instance and calling an API. This is replaced by using another extractor based on the Sweble parser [6] to decrease the runtime.

3.3 Linking

The output of the previous step is an isolated knowledge graph per wiki. It happens a lot that the same concept appears in multiple wikis, e.g., the actor Tom Cruise appears in moviepedia wiki, topgun wiki, and jack reacher wiki. To create a consolidated KG, all source KGs need to be matched together. This does not only require instance matches but also schema level correspondences which include property and class matches. Thus, the KG matching task at hand is different from entity matching because the schema in each data source can be quite different. The reason is that each attribute value pair in the infobox is transformed into a triple. Therefore, different information about a resource might be extracted e.g. eye color of Tom Cruise in jack reacher wiki and height in topgun wiki. Matching systems needs to be very precise and should be able to handle multi source KG matching. A blocking approach based on entity types is not possible because of the missing schema alignment.

Due to the fact that there are not many systems able to solve the multi source KG matching task, we implemented an approach presented in [14]. The basic idea is to use a one to one matching system over and over again in a multi source setup. The implementation of the all-pairs approach does not scale (which runs the binary matching system for all KG combinations). For n KGs this would need $\frac{n*(n-1)}{2}$ executions of the matching system e.g. $\frac{300,000*299,999}{2} = 44,999,850,000$ comparisons. Even if each comparison of two KGs only needed one second (which is overly optimistic), the whole computation would require 1,426 years without parallelization. The same complexity also applies to the resulting mapping size. If each correspondence is stored in a file, it becomes very huge even for a small number of wikis.

Thus the selected approach is called incremental merge based on similarity. Figure 2 shows an example for isolated KGs A to H. The one to one matcher is first applied to KG A and B which results in an alignment. This alignment is used to create a union KG AB out of it. It is generated by adding all triples of A and B to the result but replacing URIs in B with the corresponding one from A in case it exists. Thus more information about an entity is available at higher levels of the execution.

The order in which those KGs are merged is determined by the similarity between them. The overall tree is computed by an agglomerative hierarchical clustering (HAC) using tf-idf vectors as features to lower the impact of frequent terms, e.g. upper-level ontology terms. The vectors are generated by using all values which occur in literals and are string-valued. Creating tokens is done in a standard way by sentence splitting, tokenization, lowercasing, stopword removal, and stemming. The tokenization is adapted due to special naming conventions (like underscore between words or camel case). The output of the hierarchical clustering can be directly used as an execution plan for the matching and merging of KGs. The advantage of such an approach is that the schema and instances are matched at the same time, at least if the binary matching system is able to do so.

The hierarchical agglomerative clustering algorithm works in the following way: (1) compute the distance matrix (2) let each datapoint be a cluster (3) repeatedly merge the two closest clusters and update the distance matrix accordingly until only a single cluster remains. It turned out that many implementations like ELKI [31] and SMILE [19] do not scale to more than 65,535 examples (and in this case data sources) because the distance matrix is stored in one array which can only hold up to 2,147,483,647 values (whereas the lower triangular matrix needs $\frac{n*(n-1)}{2}$ entries). Thus we use the largest 40,000 wikis for merging. The main time is spent in part (1) where the distance matrix is computed (799,980,000 values for 40,000 input wikis). Therefore a faster way is implemented which runs in parallel and uses low-level BLAS calls to compute multiple distances at the same time.

One important parameter is the linkage criteria which heavily influences the height of the resulting tree. All executions within one level (shown in Fig. 2) can be executed in parallel because they do not depend on each other. Thus the height corresponds to the number of matching stages which need at least to be executed sequentially. When using single linkage, the corresponding height is 22,007 with 5,869 possible parallel merges in level 0. This is in contrast to complete linkage where the height is only 145 and 12,639 merges in level 0. Thus complete linkage is chosen because of the improved parallelism. In addition, an efficient algorithm for computing complete linkage in $O(n^2)$ exists which is called CLINK [5].

Choosing a one to one matching system is done by analyzing the results of OAEI 2021 [27]. Alod2Vec[26] was one of the top performing systems in the KG track which is very similar to the KGs used in this work. Furthermore, the system is able to produce schema as well as instance alignments. A postprocessing step is added to the matching system to ensure a 1 : 1 alignment which is required by the multi source matching strategy in each execution step. For this, a naive descending approach [20] is used which sorts the alignment by confidence and extracts the correspondences as long as no duplicate source or target entities appear. The overall runtime of the matching needs six days - most probably due to costly matching approaches executed by the one to one matcher. In the output alignment, no property mappings are generated. As a replacement, the property alignment of a simpler element-based matcher is used.

Even though a blocking-based approach for matching would be possible, each entity combination in the identity set needs to be analyzed and thus the information about the schema and instances of each KG needs to present.

3.4   Knowledge Graph Fusion

Given the alignment and the source KGs, this section describes the approach to generate a fused KG. The basic idea is to replace the URIs of matched entities with a canonical URI representing the set of identical resources.

Due to the fact that the alignment is generated on various levels and in each of them, only correspondences between two concepts are made, the transitive closure of the alignment needs to be computed. As an example, a concept a_1 from KG A is mapped to concept b_1. In the resulting union AB, all appearances of b_1 are replaced with a_1. In the following, the union is matched to CD which results in the correspondence $< a_1, c_1 >$. Only under the transitive closure b_1 and c_1 are also matched. For computing the transitive closure the approach of [3] (Sect. 3.3) is re-implemented to store the closure in memory which decreases the runtime.

Given the identity sets computed by the transitive closure, the next step is to calculate for each set the corresponding canonical URI which is the replacement for all URIs in this set. In contrast to Wikidata, the URIs should have speaking names. The transitive closure is thus sorted by descending size (number of elements in this set). Starting from the largest set, the following algorithm is applied: the most common URI fragment is used as the canonical URI. In case the most common fragment is already in use, a postfix with an increasing number is added. This results in URI fragments like New_York, New_York_1, New_York_2 to distinguish between different cities (but just in the case that the majority of the fragments are New_York). For new extractions of the knowledge graph, one should make sure that the URIs are stable and always pointing to the same disambiguated entity. Thus it should be checked if the URIs of the new extraction correspond to the same entity as in the older extraction (which usually requires an additional matching step).

In addition to merging the KG, different sources may provide conflicting data values for specific properties [9]. In data integration frameworks like WInte.r [17], conflict resolution functions are provided on an attribute basis e.g. for property first_name use the shortest (or longest) string. Due to the number of relations in a KG, such an approach does not scale. Currently, a union-based approach is used for all properties.

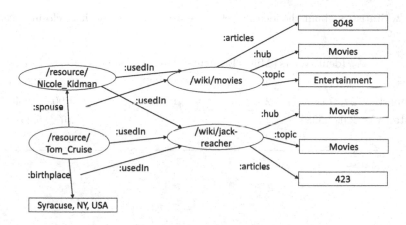

Fig. 3. Provenance information such as wiki information about resources.

3.5 Provenance Information

After the fusion of all KGs, the information which resource originates from which wiki is lost. Thus each wiki is represented in the final KG as its own resource with additional metadata attached to it. It is collected during the crawling of the wikis. Each MediaWiki has a special statistics page that lists information about the number of pages, users, active users, etc. In Fandom, additional statistics like the WAM score are created. It is a value between 0 and 100 and should indicate the activeness of a wiki[9]. The computation is based on page views, contributions, and level of user activity. All this information is added to the resource representing the corresponding wiki. Figure 3 shows an example where the property *usedIn* connects the resources Nicole Kidman and Tom Cruise to the corresponding source wikis. Note that one resource (like Nicole Kidman in this example) might be extracted from multiple wikis.

After this step, the final consolidated KG is generated. In the following two sections, we analyze the resulting alignment as well as the final KG.

4 Analysis of the Resulting Alignment

The generated alignment consists of 7,847,480 correspondences on the schema and instance level. Out of those, 5,254,001 correspondences have the same label and 2,593,479 correspondences (33.05%) link entities that do not share a label. The fraction of 33% is in line with the reference alignment of the OAEI KG track which has between 16% and 47% matches of entities with different labels. [13].

Each correspondence maps only one entity to another and thus the transitive closure needs to be computed as discussed before. It has overall 2,175,184 identity sets. The smallest identity set is of size two and the maximum size is 16,929. Figure 4 shows the number of elements in each identity set in a log-log plot. It

---

[9] https://community.fandom.com/wiki/WAM_Hall_of_Fame.

Table 2. Statistics about the identity sets generated by the transitive closure over the alignment

	#Identity sets	Min #Elements	Max #Elements	Mean	Std Dev
Classes	22,772	2	11,899	12.83	139.53
Properties	54,814	2	16,929	21.90	219.40
Instances	2,097,598	2	821	4.07	5.26

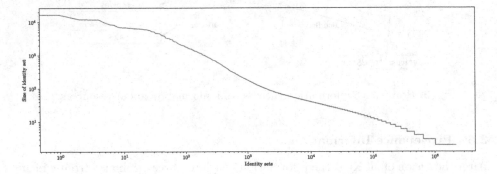

Fig. 4. Log-log plot of the number of elements in each identity set sorted by size.

shows that there are a few sets with many elements but a lot of sets with a rather low number of identical resources.

Table 2 shows the statistics divided into classes, properties, and instances. One can see that properties are matched the most and also contains the largest number of elements in one identity set. This is followed by classes which also occur in many different KGs. Due to the restriction that one entity can be matched to at most one other entity in another KG, the number of elements in an identity set also represents the number of KGs in which a concept can be found. Table 4 shows the evaluation measures for a very small subset of the alignment where a gold standard is given by the OAEI track.

After this quantitative analysis, the top matched entities (largest identity sets) are also inspected. Table 3 shows the top ten matched classes, properties, and instances together with the number of elements in the identity set (in how many wikis this concept is contained). Wiki-related entities like Main Pages are excluded and the names are the most common label appearing in the corresponding identity set (because entities with different labels can be also matched).

The class `character` is contained in 11,899 wikis. This also shows the domain of the majority of the wikis which is entertainment, movies, and games. But there are also more general classes like location, game, and film which appear in at least 1,325 wikis. The largest identity set contains the property `title` which is used in nearly 17,000 out of the 40,000 inspected wikis. It is directly followed by the property `name` which is used in 16,593 wikis. These properties are used to give a label to an instance. Gender, height, and occupation are more interesting

Table 3. Most frequently matched classes, properties, and instances

Classes		Properties		Instances	
Name	#Match	Name	#Match	Name	#Match
Character	11,899	Title	16,929	Discussion	821
Episode	5,153	Name	16,593	Jim cummings	210
Quote	4,148	Gender	12,109	Circe	198
Location	3,431	Caption	11,963	Rhea	186
Item	2,187	Image	9,696	Tara strong	178
Book	1,735	Type	8,247	Wonderland	177
Season	1,564	Height	6,931	Kevin Michael Richardson	176
Game	1,429	Status	6,860	Fred Tatasciore	162
Album	1,385	Occupation	6,827	Tom Kenny	160
Film	1,325	Species	6,655	Israel	159

Table 4. Evaluation of the multi source matching strategy using the OAEI KG track. For each category, precision (P), recall (R), and f-measure (F) are reported.

	Instance			Class			Property		
	P	R	F1	P	R	F1	P	R	F1
Alod2vec	.937	.390	.551	.842	.432	.571	.702	.549	.617

because it describes some attributes of persons. This is also in line with the top matched instances because they are mostly from the class person. One can already see that the number of matches is rather low for instances which means that fewer wikis have resources in common but rather describe different topics. The instance Jim Cummings appears in 210 wikis (e.g. disney-mickey-mouse, tarzan, sonic, and starwars) and is an American voice actor.

5 Analysis of the Resulting Knowledge Graph

After analyzing the alignment, the resulting fused KG is inspected. Overall 15,346,033 instances are contained and described by 215,273 properties. The number of classes is 109,042 but only 15,642 classes are based on infobox templates (which usually better indicate actual classes). The number of assertions depends on which triples are actually counted. When only using the extracted triples from the WikiMedia templates (main information source), then the value is 106,347,347. When adding additional information like page links, labels, comments, categories, etc the value increases to 1,032,941,819 triples. Further adding NIF files which provide the structure of wiki pages results in 4,422,408,859 triples. Table 1 shows all reported numbers in comparison to other available KGs like DBpedia, YAGO, and Babelnet. The only slight increase of instances

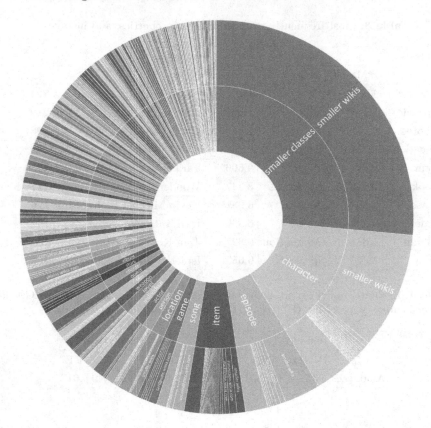

Fig. 5. Sunburst diagram of the final KG showing the relative amounts of instances per class. Inner circles: matched classes. Outer circles: Wikis containing this class.

between DBkWik and DBkWik++ can be explained by the fact that DBkWik uses the largest 12,000 wikis whereas DBkWik++ uses the largest 40,000 wikis and the fact that the size of the wikis follows a power law (meaning there are only a few large wikis but a lot of small wikis).

Figure 5 shows the relative number of instances per matched class via a sunburst diagram. The inner circle represents the matched classes and the radius corresponds to the number of instances in this class. The outer circle shows the distribution of this class over all wikis containing it. Due to displaying issues, too small classes/wikis are grouped together (indicated by smaller classes/wikis). Characters, episodes, and items are the largest classes but on the other side, there are many specific classes e.g. soap character with only a few instances.

5.1 Resource Availability

The final dataset [15] can be downloaded from https://doi.org/10.6084/m9.figshare.20407864.v1.

6 Conclusion and Outlook

In this paper, we presented DBkWik++, which is a consolidated knowledge graph generated from 40,000 wikis belonging to the Fandom wiki hosting platform. The matching approach reuses top performing binary matchers from the OAEI competition and shows that a multi source matching problem can be reduced to one to one matching tasks in a reasonable time. We analyzed the resulting alignment as well as the final KG to show which kind of information can be found in DBkWik++. Due to the fact that it does not originate from Wikipedia, it contains a lot of instances which do not appear there (long tail entities).

As an outlook, three further steps to improve and evaluate DBkWik++ are described in the following paragraphs.

In the current approach, the classes are created from infobox templates. As already discussed, those do not form a hierarchy. But there are also categories which are manually created, form a hierarchy, and are used to group wiki articles. Thus it is also possible to use them as a class replacement. [4] already clean those category trees to induce a class hierarchy. After this step, each KG has a class hierarchy which needs to be merged to create a huge consolidated hierarchy which does not need to be manually created (like in DBpedia).

For the fusion of the KGs, a union-based approach is currently applied. But for functional properties (only one value per instance) multiple sources can provide conflicting values. A first step toward this direction is to determine if properties are functional. One trivial approach is data driven and checks in each KG if the corresponding property (which is matched across multiple KGs) is functional. If this is the case, then a datatype specific resolution method can be applied (e.g. average for numeric values and shortest/longest values for strings). In the case of a voting-based approach, metadata of the sources (see Sect. 3.5) can be used to give higher weights for the voting e.g. number of editors (more editors in a wiki should increase the trustworthiness of values). A property based resolution method is not suitable because this would require knowledge on how to resolve the values for each property. This would not scale to the presented use case.

A next step is the accuracy evaluation of the resulting knowledge graph with intrinsic and extrinsic methods. For an intrinsic evaluation, random samples of triples from the knowledge graph can be assessed. The more of these triples are correct, the more accurate and better is the graph. This can be scaled to many judges in a crowdsourced survey. An important part is to show the source wiki page from which the information is extracted because the information might be very domain specific and not known to the survey participant. In an extrinsic evaluation, the knowledge graph is used as an external source in other applications such as recommender systems or classification approaches. [25] shows that the coverage of entities in DBpedia for a recommender system is not larger than 85 % for movies, 63 % for music artists, and 31 % for books. The LyricWiki[10],

---

[10] https://lyrics.fandom.com.

which is contained in DBkWik++, could help in this scenario to increase the recall for artists.

References

1. Alshammari, G., Jorro-Aragoneses, J.L., Kapetanakis, S., Petridis, M., Recio-García, J.A., Díaz-Agudo, B.: A hybrid CBR approach for the long tail problem in recommender systems. In: Aha, D.W., Lieber, J. (eds.) ICCBR 2017. LNCS (LNAI), vol. 10339, pp. 35–45. Springer, Cham (2017). https://doi.org/10.1007/978-3-319-61030-6_3
2. Auer, S., Bizer, C., Kobilarov, G., Lehmann, J., Cyganiak, R., Ives, Z.: DBpedia: a nucleus for a web of open data. In: Aberer, K., Choi, K.-S., Noy, N., Allemang, D., Lee, K.-I., Nixon, L., Golbeck, J., Mika, P., Maynard, D., Mizoguchi, R., Schreiber, G., Cudré-Mauroux, P. (eds.) ASWC/ISWC -2007. LNCS, vol. 4825, pp. 722–735. Springer, Heidelberg (2007). https://doi.org/10.1007/978-3-540-76298-0_52
3. Beek, W., Raad, J., Wielemaker, J., van Harmelen, F.: sameAs.cc: the closure of 500M owl:sameAs Statements. In: Gangemi, A., et al. (eds.) ESWC 2018. LNCS, vol. 10843, pp. 65–80. Springer, Cham (2018). https://doi.org/10.1007/978-3-319-93417-4_5
4. Chu, C.X., Razniewski, S., Weikum, G.: Tifi: taxonomy induction for fictional domains. In: The World Wide Web Conference (2019)
5. Defays, D.: An efficient algorithm for a complete link method. Comput. J. **20**(4), 364–366 (1977)
6. Dohrn, H., Riehle, D.: Design and implementation of the sweble wikitext parser: unlocking the structured data of wikipedia. In: Proceedings of the 7th International Symposium on Wikis and Open Collaboration, pp. 72–81 (2011)
7. Fabian, M., Gjergji, K., Gerhard, W., et al.: Yago: a core of semantic knowledge unifying wordnet and wikipedia. In: 16th International World Wide Web Conference, WWW (2007)
8. Faria, D., Lima, B., Silva, M.C., Couto, F.M., Pesquita, C.: AML and AMLC results for OAEI 2021. In: Ontology Matching Workshop at ISWC, vol. 2536 (2021)
9. Fensel, D., et al.: How to Build a Knowledge Graph, pp. 11–68. Springer (2020)
10. Heist, N., Hertling, S., Ringler, D., Paulheim, H.: Knowledge graphs on the web - an overview. In: Knowledge Graphs for Explainable Artificial Intelligence. IOS Press (2020)
11. Heist, N., Paulheim, H.: Uncovering the semantics of wikipedia categories. In: Ghidini, C., et al. (eds.) ISWC 2019. LNCS, vol. 11778, pp. 219–236. Springer, Cham (2019). https://doi.org/10.1007/978-3-030-30793-6_13
12. Hertling, S., Paulheim, H.: DBkWik: extracting and integrating knowledge from thousands of Wikis. Knowl. Inf. Syst. **62**(6), 2169–2190 (2019). https://doi.org/10.1007/s10115-019-01415-5
13. Hertling, S., Paulheim, H.: The knowledge graph track at OAEI. In: Harth, A., et al. (eds.) ESWC 2020. LNCS, vol. 12123, pp. 343–359. Springer, Cham (2020). https://doi.org/10.1007/978-3-030-49461-2_20
14. Hertling, S., Paulheim, H.: Order matters: matching multiple knowledge graphs. In: K-CAP 2021: Knowledge Capture Conference, Virtual Event, USA, 2–3 December 2021, pp. 113–120 (2021)
15. Hertling, S., Paulheim, H.: DBkWik Plus Plus (2022). https://doi.org/10.6084/m9.figshare.20407864.v1, https://figshare.com/articles/dataset/DBkWik_Plus_Plus/20407864

16. Köpcke, H., Rahm, E.: Frameworks for entity matching: a comparison. Data & Knowl. Eng. **69**(2), 197–210 (2010)
17. Lehmberg, O., Bizer, C., Brinkmann, A.: Winte.r - a web data integration framework. In: ISWC 2017 Posters & Demonstrations (2017)
18. Lenat, D.B.: Cyc: a large-scale investment in knowledge infrastructure. ACM Commun. **38**(11), 33–38 (1995)
19. Li, H.: Smile (2014). https://haifengl.github.io
20. Meilicke, C., Stuckenschmidt, H.: Analyzing mapping extraction approaches. In: OM (2007)
21. Miller, G.A.: Wordnet: a lexical database for English. Commun. ACM **38**(11), 39–41 (1995)
22. Mitchell, T., et al.: Never-ending learning. In: AAAI (2015)
23. Mudgal, S., et al.: Deep learning for entity matching: a design space exploration. In: SIGMOD Conference 2018, pp. 19–34 (2018)
24. Navigli, R., Ponzetto, S.P.: Babelnet: the automatic construction, evaluation and application of a wide-coverage multilingual semantic network. Artif. Intell. **193**, 217–250 (2012)
25. Noia, T.D., Ostuni, V.C., Tomeo, P., Sciascio, E.D.: Sprank: semantic path based ranking for top-n recommendations using linked open data. ACM Trans. Intell. Syst. Technol. (TIST) **8**(1), 1–34 (2016)
26. Portisch, J., Paulheim, H.: Alod2vec matcher results for OAEI 2021. In: CEUR Workshop Proceedings (2022)
27. Pour, M.A.N., et al.: Results of the ontology alignment evaluation initiative 2021. In: Ontology Matching Workshop at ISWC, vol. 3063, pp. 62–108 (2021)
28. Primpeli, A., Bizer, C.: Graph-Boosted active learning for multi-source entity resolution. In: Hotho, A., et al. (eds.) ISWC 2021. LNCS, vol. 12922, pp. 182–199. Springer, Cham (2021). https://doi.org/10.1007/978-3-030-88361-4_11
29. Saeedi, A., David, L., Rahm, E.: Matching entities from multiple sources with hierarchical agglomerative clustering. In: KEOD, pp. 40–50 (2021)
30. Saeedi, A., Peukert, E., Rahm, E.: Using link features for entity clustering in knowledge graphs. In: Gangemi, A., et al. (eds.) ESWC 2018. LNCS, vol. 10843, pp. 576–592. Springer, Cham (2018). https://doi.org/10.1007/978-3-319-93417-4_37
31. Schubert, E., Koos, A., Emrich, T., Züfle, A., Schmid, K.A., Zimek, A.: A framework for clustering uncertain data. Proc, VLDB Endow (2015)
32. Singh, R., et al.: Generating concise entity matching rules. In: SIGMOD Conference 2017, pp. 1635–1638 (2017)
33. Thor, A., Rahm, E.: MOMA - A mapping-based object matching system. In: Third Biennial Conference on Innovative Data Systems Research, CIDR 2007, Asilomar, CA, USA, January 7–10, 2007, Online Proceedings (2007)
34. Tonon, A., Felder, V., Difallah, D.E., Cudré-Mauroux, P.: VoldemortKG: mapping schema.org and web entities to linked open data. In: Groth, P., et al. (eds.) ISWC 2016. LNCS, vol. 9982, pp. 220–228. Springer, Cham (2016). https://doi.org/10.1007/978-3-319-46547-0_23
35. Volz, J., Bizer, C., Gaedke, M., Kobilarov, G.: Silk - a link discovery framework for the web of data. In: WWW2009 Workshop on Linked Data on the Web (2009)
36. Vrandečić, D., Krötzsch, M.: Wikidata: a free collaborative knowledgebase. Commun. ACM **57**(10), 78–85 (2014)

A Survey on Knowledge Graph-Based Methods for Automated Driving

Juergen Luettin[1(✉)], Sebastian Monka[1], Cory Henson[2], and Lavdim Halilaj[1]

[1] Bosch Center for AI, Renningen, Germany
{juergen.luettin,sebastian.monka,lavdim.halilaj}@bosch.com
[2] Bosch Center for AI, Pittsburgh, PA, USA

Abstract. Deep learning methods have made remarkable breakthroughs in machine learning in general and in automated driving (AD) in particular. However, there are still unsolved problems to guarantee reliability and safety of automated systems, especially to effectively incorporate all available information and knowledge in the driving task. Knowledge graphs (KG) have recently gained significant attention from both industry and academia for applications that benefit by exploiting structured, dynamic, and relational data. The complexity of graph-structured data with complex relationships and inter-dependencies between objects has posed significant challenges to existing machine learning algorithms. However, recent progress in knowledge graph embeddings and graph neural networks allows to applying machine learning to graph-structured data. Therefore, we motivate and discuss the benefit of KGs applied to AD. Then, we survey, analyze and categorize ontologies and KG-based approaches for AD. We discuss current research challenges and propose promising future research directions for KG-based solutions for AD.

Keywords: Knowledge graph · Automated driving · Automotive ontology · Knowledge representation learning · Knowledge graph embedding · Knowledge graph neural network

1 Introduction

The first successful AD vehicle was demonstrated in the 1980s [17]. However, despite remarkable progress, fully AD has not been realized to date. One unsolved problem is that AD vehicles must be able to drive safely in situations that have not been seen before in the training data. Moreover, AD systems must consider the strong safety requirements specified in ISO 26262 [48], which states that the behavior of the components needs to be fully specified and that each refinement can be verified with respect to its specification. verified.

Deep learning (DL) [40] has made remarkable breakthroughs with significant impact on the performance of AD systems. However, DL methods do not provide information to adequately understand what the network has learned and thus are hard to interpret and validate [9,38]. In safety-critical applications, this is

B. Villazón-Terrazas et al. (Eds.): KGSWC 2022, CCIS 1686, pp. 16–31, 2022.
https://doi.org/10.1007/978-3-031-21422-6_2

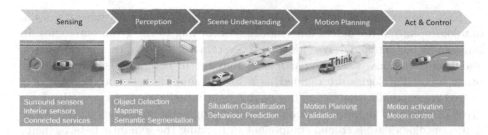

Fig. 1. Standard components of an AD system, modified from [57].

a major drawback. Moreover, the performance of DL methods is heavily dependent on the availability of suitable training data. When the testing environment deviates from the distribution of the training data, DL methods tend to produce unpredictable and critical errors. Whereas driving is governed by traffic laws and typical driver behaviors that represents a crucial knowledge source, traditional DL methods cannot easily incorporate such explicit knowledge. We argue that KGs are well suited to address all of these drawbacks.

The use of graphs to represent knowledge has been researched for a long time. The term knowledge graph (KG) was popularized with the announcement of the Google Knowledge Graph [90]. A graph-based representation has several advantages over alternative approaches to represent knowledge. Graphs represent a concise and intuitive abstraction with edges representing the relations that exist between entities. Recently, methods to apply machine learning directly on graphs have generated new opportunities to use KGs in data-based applications [101]. Figure 1 shows the standard components of an AD system together with their sub-tasks. In this survey, we address KG-based approaches for the components *Perception, Scene Understanding, and Motion Planning*.

2 Preliminaries

We first describe the basic terminology relevant in the context of this survey as well as insights of related work regarding generic principles of joint usage of knowledge graphs and machine learning pipelines.

2.1 Knowledge Graphs and Ontologies

Knowledge graphs are means for structuring facts, with entities connected via named relationships. Hogan et al. [41] define a KG as "a graph of data with the objective of accumulating and conveying real-world knowledge, where nodes represent entities and edges are relations between entities". Knowledge can be expressed in a factual triple in the form of (head, relation, tail) or (subject, predicate, object) under the Resource Description Framework (RDF), for example, (Albert Einstein, WinnerOf, Nobel Prize). A KG is a set of triples $G = H, R, T$, where H represents a set of entities E, $T \subseteq E \times L$, a set of entities and literal values, and R set of relationships connecting H and T.

KGs are essentially composed of two main components: 1) schemas a.k.a. ontologies; and 2) the actual data modeled according to the given ontologies. In philosophy, an ontology is considered as a systematic study of things, categories, and their relations within a particular domain. In computer science, on the other hand, ontologies are defined as a formal and explicit specification of a shared conceptualization [92]. They enable conceptualization of knowledge for a given domain and support common understanding between various stakeholders. Thus, ontologies are a crucial component in tackling the semantic heterogeneity problem and enabling interoperability in scenarios where different agents and services are involved.

2.2   Knowledge Representation Learning

While most symbolic knowledge is encoded in graph representation, conventional machine learning methods operate in vector space. Using a *knowledge graph embedding method* (KGE-Method), a KG can be transformed into a *knowledge graph embedding* (KGE), a representation where entities and relations of a KG are mapped into low-dimensional vectors. The KGE captures semantic meanings and relations of the graph nodes and reflects them by locality in vector space [101]. KGEs are originally used for graph-based tasks such as node classification or link prediction, but have recently been applied to tasks such as object classification, detection, or segmentation. As defined in [11], graph embedding algorithms can be clustered into unsupervised and supervised methods.

Unsupervised KGE-Methods form a KGE based on the inherent graph structure and the attributes of the KG. One of the earlier works [76] focused on statistical relational learning. Recent surveys [10,28,51,101] categorize unsupervised KGE-Methods based on their *representation space* (vector, matrix, and tensor space), the *scoring function* (distance-based, similarity-based), the *encoding model* (linear/bilinear models, factorization models, neural networks), and the *auxiliary information* (text descriptions, type constraints).

Supervised KGE-Methods learn a KGE to best predict additional node or graph labels. Forming a KGE by using task-specific labels for the attributes, the KGE can be optimized for a particular task while retaining the full expressivity of the graph. Common supervised KGE-Methods are based on *graph neural networks* (GNNs) [26], an extension of DL networks that can directly work on a KG. For scalability reasons and to overcome challenges that arise from graph irregularities, various adaptations have emerged such as *graph convolutional networks* (GCN) [56] and *graph attention networks* (GAT) [99].

Several surveys focusing on different research topics in AD have been published, including *computer vision* [49], *object detection* [1,27], *DL based scene understanding* [29,32,45,67], *DL based vehicle behavior prediction* [74], *deep reinforcement learning* [57], *lane detection* [75,94,113], *semantic segmentation* [61, 69], and *planning and decision making* [5,15,25,78,89]. More recently, ontologies and KGs have gained interest for knowledge-infused learning approaches. Monka et al. [71] provided a survey about visual transfer learning using KGs. However, we did not find a survey that cover the use of KGs applied to AD.

3   Ontologies for Automated Driving

Several ontologies have been developed to model relevant knowledge in the automotive domain. They cover elements such as vehicle, driver, route, and scenery, including their spatial and temporal relationships. The authors in [24,96] propose a consolidated definition and taxonomy for AD terminology. The goal is to establish a standard and consistent terminology and ontology. Additionally, there exist well-known ontologies such as *DBpedia* [4], *Schema.org* [31] and *SOSA* [50] which contain concepts related to the automotive domain described from a more generic perspective. Figure 2 illustrates main concepts such as *Scene, Participant* and *Trip* with sub-categories and relationships. In the following, we categorize and describe the surveyed ontologies considering their primary focus.

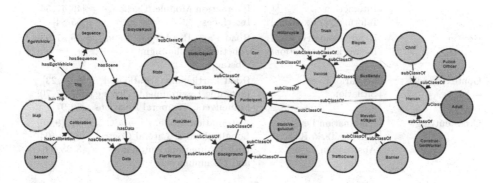

Fig. 2. Scene Ontology. Excerpt of an ontology to describe information of a driving scene [35]

Vehicle Model. An ontology modeling the main concepts of vehicles, such as vehicle type, installed components and sensors, is described in [112]. The work in [58] focuses on representing sensors, their attributes, and generated signals to increase data interoperability.

Driver Model. Combined approaches for managing driving related information by a model representing the driver, the vehicle, and the context are described in [21,39]. An ontology for modeling driver profiles based on their demographic and behavioral aspects is proposed by [85]. It is used as an input for a hybrid learning approach for categorizing drivers according to their driving styles.

Driver Assistance. An ontology for driver assistance is described in [54]. It serves as a common basis for domain understanding, decision-making, and information sharing. Feld et al. [21] presented an ontology focusing on human-machine-interfaces and inter-vehicle communication. Huelsen et al. [46,47] developed an ontology for traffic intersections to reason about the right-of-way including information about traffic signs and traffic lights. Related approaches can be found in [22,33,65,84].

Table 1. Automotive ontologies. Relevant ontologies split into different categories considering their main focus and concepts.

Focus	Scope	Main concepts	Ref.
Generic	Vehicle, Driver, Scene	Vehicle, Engine Spec., Usage Type, Driving Wheel, Configuration, Manufacturer, Sensor, Scene, Time	[4,31,50]
Vehicle	Vehicle, Categories, Parts	Vehicle, Sensors, Actuators, Signals, Taxonomy, Speed, Acceleration	[58,112]
Driver model	Driver, Characteristics, Objectives	Ability, Physiological & Emotional State, Driving Style, Preferences, Behavior	[21,39,85]
Driver assistance	Driver Machine Interactions, Information Exchange	Assistance Functions, Interaction Module, Exchange Interfaces	[21,22,33] [46,47,54] [65,84]
Routing	Road Geometry, Road Network, Intersections	Road Part, Road Type, Lane, Junction, Traffic Sign, Obstacles	[59,87,93]
Context model	Scenario, Event, Situation	Road Participants, Static and Dynamic Environment, Driving Maneuvers, Temporal Relations	[8,21,23] [34,37,70] [79,97]
Cross-cutting	Automation Level, Driving Task, Risk, Abstraction Layer	Automation Mode, Longitudinal & Lateral Control, Risk Estimation, Autonomy and Communication	[62,79,86] [2,88,104] [3,35,98]

Route Planning. Schlenoff et al. [87] explored the use of ontologies to improve the capabilities and performance of on-board route-planning. An ontology-based method for modeling and processing map data in cars was presented by Suryawanshi et al. [93]. Kohlhaas et al. [59] introduced a modeling approach for the semantic state space for maneuver planning.

Context Model. Ulbrich et al. [97] described an ontology for context representation and environmental modelling for AD. It contains different layers including a metric layer of lane context, traffic rules, object, and situation description. Buechel et al. [8] proposed a modular framework for traffic regulation based decision-making in AD. The traffic scene is represented by an ontology and includes knowledge about traffic regulations. Henson et al. [37] presented an ontology-based method for searching scenes in AD datasets. Similar approaches representing the context in a driving scenario are shown in [21,23,34,70]. Ontologies have also been used for context-dependent recommendation tasks [36,103].

Cross-cutting. An ontology describing the levels of automated driving systems, ranging from fully manual to fully automated was presented in [79]. In line with this, [86] analyses crucial questions of driving tasks and maps them to the level of driving automation. The interaction risks between human driven and automated vehicles investigated in [62] are based on five main components: obstacle, road, ego-vehicle, environment, and driver. An informal but comprehensive 6-layer model for a structured description and categorization of urban scenes was describe in [88]. This was further developed into the Automotive Urban Traffic Ontology (A.U.T.O.) [104]. The Automotive Global Ontology(AGO) [98] focuses on semantic labeling and testing use cases. Ontologies representing data structures of AD datasets have been presented in [35]. Several standards aim to develop ontologies to provide a foundation of common definitions, properties, and relations for central concepts, e.g. ASAM OpenScenario [2] and ASAM OpenX [3].

A summary of automotive ontologies with scope and main concepts is shown in Table 1. It can be seen that many ontologies cover only specific concepts and use cases but other are more complete and provide a more comprehensive coverage for all AD components and tasks. Since one of the main benefits of ontologies is re-usability, these ontologies can be re-used and extended for various scenarios.

4 Knowledge Graphs Applied to Automated Driving

In this section, we provide an overview of KG-based approaches and categorize them based on their respective components and tasks in the AD pipeline as shown in Fig. 1. We consider approaches that use ontologies or KGs as well as approaches that combine KGs with machine learning for AD tasks.

4.1 Object Detection

Object detection in AD includes the detection of traffic participants such as vehicles, pedestrians, road markings, traffic signs, and others. Typical sensors used are video cameras, RADAR and LiDAR.

Scene graphs are a relatively recent technique to semantically describe and represent objects and relations in a scene [52]. Much research is targeting the generation of scene graphs [12] which can be divided into two types. The first, bottom up approach, consists of object detection followed by pairwise relationship recognition. The second, top-down approach, consists of joint detection and recognition of objects and their relationships. Current research in scene graphs focuses on the integration of prior knowledge. The use of KGs to provide background knowledge and generate scene graphs is recently proposed [109] with Graph Bridging Networks (GB-Net). The GB-Net regards the scene graph as the image conditioned instantiation of the common sense knowledge graph. ConceptNet [66] is used as the knowledge graph. Gu et al. [30] presented KB-GAN, a knowledge base and auxiliary image generation approach based on external knowledge and image reconstruction loss to overcome the problems found in datasets. We found no applications of these methods for object detection in AD.

Wickramarachchi et al. [106] generated a KGE from a scene knowledge graph, and use the embedding to predict missing objects in the scene with high accuracy. This is accomplished with a novel mapping and formalization of object detection as a KG link prediction problem. Several KGE algorithms are evaluated and compared [105]. Chowdhury et al. [14] extended this work by integrating common-sense knowledge into the scene knowledge graph. Woo et al. [107] presented a method to embed relations by jointly learning connections between objects. It contains a global context encoding and a geometrical layout encoding witch extract global context information and spatial information.

Road-Sign Detection. Several approaches focus on using knowledge-graphs for road sign detection [55,72]. Kim et al. [55] described a method to assist human annotation of road signs by combining KGs and machine learning. Monka et al. [72] proposed an approach for object recognition based on a knowledge graph neural network (KG-NN) that supervises the training using image-data-invariant auxiliary knowledge. The auxiliary knowledge is encoded in a KG with respective concepts and their relationships. These are transformed into a dense vector representation by an embedding method. The KG-NN learns to adapt its visual embedding space by a contrastive loss function.

Lane Detection. Lane detection approaches can be divided into two categories: (1) data from high definition (HD) maps; and (2) data from the vehicle perception sensors (e.g. cameras). The drawback of HD maps is that such data is not always available and not always up to date. Traditional methods for lane detection usually perform first edge detection and then model fitting [113]. A graph-embedding-based approach for lane detection [68] can robustly detect parallel, non-parallel (merging or splitting), varying lane width, and partially blocked lane markings. A novel graph structure is used to represent lane features, lane geometry, and lane topology. Homayounfar et al. [42] focused on discovering lane boundaries of complex highways with many lanes that contain topology changes due to forks and merges. They formulate the problem as inference in a directed *acyclic graph model* (DAG), where the nodes of the graph encode geometric and topological properties of the local regions associated with the lane boundaries.

4.2   Semantic Segmentation

The goal of semantic segmentation is to assign a semantic meaningful class label (e.g. road, sidewalk, pedestrian, road sign, vehicle) to each pixel in a given image. A KG-based approach for scene segmentation is described by Kunze et al. [60]. A scene graph is generated from a set of segmented entities that models the structure of the road using an abstract, logical representation to enable the incorporation of background knowledge. Similar approaches based on the (non-semantic) graph representation for scene segmentation can be found in [7,18,91, 95,100].

4.3   Mapping

Automated vehicles often use digital maps as a virtual sensor to retrieve information about the road network for understanding, navigating, and making decisions

Table 2. Knowledge graphs applied to automated driving.

Perception	Mapping, Understanding	Plan & Validate
Object detection	**Mapping**	**Motion Planning**
Scene graphs [12,52]	Map representation [93]	Decision making [83]
Context learning [107]	Map integration [80]	Rules [110–112]
KG-scene-graphs [30,109]	Map updating [81]	Reasoning [44,108]
KG-based detection [105]	Quality of maps [82]	Rule learning [16,43,73]
Common-sense [14]	**Scene understanding**	KG from text [19]
Road sign recog. [55,72]	Context model [102]	**Validation**
Lane detection [42,68]	Situation understanding [34]	Risk assessm. [6,77,104]
Segmentation	**Behavior Prediction**	Test gener. [13,23,64]
Scene segm. [60]	Motion prediction [20,114]	Verification [53]

about the driving path. Qui et al. [80] proposed a knowledge architecture with two levels of abstraction to solve the map data integration problem. How to use different types of rules to achieve two-dimensional reasoning is detailed in [81]. Qiu et al. [82] addressed the issue of quality assurance in ontology- based map data for AD, specifically the detection and rectification of map errors.

4.4   Scene Understanding

Scene understanding aims to understand what is happening in the scene, the relations between the objects in order to obtain a comprehension for further steps in automated driving that deal with motion planning and vehicle control. An approach for KG-based scene understanding was described by Werner et al. [102]. The paper proposes a KG to model temporally contextualized observations and Recurrent Transformers (RETRA), a neural encoder stack with a feedback loop and constrained multi-headed self-attention layers. RETRA enables transformation of global KG-embeddings into custom embeddings, given the situation-specific factors of the relation and the subjective history of the entity. Halilaj et al. [34] presented a KG-based approach for fusing and organizing heterogeneous types and sources of information for driving assistance and automated driving tasks. The model builds on the terminology of [96] and uses existing ontologies such as SOSA [50]. A KGE based on graph neural networks [101] is then used for the task to classify driving situations.

4.5   Object Behavior Prediction

Fang et al. [20] described an ontology-based reasoning approach for long-term behavior prediction of road users. Long-term behavior is predicted with estimated probabilities based on semantic reasoning that considers interactions among various players. Li et al. [63] present a graph based interaction-aware trajectory prediction (GRIP) approach. This is based on a GCN and graph operations that model the interaction between the vehicles. Relation-based traffic motion prediction using traffic scene graphs and GNN is described in [114].

4.6 Motion Planing

The aim of motion planning is to plan and execute driving actions such as steering, accelerating, and braking taking all information of previous steps into account. Regele [83] uses an ontology-based high-level abstract world model to support the decision-making process. It consists of a low-level model for trajectory planning and a high-level model for solving traffic coordination problems. Zhao et al. [110–112] presented core ontologies to enable safe automated driving. They are combined with rules for ontology-based decision-making on uncontrolled intersections and narrow roads. Another approach for ontology development, focusing on vehicle context is described in [108]. Huang et al. [44] use ontologies for scene-modeling, situation assessment and decision-making in urban environments and a knowledge base of driving knowledge and driving experience. Using ontologies to generate semantic rules from a decision tree classifier trained on driving test descriptions of a driving school is outlined in [16]. Khan et al. [19] retrieve human knowledge from natural text descriptions of traffic scenes with pedestrian-vehicle encounters. Hovi [43] presents how rules for ontology-based decision-making systems can be learned through machine learning. Morignot [73] presents an ontology-based approach for decision-making and relaxing traffic regulations in situations when this is a preferred scenario.

4.7 Validation

In the following, we list KG-based approaches that support the validation of AD systems including requirements verification, test case generation, and risk assessment. Bagschik et al. [6] introduced a concept of ontology-based scene creation, that can serve in a scenario-based design paradigm to analyze a system from multiple viewpoints. In particular, they propose the application for hazard and risk assessment. A similar approach for use case generation is described in [13,23]. Paardekooper et al. [77] presents a hybrid-AI approach to situational awareness. A data-driven method is coupled with a KG along with the reasoning capabilities to increase the safety of AD systems. Criticality recognition using the A.U.T.O ontology is demonstrated in [104]. Li et al. [64] outlined a framework for testing, verification, and validation for AD. It is based on ontologies for describing the environment and converted to input models for combinatorial testing. Kaleeswaran et al. [53] presented an approach for verification of requirements in AD. A semantic model translates requirements using world knowledge into formal representation, which is then used to check plausibility and consistency of requirements.

A summary of KG-based methods for AD-tasks is shown in Table 2. It can be seen, that for every component and task only very few KG-based approaches have been developed, suggesting much potential for further research in this area.

5 Conclusions

While AD has made tremendous progress over the last few years, many questions remain still unanswered. Among these are verifiability, explainability, and safety. AD systems that operate in complex, dynamic, and interactive environments require approaches that generalize to unpredictable situations and that can reason about the interactions with multiple participants and variable contexts.

Recent progress on KGs and KG-based representation learning has opened new possibilities in addressing these open questions. This is motivated by two properties of KGs. First, the ability of KGs to represent complex structured information and in particular, relational information between entities; second, the ability to combine heterogeneous sources of knowledge, such as common sense knowledge, rules, or crowd-sourced knowledge, into a unified graph-based representation. Recent progress in KG-based representation learning opened a new research direction in using KG-based data for machine learning applications such as AD.

We have surveyed ontologies and KG-based approaches for AD. A few automotive ontologies have been developed. However, harmonization of automotive ontologies and AD dataset structures is an important step towards enabling KG construction and usage for AD tasks. Only a few KG-based methods have been developed for AD. Given the benefits and improvements of KG-based methods in other domains indicates great potential of KG-based methods in the AD domain.

Future topics include but are not limited to (1) the inclusion of additional knowledge sources; (2) task-oriented knowledge representation and knowledge embedding; (3) temporal representation of KG-based approaches; and (4) rule extraction and verification for explainability. While KGs have already been used in the areas of perception, scene understanding, and motion planning, the use of the technology for the tasks of sensing and act & control will bring further advances for AD. The work indicates that knowledge graphs will play an important role in making automated driving better, safer, and ultimately feasible for real-world use.

References

1. Arnold, E., Al-Jarrah, O.Y., et al.: A survey on 3d object detection methods for autonomous driving applications. T-ITS **20**(10), 3782–3795 (2019)
2. ASAM: ASAM OpenSCENARIO V2.0 (2022)
3. ASAM: ASAM OpenX, proposal (2022)
4. Auer, S., Bizer, C., Kobilarov, G., Lehmann, J., Cyganiak, R., Ives, Z.: DBpedia: a nucleus for a web of open data. In: Aberer, K., et al. (eds.) ASWC/ISWC -2007. LNCS, vol. 4825, pp. 722–735. Springer, Heidelberg (2007). https://doi.org/10.1007/978-3-540-76298-0_52
5. Badue, C., Guidolini, R., Carneiro, R.V., et al.: Self-driving cars: a survey. Expert Syst. Appl. **165**, 113816 (2021)

6. Bagschik, G., Menzel, T., Maurer, M.: Ontology based scene creation for the development of automated vehicles. In: 2018 IEEE Intelligent Vehicles Symposium (IV), pp. 1813-1820. IEEE (2018)
7. Bordes, J.B., Davoine, F., Xu, P., Denoeux, T.: Evidential grammars: a compositional approach for scene understanding. application to multimodal street data. Appl. Soft Comput. **61**, 1173–1185 (2017)
8. Buechel, M., Hinz, G., Ruehl, F., et al.: Ontology-based traffic scene modeling, traffic regulations dependent situational awareness and decision-making for automated vehicles. In: IEEE Intelligent Vehicles Symposium, IV (2017)
9. Burton, S., Gauerhof, L., Heinzemann, C.: Making the case for safety of machine learning in highly automated driving. In: SAFECOMP Workshops (2017)
10. Cai, H., Zheng, V., Chang, K.: A comprehensive survey of graph embedding: problems, techniques, and applications. In: IEEE TKDE (2018)
11. Chami, I., Abu-El-Haija, S., Perozzi, B., Ré, C., Murphy, K.: Machine learning on graphs: a model and comprehensive taxonomy. CoRR abs/2005.03675 (2020)
12. Chang, X., Ren, P., Xu, P., Li, Chen, X., Hauptmann, A.: Scene graphs: a survey of generations and applications. arXiv abs/2104.01111 (2021)
13. Chen, W., Kloul, L.: An ontology-based approach to generate the advanced driver assistance use cases of highway traffic. In: Proceedings of the 10th IC3K (2018)
14. Chowdhury, S.N., Wickramarachchi, R., Gad-Elrab, M.H., Stepanova, D., Henson, C.: Towards leveraging commonsense knowledge for autonomous driving. In: ISWC (2021)
15. Claussmann, L., Revilloud, M., Glaser, S., Gruyer, D.: A study on al-based approaches for high-level decision making in highway autonomous driving. In: SMC (2017)
16. Dianov, I., Ramírez-Amaro, K., Cheng, G.: Generating compact models for traffic scenarios to estimate driver behavior using semantic reasoning. In: IROS (2015)
17. Dickmanns, E., Graefe, V.: Dynamic monocular machine vision. Mach. Vis. Appl. (2005)
18. Dierkes, F., Raaijmakers, M., Schmidt, M., Bouzouraa, M.E., Hofmann, U., Maurer, M.: Towards a multi-hypothesis road representation for automated driving. In: IEEE ITSC (2015)
19. Elahi, M.F., Luo, X., Tian, R.: A framework for modeling knowledge graphs via processing natural descriptions of vehicle-pedestrian interactions. In: Stephanidis, C., Duffy, V.G., Streitz, N., Konomi, S., Krömker, H. (eds.) HCII 2020. LNCS, vol. 12429, pp. 40–50. Springer, Cham (2020). https://doi.org/10.1007/978-3-030-59987-4_4
20. Fang, F., Yamaguchi, S., Khiat, A.: Ontology-based reasoning approach for longterm behavior prediction of road users. In: IEEE ITSC (2019)
21. Feld, M., Müller, C.A.: The automotive ontology: managing knowledge inside the vehicle and sharing it between cars. In: AutomotiveUI (2011)
22. Fuchs, S., Rass, S., Lamprecht, B., Kyamakya, K.: A model for ontology-based scene description for context-aware driver assistance systems. In: ICST AMBISYS (2008)
23. de Gelder, E., et al.: Ontology for scenarios for the assessment of automated vehicles. CoRR abs/2001.11507 (2020)
24. Geyer, S., Baltzer, M., Franz, B., Hakuli, S., Kauer, M., Kienle, M., et al.: Concept and development of a unified ontology for generating test and use-case catalogues for assisted and automated vehicle guidance. In: IET ITS (2014)
25. González, D., Pérez, J., Montero, V.M., Nashashibi, F.: A review of motion planning techniques for automated vehicles. In: T-ITS (2016)

26. Gori, M., Monfardini, G., Scarselli, F.: A new model for learning in graph domains. In: IJCNN (2005)
27. Gouidis, F., Vassiliades, A., Patkos, T., et al.: A review on intelligent object perception methods combining knowledge-based reasoning and machine learning. In: Proceedings of the AAAI-MAKE Symposium (2020)
28. Goyal, P., Ferrara, E.: Graph embedding techniques, applications, and performance: a survey. Knowl. Based Syst. **151**, 78–94 (2018)
29. Grigorescu, S.M., Trasnea, B., Cocias, T.T., Macesanu, G.: A survey of deep learning techniques for autonomous driving. J. Field Robot. **37**(3), 362–386 (2020)
30. Gu, J., Zhao, H., Lin, Z., Li, S., Cai, J., Ling, M.: Scene graph generation with external knowledge and image reconstruction. In: CVPR (2019)
31. Guha, R.V., Brickley, D., Macbeth, S.: Schema.org: evolution of structured data on the web. Commun. ACM (2016)
32. Guo, Z., Huang, Y., Hu, X., Wei, H., Zhao, B.: A survey on deep learning based approaches for scene understanding in autonomous driving. Electronics **10**(4), 471 (2021)
33. Gutiérrez, G., Iglesias, J.A., Ordóñez, F.J., Ledezma, A., Sanchis, A.: Agent-based framework for advanced driver assistance systems in urban environments. In: FUSION (2014)
34. Halilaj, L., Dindorkar, I., Lüttin, J., Rothermel, S.: A knowledge graph-based approach for situation comprehension in driving scenarios. In: Verborgh, R., et al. (eds.) ESWC 2021. LNCS, vol. 12731, pp. 699–716. Springer, Cham (2021). https://doi.org/10.1007/978-3-030-77385-4_42
35. Halilaj, L., Luettin, J., Henson, C., Monka, S.: Knowledge graphs for automated driving. In: IEEE AIKE-Artificial Intelligence and Knowledge Engineering (2022)
36. Halilaj, L., Luettin, J., Rothermel, S., Arumugam, S.K., Dindorkar, I.: Towards a knowledge graph-based approach for context-aware points-of-interest recommendations. In: ACM/SIGAPP SAC, pp. 1846–1854 (2021)
37. Henson, C., Schmid, S., Tran, A.T., Karatzoglou, A.: Using a knowledge graph of scenes to enable search of autonomous driving data. In: ISWC (2019)
38. Herrmann, M., Witt, C., Lake, L., Guneshka, S., Heinzemann, C., Bonarens, F., et al.: Using ontologies for dataset engineering in automotive AI applications. In: 2022 Design, Automation & Test in Europe Conference & Exhibition (DATE) (2022)
39. Hina, M.D., Thierry, C., Soukane, A., Ramdane-Cherif, A.: Ontological and machine learning approaches for managing driving context in intelligent transportation. In: IC3K (2017)
40. Hinton, G.E., Osindero, S., Teh, Y.: A fast learning algorithm for deep belief nets. Neural Comput. **18**(7), 1527–1554 (2006)
41. Hogan, A., Blomqvist, E., Cochez, M., et al.: Knowledge graphs. In: ACM Computing Surveys (2021)
42. Homayounfar, N., Liang, J., Ma, W.C., Fan, J., Wu, X., Urtasun, R.: Dagmapper: learning to map by discovering lane topology. In: ICCV (2019)
43. Hovi, J., Ichise, R.: Feasibility study: rule generation for ontology-based decision-making systems. In: Wang, X., Lisi, F.A., Xiao, G., Botoeva, E. (eds.) JIST 2019. CCIS, vol. 1157, pp. 88–99. Springer, Singapore (2020). https://doi.org/10.1007/978-981-15-3412-6_9
44. Huang, L., Liang, H., Yu, B., Li, B., Zhu, H.: Ontology-based driving scene modeling, situation assessment and decision making for autonomous vehicles. In: (ACIRS) (2019)

45. Huang, Y., Chen, Y.: Survey of state-of-art autonomous driving technologies with deep learning. In: IEEE QRS-C (2020)
46. Hülsen, M., Zöllner, J.M., Weiss, C.: Traffic intersection situation description ontology for advanced driver assistance. IEEE IV Symposium (2011)
47. Hülsen, M., Zöllner, J.M., Haeberlen, N., Weiss, C.: Asynchronous real-time framework for knowledge-based intersection assistance. In: IEEE ITSC (2011)
48. ISO: ISO 26262–1:2018: Road vehicles - functional safety (2018)
49. Janai, J., Güney, F., Behl, A., Geiger, A.: Computer vision for autonomous vehicles: Problems, datasets and state of the art. Found. Trends Comput. Graph. Vis. **12**(1–3), 1–308 (2020)
50. Janowicz, K., Haller, A., Cox, S.J.D., Phuoc, D.L., Lefrançois, M.: SOSA: a lightweight ontology for sensors, observations, samples, and actuators. J. Web Semant. **56**, 1–10 (2019)
51. Ji, S., Pan, S., Cambria, E., Marttinen, P., Yu, P.S.: A survey on knowledge graphs: representation, acquisition and applications. In: IEEE Transactions on Neural Networks and Learning Systems (2021)
52. Johnson, J., et al.: Image retrieval using scene graphs. In: CVPR (2015)
53. Kaleeswaran, A., Nordmann, A., Mehdi, A.: Towards integrating ontologies into verification for autonomous driving. In: ISWC Satellites (2019)
54. Kannan, S., Thangavelu, A., Kalivaradhan, R.: An intelligent driver assistance system (i-das) for vehicle safety modelling using ontology approach. Int. J. Ubi-Comp **1**(3), 15–29 (2010)
55. Kim, J.E., Henson, C., Huang, K., Tran, T.A., Lin, W.Y.: Accelerating road sign ground truth construction with knowledge graph and machine learning. arXiv abs/2012.02672 (2020)
56. Kipf, T.N., Welling, M.: Semi-supervised classification with graph convolutional networks. In: ICLR (2017)
57. Kiran, B.R., et al.: Deep reinforcement learning for autonomous driving: a survey. IEEE Trans. Intell. Transp. Syst. **23**, 4909–4926 (2022)
58. Klotz, B., Troncy, R., Wilms, D., Bonnet, C.: Vsso: the vehicle signal and attribute ontology. In: SSN@ISWC (2018)
59. Kohlhaas, R., Bittner, T., Schamm, T., Zöllner, J.M.: Semantic state space for high-level maneuver planning in structured traffic scenes. In: ITSC (2014)
60. Kunze, L., Bruls, T., Suleymanov, T., Newman, P.: Reading between the lanes: road layout reconstruction from partially segmented scenes. In: ITSC (2018)
61. Lateef, F., Ruichek, Y.: Survey on semantic segmentation using deep learning techniques. Neurocomputing **338**, 321–348 (2019)
62. Leroy, J., Gruyer, D., Orfila, O., Faouzi, N.E.E.: Five key components based risk indicators ontology for the modelling and identification of critical interaction between human driven and automated vehicles. In: IFAC (2020)
63. Li, X., Ying, X., Chuah, M.C.: Grip: graph-based interaction-aware trajectory prediction. In: 2019 IEEE ITSC, pp. 3960–3966 (2019)
64. Li, Y., Tao, J., Wotawa, F.: Ontology-based test generation for automated and autonomous driving functions. In: IST (2020)
65. Lilis, Y., Zidianakis, E., Partarakis, N., Antona, M., Stephanidis, C.: Personalizing HMI elements in ADAS using ontology meta-models and rule based reasoning. In: Antona, M., Stephanidis, C. (eds.) UAHCI 2017. LNCS, vol. 10277, pp. 383–401. Springer, Cham (2017). https://doi.org/10.1007/978-3-319-58706-6_31
66. Liu, H., Singh, P.: Conceptnet - a practical commonsense reasoning tool-kit. BT Technol. J. **22**, 211–226 (2004). https://doi.org/10.1023/B:BTTJ.0000047600.45421.6d

67. Liu, L., et al.: Deep learning for generic object detection: a survey. Int. J. Comput. Vis. **128**(2), 261–318 (2019). https://doi.org/10.1007/s11263-019-01247-4
68. Lu, P., Xu, S., Peng, H.: Graph-embedded lane detection. In: IEEE Transactions on Image Processing (2021)
69. Minaee, S., Boykov, Y., Porikli, F., Plaza, A., Kehtarnavaz, N., Terzopoulos, D.: Image segmentation using deep learning: a survey. In: IEEE Transactions PAMI (2021)
70. Mohammad, M.A., Kaloskampis, I., Hicks, Y., Setchi, R.: Ontology-based framework for risk assessment in road scenes using videos. In: International Conference KES (2015)
71. Monka, S., Halilaj, L., Rettinger, A.: A survey on visual transfer learning using knowledge graphs. Semant. Web **13**, 477–510 (2022)
72. Monka, S., Halilaj, L., Schmid, S., Rettinger, A.: Learning visual models using a knowledge graph as a trainer. In: Hotho, A., et al. (eds.) ISWC 2021. LNCS, vol. 12922, pp. 357–373. Springer, Cham (2021). https://doi.org/10.1007/978-3-030-88361-4_21
73. Morignot, P., Nashashibi, F.: An ontology-based approach to relax traffic regulation for autonomous vehicle assistance. arXiv abs/1212.0768 (2012)
74. Mozaffari, S., Al-Jarrah, O.Y., Dianati, M., Jennings, P., Mouzakitis, A.: Deep learning-based vehicle behaviour prediction for autonomous driving applications: a review. arXiv abs/1912.11676 (2019)
75. Narote, S.P., Bhujbal, P.N., Narote, A.S., Dhane, D.M.: A review of recent advances in lane detection and departure warning system. Pattern Recognit. **73**, 216–234 (2018)
76. Nickel, M., Murphy, K., Tresp, V., Gabrilovich, E.: A review of relational machine learning for knowledge graphs. In: Proceedings of the IEEE (2016)
77. Paardekooper, J.P., Comi, M., et al.: A hybrid-ai approach for competence assessment of automated driving functions. In: SafeAI@AAAI (2021)
78. Paden, B., Cáp, M., Yong, S.Z., Yershov, D.S., Frazzoli, E.: A survey of motion planning and control techniques for self-driving urban vehicles. In: IEEE Transactions on Intelligent Vehicles (2016)
79. Pollard, E., Morignot, P., Nashashibi, F.: An ontology-based model to determine the automation level of an automated vehicle for co-driving. In: Proceedings of the FUSION (2013)
80. Qiu, H., Ayara, A., Glimm, B.: A knowledge architecture layer for map data in autonomous vehicles. In: ITSC (2020)
81. Qiu, H., Ayara, A., Glimm, B.: Ontology-based processing of dynamic maps in automated driving. In: KEOD (2020)
82. Qiu, H., Ayara, A., Glimm, B.: Ontology-based map data quality assurance. In: Verborgh, R., et al. (eds.) ESWC 2021. LNCS, vol. 12731, pp. 73–89. Springer, Cham (2021). https://doi.org/10.1007/978-3-030-77385-4_5
83. Regele, R.: Using ontology-based traffic models for more efficient decision making of autonomous vehicles. In: ICAS (2008)
84. Ryu, M.W., Cha, S.H.: Context-awareness based driving assistance system for autonomous vehicles. Int. J. Control Autom. **1**(1), 153–162 (2018)
85. Sarwar, S., Zia, S., ul Qayyum, Z., Iqbal, M., Safyan, M., Mumtaz, S., et al.: Context aware ontology-based hybrid intelligent framework for vehicle driver categorization. In: Transactions on Emerging Telecommunications Technologies (2019)
86. Schafer, F., Kriesten, R., Chrenko, D., Gechter, F.: No need to learn from each other? - potentials of knowledge modeling in autonomous vehicle systems engineering towards new methods in multidisciplinary contexts. In: ICE/ITMC (2017)

87. Schlenoff, C., Balakirsky, S., Uschold, M., Provine, R., Smith, S.: Using ontologies to aid navigation planning in autonomous vehicles. Knowl. Eng. Rev. **18**(3), 243–255 (2003)
88. Scholtes, M., Westhofen, L., Turner, L.R., Lotto, K., Schuldes, M., Weber, H., et al.: 6-layer model for a structured description and categorization of urban traffic and environment. IEEE Access **9**, 59131–59147 (2021)
89. Schwarting, W., Pierson, A., et al.: Social behavior for autonomous vehicles. In: Proceedings of the National Academy of Sciences, USA (2019)
90. Singhal, A.: Introducing the knowledge graph: things, not strings. https://blog.google/products/search/introducing-knowledge-graph-things-not/ (2012). 07 May 2021
91. Spehr, J., Rosebrock, D., Mossau, D., Auer, R., Brosig, S., Wahl, F.: Hierarchical scene understanding for intelligent vehicles. In: 2011 IEEE Intelligent Vehicles Symposium (IV) (2011)
92. Studer, R., Benjamins, V.R., Fensel, D.: Knowledge engineering: Principles and methods. Data Knowl. Eng. **25**(1–2), 161–197 (1998)
93. Suryawanshi, Y., Qiu, H., Ayara, A., Glimm, B.: An ontological model for map data in automotive systems. In: IEEE AIKE (2019)
94. Tang, J., Li, S., Liu, P.: A review of lane detection methods based on deep learning. Pattern Recognit. **11**, 107623 (2021)
95. Töpfer, D., Spehr, J., Effertz, J., Stiller, C.: Efficient road scene understanding for intelligent vehicles using compositional hierarchical models. In: T-ITS (2015)
96. Ulbrich, S., Menzel, T., Reschka, A., Schuldt, F., Maurer, M.: Defining and substantiating the terms scene, situation, and scenario for automated driving. In: ITSC (2015)
97. Ulbrich, S., Nothdurft, T., Maurer, M., Hecker, P.: Graph-based context representation, environment modeling and information aggregation for automated driving. In: IEEE Intelligent Vehicles Symposium Proceedings (2014)
98. Urbieta, I.R., Nieto, M., García, M., Otaegui, O.: Design and implementation of an ontology for semantic labeling and testing: automotive global ontology (AGO). Appli. Sci. **11**(17), 7782 (2021)
99. Velickovic, P., Cucurull, G., Casanova, A., Romero, A., Liò, P., Bengio, Y.: Graph attention networks. In: 6th International Conference on Learning Representations, ICLR (2018)
100. Venkateshkumar, S., Sridhar, M., Ott, P.: Latent hierarchical part based models for road scene understanding. In: ICCVW (2015)
101. Wang, Q., Mao, Z., Wang, B., Guo, L.: Knowledge graph embedding: a survey of approaches and applications. In: IEEE Transactions on Knowledge and Data Engineering (2017)
102. Werner, S., Rettinger, A., Halilaj, L., Luettin, J.: Embedding Taxonomical, Situational or Sequential Knowledge Graph Context for Recommendation Tasks. In: Further with Knowledge Graphs (2021)
103. Werner, S., Rettinger, A., Halilaj, L., Lüttin, J.: RETRA: recurrent transformers for learning temporally contextualized knowledge graph embeddings. In: Verborgh, R., et al. (eds.) ESWC 2021. LNCS, vol. 12731, pp. 425–440. Springer, Cham (2021). https://doi.org/10.1007/978-3-030-77385-4_25
104. Westhofen, L., Neurohr, C., Butz, M., Scholtes, M., Schuldes, M.: Using ontologies for the formalization and recognition of criticality for automated driving. IEEE Open J. Intell. Transp. Syst. **3**, 519–538 (2022)

105. Wickramarachchi, R., Henson, C., Sheth, A.: An evaluation of knowledge graph embeddings for autonomous driving data: experience and practice. In: AAAI-MAKE (2020)
106. Wickramarachchi, R., Henson, C., Sheth, A.: Knowledge-infused learning for entity prediction in driving scenes. Frontiers in Big Data (2021)
107. Woo, S., Kim, D., Cho, D., Kweon, I.S.: Linknet: relational embedding for scene graph. In: NeurIPS (2018)
108. Xiong, Z., Dixit, V., Waller, S.: The development of an ontology for driving context modelling and reasoning. In: ITSC (2016)
109. Zareian, A., Karaman, S., Chang, S.-F.: Bridging knowledge graphs to generate scene graphs. In: Vedaldi, A., Bischof, H., Brox, T., Frahm, J.-M. (eds.) ECCV 2020. LNCS, vol. 12368, pp. 606–623. Springer, Cham (2020). https://doi.org/10.1007/978-3-030-58592-1_36
110. Zhao, L., Ichise, R., et al., T.Y.: Ontology-based decision making on uncontrolled intersections and narrow roads. In: 2015 IEEE Intelligent Vehicles Symposium (IV) (2015)
111. Zhao, L., Ichise, R., Liu, Z., Mita, S., Sasaki, Y.: Ontology-based driving decision making: a feasibility study at uncontrolled intersections. In: IEICE (2017)
112. Zhao, L., Ichise, R., Mita, S., Sasaki, Y.: Core ontologies for safe autonomous driving. In: ISWC (2015)
113. Zhu, H., Yuen, K., Mihaylova, L., Leung, H.: Overview of environment perception for intelligent vehicles. In: T-ITS (2017)
114. Zipfl, M., et al.: Relation-based motion prediction using traffic scene graphs. In: IEEE ITSC (2022)

Physicians' Brain Digital Twin: Holistic Clinical & Biomedical Knowledge Graphs for Patient Safety and Value-Based Care to Prevent the Post-pandemic Healthcare Ecosystem Crisis

Asoke K. Talukder[1,2]([✉]), Erwin Selg[3], and Roland E. Haas[4]

[1] SRIT, 113/1B ITPL Road, Brookfield, Bangalore 560037, India
asoke.talukder@sritindia.com
[2] National Institute of Technology Karnataka, Surathkal, India
[3] SRH Fernhochschule GmbH, Kirchstraße 26, 88499 Riedlingen, Germany
erwin.selg@mobile-university.de
[4] IIITB, 26/C, Electronics City, Hosur Road, Bangalore 560100, India
roland.haas@iiitb.ac.in

Abstract. The 'reading to cognition gaps' and the 'knowledge to action gaps' for a physician or a care provider are the root causes of patient harm and the low-value healthcare. Rule-based symptom-checkers often fail when there are multiple co-occurring symptoms. To ensure patient safety and value-based care we have constructed nine AI-driven and evidence based interconnected holistic knowledge graphs covering the entire spectrum of medical knowledge starting from symptoms to therapeutics. These knowledge graphs are in fact the digital twin of all physicians' brains. These nine knowledge graphs are Symptomatomics, Diseasomics, SNOMED CT, Disease-Gene Network, Multimorbidity, Resistomics, Patholomics, Oncolomics, and Drugomics. These knowledge graphs are constructed from semantic integration of biomedical ontologies like Disease Ontology, Symptom Ontology, Gene Ontology, Drug Ontology, NCI Thesaurus, DisGenomics Network, PharmGKB, ChEBI, WHO AWaRe, and WHOCC. This is further enhanced through thematic integration of the knowledge mined from PubMed, DailyMed, FAERS, Wikipedia and patient data (EHR) from hospitals and cancer registry. These knowledge graphs are interconnected through common vocabularies like SNOMED CT, ICD10, ICDO, UMLS, NCIT, DOID, HGNC, GO, LOINC, ATC, RXCUI, and RxNORM codes that helped us to construct a complete clinical, medical, therapeutic, and conflicting medication knowledge graph with 723,801 nodes and 10,657,694 edges. This knowledge graph is stored in a Neo4j property graph database which is deployed in the cloud accessible 24×7 through REST/JSON-RPC and AIoT API. On top of this integrated knowledge graph we used node2vec to construct digital triplet discovering many unknown and hidden knowledge. This integrated clinical & biomedical knowledge functions as the digital twin of all physicians' brains.

Keywords: Physicians' brain digital twin · Digital triplet · Symptomatomics · Diseasomics · Resistomics · Patholomics · DisGenomics · Oncolomics · Drugomics · Knowledge graph · Healthcare ecosystem crisis · Patient safety

1 Introduction

To offer the best possible care, each care provider be it a nurse, paramedic, a primary care doctor, or a specialist clinician – is expected to know every piece of latest and accurate holistic clinical, medical, and therapeutic knowledge, which is humanly impossible. To avoid even the smallest risk of missing a serious condition, the clinician with his or her incomplete knowledge orders multiple medical services or even procedures. These unnecessary services often cause patient harm that results into a cascade of further care and medical services. This type of care is known as low-value care. Low-value care is defined as "services that provide little or no benefit to patients, have potential to cause harm, incur unnecessary cost to patients, or waste limited healthcare resources" [1]. In 2006, the Institute of Medicine (IOM) Evidence-Based Medicine Roundtable experts noted that "care that is important is often not delivered. Care that is delivered is often not important" [2]. We do not have quantifiable numbers for the burden of low-value medical care in India or globally. In the US however, the low-value care contributes to over $345 billion annually in wasteful health spending [1].

The root cause of the low-value care are the knowledge gaps. These knowledge gaps occur in two distinct domains, namely, 'reading to cognition gaps' and 'knowledge to action gaps' [3]. Reading to cognition gap is the gap during the knowledge acquisition – the gap between the true knowledge and the ingested knowledge. Whereas knowledge to action gap is the gap during the decision making or problem solving. Both these gaps result into erroneous outcomes or low-value care.

Healthcare professionals are humans with uneven knowledge levels; therefore, they make mistakes and errors. Globally medical error is one of the leading causes of death. In 2000, IOM highlighted this through the publication of "To Err is Human: Building a Safer Health System" [4]. Medical errors are the third leading cause of death in the US after heart disease and cancer [5]. The preventable adverse drug reactions (ADR) or the conflicting medications are the fourth leading cause of death in the US [6]. Frost & Sullivan estimates that the cost burden of patient harm in 2022 is expected to increase at a CAGR (Compound Annual Growth rate) of 3.2% to reach USD $383.7 billion in USA and Western Europe [7]. The accelerating antibiotic abuse during the COVID-19 pandemic increased patient harm further [8]. A study found that COVID-19 likely contributed to 216.4 million excess doses of antibiotics and 38 million excess doses of azithromycin from June 2020 through September 2020, a four-month period of peak COVID-19 activity in India [9]. It is estimated that the burden of antibiotic resistance alone will be about $100 trillion by 2050 [10]. The only way patient harm can be eliminated is by reducing the 'reading to cognition gaps' and the 'knowledge to action gaps' through a learning healthcare system as suggested by IOM [2]. For a long time, rule-based symptom-checkers were used for triaging and patient stratification. Such symptom-checkers often fail in multiple co-occurring symptoms. Therefore, for the first time we constructed AI-driven evidence based machine understandable clinical, medical, and therapeutic knowledge. We made this holistic medical knowledge available to the caregiver at any point-of-care over the Internet and cloud.

A WHO 2021 report estimated that 71% of all deaths globally are caused by non-communicable diseases (NCD) [11]. During the COVID-19 pandemic focus shifted from NCD to infectious diseases with delays and avoidance in elective and critical NCD care – though the most vulnerable population during pandemic were people with NCD comorbidities – in fact diabetes NCD made up 40% of COVID-19 deaths in the US. [12]. Due to high cost, increasing patient harm [5–7], accelerating antibiotic abuse [8–10], increasing non-communicable disease burden [11], growing population of geriatric, aging, & chronic diseases [13], exponential growth in medical knowledge [14], 17 years of study to practice gap [15], increased antibiotic misuse [16], and high workforce turnover intention [17], the post-pandemic healthcare is heading for a crisis, which we call Healthcare Ecosystem Crisis (HEC).

To prevent the post-pandemic healthcare ecosystem crisis, we present in this paper a learning healthcare system. This learning healthcare system will reduce the 'reading to cognition gaps' and the 'knowledge to action gaps' through the democratization of biomedical knowledge. Our ultimate goal is to transform Healthcare Science from reductionism to the paradigm of holism through holistic clinical, medical, biological, and therapeutic knowledge in knowledge graphs. These knowledge graphs work like all physicians' brain digital twins. Our approach utilizes methods from Digital Twins [18], Digital Triplets [18], Neo4j Property Graph database, Strong AI, Symbolic AI, Ontologies, Machine Learning, Digital Health, NLP (Natural Language Processing), transformers, and Web Technologies. This latest validated clinical & medical knowledge will supplement the working knowledge used by the healthcare providers for making accurate clinical decision at the point-of-care. We did semantic integration of Disease Ontology, Symptom Ontology, Gene Ontology, Drug Ontology, NCI Thesaurus, DisGenomics Network, PubMed, PharmGKB, WHO AWaRe, WHOCC, FAERS, DailyMed, and Wikipedia to create the Physician Digital Twin. We constructed the patient spatial digital twin from statistically significant knowledge from Emergency Room, and cross-sectional EHR (Electronic Health Record) data. We created the patient temporal digital twin or disease trajectories from longitudinal EHR data [19], and cancer registry data [20]. From these digital twins we constructed nine knowledge graphs namely, Symptomatomics, Diseasomics [21], SNOMED CT [22], DisGenomics [23], Multimorbidity [24], Resistomics [25], Patholomics [24], Oncolomics [18], and Drugomics and stored in the Neo4j property graph [26].

This paper is organized as follows: Sect. 1 is the introduction. In Sect. 2 we describe the nine biomedical knowledge graphs that we constructed. In Sect. 3 we explain how these nine biomedical knowledge graphs are integrated to provide a holistic and complete biomedical knowledge starting from symptoms to therapeutic drugs. In Sect. 4 we discuss the validation techniques we used in the knowledge graph. In Sect. 5 we describe the architecture of the system and a deployment use case that delivers this integrated holistic knowledge at the point-of-care through smartphone and Web to ensure patient safety and value-based care. We conclude the paper in Sect. 6.

2 Healthcare Knowledge Graphs

We have constructed nine knowledge graphs namely, Symptomatomics, Diseasomics, SNOMED, DisGenomics, Multimorbidity, Resistomics, Patholomics, Oncolomics, and Drugomics as described below. These nine knowledge graphs are interconnected through common controlled vocabularies and machine interpretable normalized codes.

2.1 Symptomatomics

Symptom is a generic term used to describe signs, symptoms, and findings of a medical condition exhibited by a patient as indications of illness. Symptoms are subjective and can be perceived only by the person affected. Signs are objective findings that can be seen or measured by anybody. A finding is the result of a clinical investigation. A clinical finding represents the patient diagnosis and the signs and symptoms presented by the patient. We used symptom ontology, symptomatology textbooks, and manual curation to identify known signs, and symptoms. We took UMLS and SNOMED raw files to extract the findings. All these have been combined to construct the symptom nodes. We then looked at PubMed, Wikipedia, textbooks, and manual curation to construct the symptom disease relationships.

2.2 Diseasomics

Diseasomics supported by Symptomatomics is at the center of our nine knowledge graphs. There are many rule-based symptom-checkers available today. Rule based symptom-checkers have many limitations. These systems fail for uncommon symptom combinations outside of the defined rule. Diseasomics overcomes this limitation through knowledge based AI driven differential diagnosis (DDx) constructed from best of the best knowledge bodies. Diseasomics accepts plain text signs, symptoms, and findings to arrive at a disease.

Diseasomics includes diseases, genes, pharmacogenomics and their associations [21]. A patient's journey starts from some signs or symptoms of an illness. Physicians establish diagnosis by assessing a patient's signs, symptoms, findings, age, sex, disease history, and physical examination. All these tasks must be completed in limited time and against the backdrop of an increasing overall workload. A 2017 study found that the patient-physician interaction time in primary care varies from country to country. The study found that it ranges on average from 48 s in Bangladesh to 22.5 min in Sweden [27]. This has worsened in the post-pandemic healthcare. In the era of evidence-based medicine it is of utmost importance for a clinician to be abreast of the latest guidelines and treatment protocols which are changing rapidly. In resource limited settings, the updated knowledge often does not reach the point-of-care.

In diseasomics, we integrated different disease-related knowledge bodies to construct a comprehensive, machine interpretable knowledge-graph. The resulting disease-symptom network comprises knowledge from the Symptom Ontology, Disease Ontology, Wikipedia, PubMed, textbooks, and various symptomatology knowledge sources. We use node2vec [28] as digital triplet for link prediction in disease-symptom networks to identify missing associations. Figure 1 shows the diseasomics results for symptoms of SOB, shortness of breath, breathlessness, fever, coughing blood, and hemoptysis.

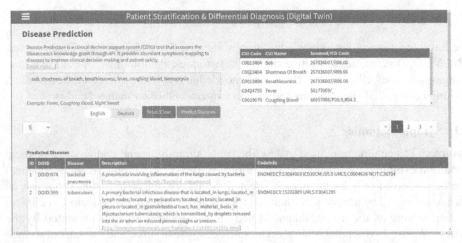

Fig. 1. The bilingual (English & German) Differential Diagnosis (DDx) module accesses the diseasomics knowledge graph through an API. In DDx human understandable signs, symptoms, findings are entered in the upper part of the DDx user interface (Chief Complaints). The Natural language processing (NLP) AI module converts patient symptoms into UMLS vocabularies. It can be seen that "sob, shortness of breath, breathlessness" all have been normalized into UMLS code "C0013404", SNOMED code "267036007" and ICD10 code "R06.00". These normalized vocabularies are used to fetch all diseases in the knowledge graph where all these conditions co-occur. The differential diagnosis is shown in the "Predicted Diseases" part of the screen. The provenance and details are provided through Web links. Diseases are converted into vocabularies like UMLS, SNOMED, ICD10, NCIT etc. for automated documentation and AI ready health data.

2.3 SNOMED CT Knowledge Graph

SNOMED CT or SNOMED Clinical Terms is a systematically organized computer-processable collection of medical terms. SNOMED CT is considered to be the most comprehensive, multilingual clinical healthcare terminology in the world. SNOMED CT includes clinical findings, symptoms, diagnoses, procedures, body structures, organisms and other etiologies, substances, pharmaceuticals, devices and specimens. We used the 2020 SNOMED CT and loaded it in the Neo4j database using the technique described by Campbell et al. in [22]. The diseasomics is semantically integrated with the SNOMED knowledge graph through the xref (cross reference) SNOMED codes.

2.4 Disease-Gene Network

Disease-Gene knowledge graph (DisGenomics) is a collection of genes and variants associated to human diseases. DisGenomics integrates data from expert curated repositories, GWAS catalogues, animal models and the scientific literature from DisGeNET [23]. DisGeNET contains associations between 21,671 genes and 30,170 diseases, disorders, traits, and clinical or abnormal human phenotypes. In DisGenomics, genes are represented as gene symbols whereas diseases are encoded in UMLS CUI. We converted the gene symbols into HGNC (HUGO Gene Nomenclature Committee) [29]. We connected the diseasomics with DisGenomics through UMLS cross references. Figure 2 shows the

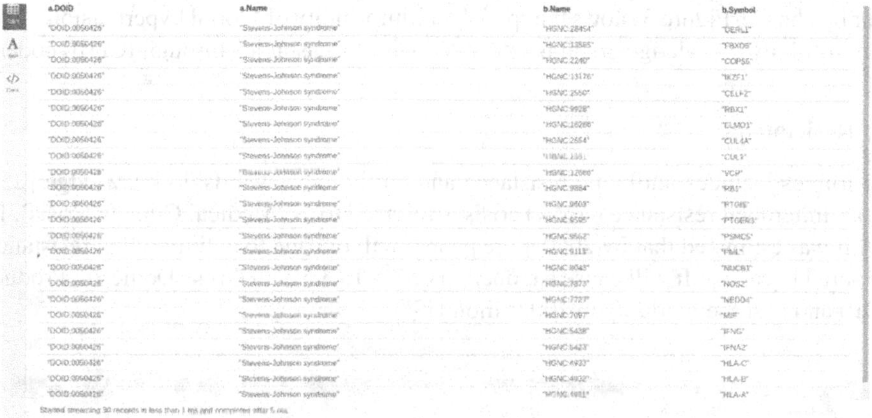

a.DOID	a.Name	b.Name	b.Symbol
"DOID:0050426"	"Stevens-Johnson syndrome"	"HGNC:28454"	"DERL1"
"DOID:0050426"	"Stevens-Johnson syndrome"	"HGNC:13585"	"FBXO6"
"DOID:0050426"	"Stevens-Johnson syndrome"	"HGNC:2240"	"COPS5"
"DOID:0050426"	"Stevens-Johnson syndrome"	"HGNC:13176"	"IKZF1"
"DOID:0050426"	"Stevens-Johnson syndrome"	"HGNC:2550"	"CELF2"
"DOID:0050426"	"Stevens-Johnson syndrome"	"HGNC:9928"	"RBX1"
"DOID:0050426"	"Stevens-Johnson syndrome"	"HGNC:16288"	"ELMO1"
"DOID:0050426"	"Stevens-Johnson syndrome"	"HGNC:2554"	"CUL4A"
"DOID:0050426"	"Stevens-Johnson syndrome"	"HGNC:2551"	"CUL1"
"DOID:0050426"	"Stevens-Johnson syndrome"	"HGNC:12666"	"VCP"
"DOID:0050426"	"Stevens-Johnson syndrome"	"HGNC:9884"	"RB1"
"DOID:0050426"	"Stevens-Johnson syndrome"	"HGNC:9603"	"PTGIS"
"DOID:0050426"	"Stevens-Johnson syndrome"	"HGNC:9595"	"PTGER3"
"DOID:0050426"	"Stevens-Johnson syndrome"	"HGNC:9562"	"PSMC5"
"DOID:0050426"	"Stevens-Johnson syndrome"	"HGNC:9113"	"PML"
"DOID:0050426"	"Stevens-Johnson syndrome"	"HGNC:8043"	"NUCB1"
"DOID:0050426"	"Stevens-Johnson syndrome"	"HGNC:7873"	"NOS2"
"DOID:0050426"	"Stevens-Johnson syndrome"	"HGNC:7727"	"NEDD4"
"DOID:0050426"	"Stevens-Johnson syndrome"	"HGNC:7097"	"MIF"
"DOID:0050426"	"Stevens-Johnson syndrome"	"HGNC:5438"	"IFNG"
"DOID:0050426"	"Stevens-Johnson syndrome"	"HGNC:5423"	"IFNA2"
"DOID:0050426"	"Stevens-Johnson syndrome"	"HGNC:4933"	"HLA-C"
"DOID:0050426"	"Stevens-Johnson syndrome"	"HGNC:4932"	"HLA-B"
"DOID:0050426"	"Stevens-Johnson syndrome"	"HGNC:4931"	"HLA-A"

Started streaming 30 records in less than 1 ms and completed after 5 ms.

Fig. 2. The Disease-Gene association for 'Stevens Johnson syndrome' (DOID:0050426). Third and fourth column of this screenshot are the HGNC and Gene symbols.

disease gene association between 'Stevens Johnson syndrome' (DOID:0050426) and associated genes.

2.5 Multimorbidity

The Multimorbidity knowledge graph is a combination of two subgraphs – spatial multimorbidity and temporal multimorbidity. Spatial multimorbidity is defined as the co-occurring diseases of a patient at 'a point in time'. Temporal multimorbidity is the set of diseases that occur in a patient's body 'over a period of time'. Temporal multimorbidity is in fact a disease trajectory of spatial multimorbidity of a patient. Spatial multimorbidity is derived from cross-sectional data, whereas temporal multimorbidity is derived from longitudinal patient data. We used statistical techniques to remove all associations that

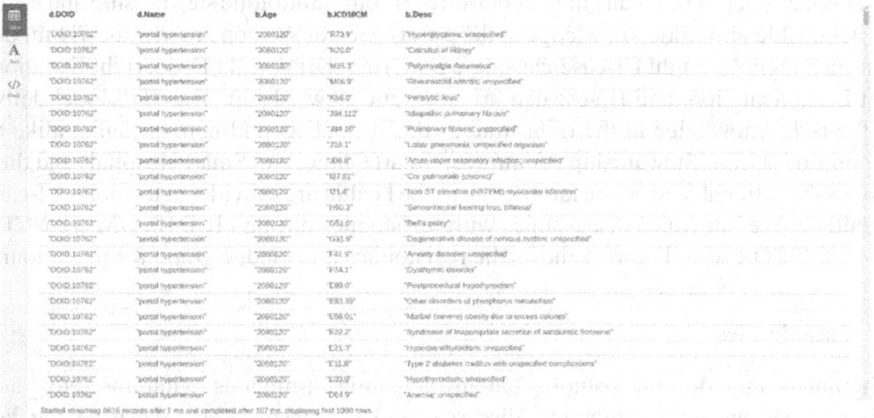

d.DOID	d.Name	b.Age	b.ICD10CM	b.Desc
"DOID:10762"	"portal hypertension"	"2080120"	"R73.9"	"Hyperglycemia, unspecified"
"DOID:10762"	"portal hypertension"	"2080120"	"N20.0"	"Calculus of kidney"
"DOID:10762"	"portal hypertension"	"2080120"	"M35.3"	"Polymyalgia rheumatica"
"DOID:10762"	"portal hypertension"	"2080120"	"M06.9"	"Rheumatoid arthritis, unspecified"
"DOID:10762"	"portal hypertension"	"2080120"	"K56.0"	"Paralytic ileus"
"DOID:10762"	"portal hypertension"	"2080120"	"J84.112"	"Idiopathic pulmonary fibrosis"
"DOID:10762"	"portal hypertension"	"2080120"	"J84.10"	"Pulmonary fibrosis, unspecified"
"DOID:10762"	"portal hypertension"	"2080120"	"J18.1"	"Lobar pneumonia, unspecified organism"
"DOID:10762"	"portal hypertension"	"2080120"	"J06.9"	"Acute upper respiratory infection, unspecified"
"DOID:10762"	"portal hypertension"	"2080120"	"I27.81"	"Cor pulmonale (chronic)"
"DOID:10762"	"portal hypertension"	"2080120"	"I21.4"	"Non-ST elevation (NSTEMI) myocardial infarction"
"DOID:10762"	"portal hypertension"	"2080120"	"H90.3"	"Sensorineural hearing loss, bilateral"
"DOID:10762"	"portal hypertension"	"2080120"	"G51.0"	"Bell's palsy"
"DOID:10762"	"portal hypertension"	"2080120"	"G31.9"	"Degenerative disease of nervous system, unspecified"
"DOID:10762"	"portal hypertension"	"2080120"	"F41.9"	"Anxiety disorder, unspecified"
"DOID:10762"	"portal hypertension"	"2080120"	"F34.1"	"Dysthymic disorder"
"DOID:10762"	"portal hypertension"	"2080120"	"E89.0"	"Postprocedural hypothyroidism"
"DOID:10762"	"portal hypertension"	"2080120"	"E83.39"	"Other disorders of phosphorus metabolism"
"DOID:10762"	"portal hypertension"	"2080120"	"E66.01"	"Morbid (severe) obesity due to excess calories"
"DOID:10762"	"portal hypertension"	"2080120"	"E22.2"	"Syndrome of inappropriate secretion of antidiuretic hormone"
"DOID:10762"	"portal hypertension"	"2080120"	"E21.3"	"Hyperparathyroidism, unspecified"
"DOID:10762"	"portal hypertension"	"2080120"	"E11.8"	"Type 2 diabetes mellitus with unspecified complications"
"DOID:10762"	"portal hypertension"	"2080120"	"E03.9"	"Hypothyroidism, unspecified"
"DOID:10762"	"portal hypertension"	"2080120"	"D64.9"	"Anemia, unspecified"

Started streaming 8616 records after 1 ms and completed after 107 ms, displaying first 1000 rows.

Fig. 3. The special multimorbidity for 'portal hypertension' (DOID:10762). Here the third column shows the ICD10 code of the spatial multimorbidity diseases with their description in column 5.

occur by chance. Figure 3 shows the spatial multimorbidity of 'portal hypertension'. The multimorbidity knowledge graph is integrated with diseasomics through ICD10 codes.

2.6 Resistomics

Resistomics includes antibiotic resistance and antibiotic stewardship knowledge [25]. The antimicrobial resistance (AMR) crisis is referred to as 'Medical Climate Crisis'. In 2014 it was estimated that by 2050 more people will die due to antimicrobial resistance compared to cancer. It will cause a reduction of 2% to 3.5% in Gross Domestic Product (GDP) and cost the world up to 100 trillion USD.

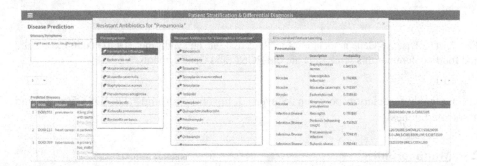

Fig. 4. The resistomics integration with the diseasomics knowledge graph. Pneumonia is shown as one of the differential diagnoses for co-occurring symptoms fever, night sweat, and coughing blood. When we click pneumonia in the Disease column a resistomics screen pops up as shown here. First column in the pop-up resistomics screen shows all disease causing bacteria for bacterial pneumonia. Second column shows the antibiotics that are not effective to treat an infection with the selected bacteria, Escherichia coli in this case, because the bacteria have become resistant. The third column is the digital triplet of pneumonia. The Digital Triplet is constructed using node2vec.

This man-made crisis can only be controlled through antibiotic stewardship and accurate actionable antibiotic knowledge at the point-of-care. We constructed the 7D model of (right Diagnosis, right Disease-causing-agent, right Drug, right Dose, right Duration, right Documentation, and De-escalation) antibiotic stewardship. The 7D Model delivers the right knowledge at the right time to the specialists and non-specialist alike at the point-of-action (Stewardship committee, Smart Clinic, and Smart Hospital) and then delivers the actionable accurate knowledge to the healthcare provider at the point-of-care in realtime. We integrated diseasomics with resistomics through ICD10, UMLS, NCIT, and SNOMED codes. Figure 4 shows the resistomics knowledge graph for pneumonia.

2.7 Patholomics

Patholomics includes the pathological investigations (minerals, vitamins, etc.) and their associations with symptoms, diseases, and neoplasms [24]. We used the latest MEDLINE/PubMed 2022 Baseline data. This release contains 1114 archives of total 33,405,863 PubMed articles. We extracted 3,662,840 relevant article abstracts to determine the association between a vitamin or mineral and the conditions like signs, symptoms, findings, diseases, syndromes, or neoplasms.

L.Test	b.Desc	b.Age	type(r)
"PHOSPHORUS INORGANIC:"	"End stage renal disease"	"2080120"	"hyper"
"PHOSPHORUS INORGANIC:"	"End stage renal disease"	"2070079"	"hyper"
"PHOSPHORUS INORGANIC:"	"End stage renal disease"	"2060069"	"hyper"
"PHOSPHORUS INORGANIC:"	"End stage renal disease"	"2050059"	"hyper"
"PHOSPHORUS INORGANIC:"	"End stage renal disease"	"2040049"	"hyper"
"PHOSPHORUS INORGANIC:"	"End stage renal disease"	"2030039"	"hyper"
"PHOSPHORUS INORGANIC:"	"End stage renal disease"	"1080120"	"hyper"
"PHOSPHORUS INORGANIC:"	"End stage renal disease"	"1070079"	"hyper"
"PHOSPHORUS INORGANIC:"	"End stage renal disease"	"1060069"	"hyper"
"PHOSPHORUS INORGANIC:"	"End stage renal disease"	"1050059"	"hyper"
"PHOSPHORUS INORGANIC:"	"End stage renal disease"	"1040049"	"hyper"
"PHOSPHORUS INORGANIC:"	"End stage renal disease"	"1030039"	"hyper"
"PHOSPHORUS INORGANIC:"	"Mild intermittent asthma with (acute) exacerbation"	"2040049"	"hypo"
"PHOSPHORUS INORGANIC:"	"Mild intermittent asthma with (acute) exacerbation"	"2030039"	"hypo"
"PHOSPHORUS INORGANIC:"	"Mild intermittent asthma with (acute) exacerbation"	"1050059"	"hypo"
"PHOSPHORUS INORGANIC:"	"Mild intermittent asthma with (acute) exacerbation"	"1040049"	"hypo"
"PHOSPHORUS INORGANIC:"	"Mild intermittent asthma with (acute) exacerbation"	"1030039"	"hypo"
"PHOSPHORUS INORGANIC:"	"Mild intermittent asthma with (acute) exacerbation"	"1020029"	"hypo"
"PHOSPHORUS INORGANIC:"	"Mild intermittent asthma with (acute) exacerbation"	"1000009"	"hypo"

Started streaming 21 records after 1 ms and completed after 2 ms.

Fig. 5. Semantic integration of patholomics with diseasomics. In this figure we show the pathology test with the disease description and the statistically significant age group. Here we show the hypophosphatemia and hyperphosphatemia for different age groups with associated diseases. The Age column contains 7 digits that indites the sex and age group of patients. First digit 1 indicates male and 2 indicates female with last 6 digits being the age group. Column 2 indicates the disease association whereas column four indicates whether the phosphorus is hypo (less than normal) and hyper (above normal range)

The vitamin/mineral investigation has been converted into controlled vocabulary LOINC codes. The conditions are converted into UMLS CUI codes. Figure 5 shows the patholomics result for Neo4j Cypher command in the Neo4j browser for 'inorganic phosphorus'. The patholomics is connected with other knowledge graphs through common UMLS, SNOMED, and ICD10 codes.

2.8 Oncolomics

Oncolomics includes cancer related knowledge along with association of genes, mutations, regimens [18]. To address the complex challenges in cancer care we integrated two digital twins namely, (A) Digital twin of oncologists' brain, and (B) Digital twin of the patients' physical state. The oncologist's twin is realized through the semantic integration of (1) NCI Thesaurus (NCIt), (2) Gene Ontology (GO), and (3) Disease-Gene Network (DisGeNET). In Oncolomics we used the digital triplet to discover missing links (link prediction) or the unknown knowledge in NCIt and enhance the accuracy of the knowledge graph. Figure 6 shows the cancer trajectory and NCI Thesaurus interactions.

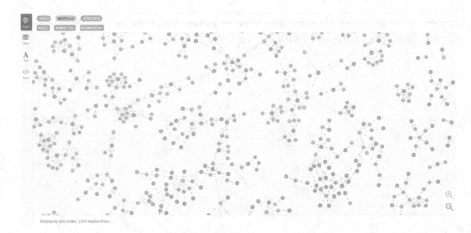

Fig. 6. The Oncolomics knowledge in the Neo4j browser for NCIt codes with ICD-O (ICD-O) codes from SEER cancer registry. We have taken all patients that had minimum 4 years of survivability with p-value 0.05. The ICD-O codes are the oncology related enhancement of ICD (The International Classification of Diseases). ICD for Oncology, third edition (ICD-O-3), is designed to categorize tumors. Unlike other diseases, ICD for Oncology code is used primarily in tumor or cancer registries for coding the site (topography) and the histology (morphology) of neoplasms, usually obtained from a pathology report.

2.9 Drugomics

Drugomics includes the knowledge about drugs [30] and their indications, contraindications, and conflicting medication or drug adverse reactions (DAR) [31]. Drug indication is the knowledge about the right use of a drug. Contraindication is about the side effect

Fig. 7. The doctor prescribed the diabetes drug 'metformin' (RXCUI:6809). The pop-up window shows the drug-drug interaction of metformin in the left side. The drug-disease interaction of metformin is shown on the right side.

or other conditions caused by the inappropriate use of a drug or the conflict between drugs. Chronic patients take multiple drugs for various conditions. These drugs often interact with each other causing adverse effects or undesirable conditions. The DAR knowledge tells us the reactions the prescribed drug many have with other drugs. In the knowledge base we also indicate the alternate drug that are unlikely to react. Figure 7 is how a clinician, or a surgeon interacts with drugomics.

3 Integration of Knowledge Graphs

Knowledge is the outcome of learning. In the human brain learning causes new connections among neurons. As the knowledge grows, the network connections among neurons grow and become a dense network. In similar fashion, as the connections (relationships) between nodes in the knowledge graph increase, the knowledge increases. This helps discover hidden knowledge.

We have presented nine holistic medical knowledge graphs. These knowledge graphs use controlled vocabularies like SNOMED CT, ICD10, ICDO, UMLS, NCIT, DOID, HGNC, GO, LOINC, ATC, RXCUI, and RxNORM. In antireductionistic fashion we integrate these nine knowledge graphs. These nine knowledge graphs are semantically and thematically connected through common vocabularies and helped us to construct a complete Clinical & Medical Knowledge Graph with 723,801 nodes and 10,657,694 edges. For example, Symptomatomics and Diseasomics are connected through UMLS codes. Diseasomics and SNOMED are connected through SNOMED codes. Diseasomics and DisGenomics are connected using DOID and HGNC codes. Diseasomics and Multimorbidity are connected using ICD10 codes. This holistic knowledge graphs are stored in a Neo4j property graph database which is deployed in the cloud accessible 24×7 through REST/JSON-RPC API.

We also used node2vec for digital triplet [28] to discover unknown knowledge. Digital triplet reduces the dimensionality of the data without changing the structure of the data. Reduced dimension helps us to discover both hidden and unknown knowledge. Utilizing a Neo4j property graph database, we can service any type of quarry or traversals instantly.

4 Validation of the Integrated Knowledge Graphs

The accuracy of the knowledge graph was validated using multiple methodologies.

4.1 Ontology Validation

Ontologies are the machine representation of knowledge. We have used many ontologies like Symptom Ontology, Disease Ontology, Gene Ontology, Gene-Disease network, Pharmacogenomics, SNOMED, NCI Thesaurus etc. These Ontologies are manually curated and updated by domain experts primarily for researchers. We have used this expert knowledge in our holistic medical knowledge graphs. These ontologies are periodically reviewed and revised. However, when these ontologies are semantically or thematically integrated with other knowledge, there might be missing links. We used digital triplet node2vec to learn the features in the holistic knowledge graphs [28]. This helped us to discover missing links between nodes at the interconnections. We validated these predicted unknown links through literature and added the links through whitelists.

4.2 Diseasomics Validation

Symptomatomics and Diseasomics knowledge graphs are at the center of our holistic medical knowledge graph. Therefore we validated diseasomics through random selection of diseases and symptom combination. We randomly selected a combination of symptoms (two symptoms and three symptoms) to fetch the associated disease from the knowledge graph. These results were validated by expert clinicians. We defined 'Positive' cases as these combinations of symptoms that fetched some knowledge from the knowledge graph that may be either correct or wrong. We also defined 'Negative' cases as these cases that did not fetch any result from the knowledge base, this in other words means that there is no disease association in the knowledge graph with this symptom combination. The negative association may be either right or wrong. Table 1 summarizes the results.

Table 1. Confusion Matrix

	Positive	Negative
Positive	805 (TP)	85 (FP)
Negative	64 (FN)	11 (TN)

We computed the F1 score to measure the model performance as following.
F1 score = 2*TP /(2*TP + FP + FN) = 0.9153.

4.3 Multimorbidity Validation

We used the EHR data for knowledge extraction from patient data from hospitals and registries. These data have biases towards geography, ethnicity, age, sex etc., We used big-data and mathematical models to eliminate these biases. We used statistical techniques and used p-value 0.05 to remove all data points that may be due to chance.

5 Knowledge at the Point-of-Care

The actionable biomedical knowledge presented in this paper is deployed in a cloud accessible through REST/JSON-RPC API. It is currently in the process of integration with the 104/108 Helpline Service in one of the states in Eastern India. The health helpline number 104 is aimed at providing free medical assistance to the needy suffering from various illnesses, whereas 108 is the ambulatory service. The knowledge graph presented here will be used for triaging and will be used as guide and second opinion.

Figure 8 shows the architecture of the entire system. The data between the client and the knowledge graph is managed through JSON data presentations. The client application uses the MERN (MongoDB, Express, React, Node) stack. The smartphone client however uses Flutter such that the knowledge is accessible through both Android and iPhone. We deliver this integrated democratized clinical and medical knowledge to a non-MD health worker empowering them to function similarly to an MD (Doctor of Medicine) specialists anywhere at any point-of-care through AIoT [32] and WebRTC. Figure 8 shows the architecture of complete system that will eliminate the 'reading to cognition' and 'knowledge to action' gaps. This architecture and the knowledge graphs will help to counter the healthcare ecosystem crisis.

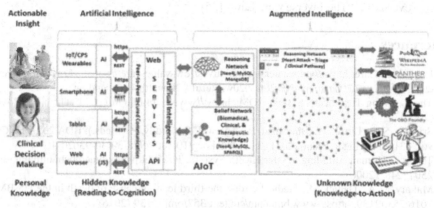

Fig. 8. The architecture of making the knowledge graphs available at the point-of-care.

6 Conclusion

The domain of clinical decision support (CDS) systems is fragmented. One CDS system does one specific function like symptom checking, drug dosing, antibiotic resistance, drug interactions etc. These techniques follow the reductionism mechanism. To overcome this limitation, we have constructed nine clinical, biomedical, and therapeutic knowledge graphs. These individual knowledge graphs are constructed from the latest and validated vast knowledge from various sources that include ontologies, literature, and patient data. The final multilingual holistic knowledge graph contains both unstructured plain text annotations for human consumption and machine understandable controlled vocabularies like SNOMED CT, ICD10, ICDO, UMLS, NCIT, DOID, HGNC, GO, LOINC, RxNORM, ATC, etc. We plan to update these knowledge bases twice a year synchronizing with the National Library of Medicine UMLS Metathesaurus release [33].

These knowledge graphs have been semantically and thematically integrated into one comprehensive antireductionistic medical knowledge graph. The antireductionism combined with digital triplets induce the 'common sense' into our physicians' brain digital twin. The commonsense knowledge is the prerequisite for an explainable AI (XAI). Through antireductionism we achieved Systems Biomedicine as well.

The knowledge-graph based systems are currently undergoing trials in India and in Germany. A German-version of the NLP engine is developed that accepts symptoms in German language and delivers the differential diagnosis in German language. We aim to get EMA and FDA approval for deployment as comprehensive clinical decision support systems that covers the entire lifecycle of a patient's journey.

Our continuous learning approach of integrating the latest and the best of the best knowledge sources will make complete interoperable machine interpretable medical knowledge available at the fingertip of expert or non-expert health workers alike. This will reduce medical errors, reduce cost of healthcare, and eliminate the power hierarchy [34]. It will offer timely, high-quality, universal value-based health care with limited resources. It will increase the patient safety in post-pandemic healthcare and ultimately help to reverse the "The Inverse Care Law" [35].

References

1. Low Value Care. https://vbidcenter.org/initiatives/low-value-care/#:~:text=Low%2Dvalue% 20care%20can%20be,annually%20in%20wasteful%20health%20spending
2. The Learning Healthcare System. IOM. https://www.ncbi.nlm.nih.gov/books/NBK53494/ (2006)
3. Srirama, S.N., Lin, J.-W., Bhatnagar, R., Agarwal, S., Reddy, P.K. (eds.): BDA 2021. LNCS, vol. 13147. Springer, Cham (2021). https://doi.org/10.1007/978-3-030-93620-4
4. To Err is Human: Building a Safer Health System. IOM. https://pubmed.ncbi.nlm.nih.gov/ 25077248/ (2000)
5. Makary, M.A., Daniel, M.: Medical error—the third leading cause of death in the US BMJ 2016;353:i2139. https://www.bmj.com/content/353/bmj.i2139 (2016)
6. Preventable Adverse Drug Reactions: A Focus on Drug Interactions. https://www.fda. gov/drugs/drug-interactions-labeling/preventable-adverse-drug-reactions-focus-drug-intera ctions
7. Patient Safety in Healthcare, Forecast to 2022. https://store.frost.com/patient-safety-in-hea lthcare-forecast-to-2022.html
8. Americas report surge in drug-resistant infections due to misuse of antimicrobials during pandemic. https://www.paho.org/en/news/17-11-2021-americas-report-surge-drug-resistant-infections-due-misuse-antimicrobials-during (2021)
9. COVID-19 aggravates antibiotic misuse in India. https://medicine.wustl.edu/news/covid-19-aggravates-antibiotic-misuse-in-india/#:~:text=After%20statistically%20adjusting%20for% 20seasonality,period%20of%20peak%20COVID%2D19 (2021)
10. O'Neill, J.: (Chair). Antimicrobial resistance: tackling a crisis for the health and wealth of nations. https://amr-review.org/sites/default/files/AMR%20Review%20Paper%20-%20Tack ling%20a%20crisis%20for%20the%20health%20and%20wealth%20of%20nations_1.pdf (2014)
11. World Health Organization, Noncommunicable Diseases (2021). https://www.who.int/news-room/fact-sheets/detail/noncommunicable-diseases
12. Diabetics make up 40% of COVID deaths in US, experts say: https://nypost.com/2021/07/ 16/diabetics-make-up-40-of-covid-deaths-in-us-experts-say/amp/
13. World's older population grows dramatically. https://www.nih.gov/news-events/news-rel eases/worlds-older-population-grows-dramatically
14. Medical knowledge doubles every few months; how can clinicians keep up?. https://www.els evier.com/connect/medical-knowledge-doubles-every-few-months-how-can-clinicians-kee p-up (2018)

15. The answer is 17 years, what is the question: understanding time lags in translational research. https://www.ncbi.nlm.nih.gov/pmc/articles/PMC3241518/ (2011)
16. Covid-19: Antimicrobial misuse in Americas sees drug resistant infections surge, says WHO. https://www.bmj.com/content/375/bmj.n2845 (2021)
17. Hou, H., et al.: Factors Associated with Turnover Intention Among Healthcare Workers During the Coronavirus Disease 2019 (COVID-19) Pandemic in China. https://www.dovepress.com/factors-associated-with-turnover-intention-among-healthcare-workers-du-peer-reviewed-fulltext-article-RMHP (2021)
18. Talukder, A.K., Haas, R.E.: Oncolomics: digital twins & digital triplets in cancer care. In: 7th Computational Approaches for Cancer Workshop (CAFCW21), Supercomputing Conference (SC21). https://sc21.supercomputing.org/presentation/?id=ws_cafcw103&sess=sess434 (2021)
19. Jensen, A., Moseley, P., Oprea, T., et al.: Temporal disease trajectories condensed from population-wide registry data covering 6.2 million patients. Nat Commun 5 (2014). https://doi.org/10.1038/ncomms5022
20. Surveillance, Epidemiology, and End Results https://seer.cancer.gov/data/
21. Talukder, A.K., Schriml, L., Ghosh, A., Biswas, R., Chakrabarti, P., Haas, R.E.: Diseasomics: actionable machine interpretable disease knowledge at the point-of-care. PLoS Digit. Health 1(10), e0000128 (2022). https://doi.org/10.1371/journal.pdig.0000128
22. Campbell, W.S., Pedersen, J., McClay, J.C., Rao, P., Bastola, D., Campbell, J.R.: An alternative database approach for management of SNOMED CT and improved patient data queries. J. Biomed. Inform. https://doi.org/10.1016/j.jbi.2015.08.016. https://www.sciencedirect.com/science/article/pii/S1532046415001847 (2015)
23. DisGeNET: https://www.disgenet.org/
24. Talukder, A.K., Sanz, J.B., Samajpati, J.: 'Precision health': balancing reactive care and proactive care through the evidence based knowledge graph constructed from real-world electronic health records, disease trajectories, diseasome, and patholome. Springer, Cham, LNCS 12581. https://www.springerprofessional.de/en/precision-health-balancing-reactive-care-and-proactive-care-thro/18718294 (2020)
25. Talukder, A.K., Chakrabarti, P., Chaudhuri, B.N., Sethi, T., Lodha, R., Haas, R.E.: 2AI&7D model of resistomics to counter the accelerating antibiotic resistance and the medical climate crisis. In: Big Data Analytics 2021. Springer, Cham, LNCS, volume 13147. https://www.springerprofessional.de/2ai-7d-model-of-resistomics-to-counter-the-accelerating-antibiot/19984472 (2021)
26. Neo4j Graph database: https://neo4j.com/
27. Irving, G., Neves, A.L., Dambha-Miller, H., et al.: International variations in primary care physician consultation time: a systematic review of 67 countries. BMJ Open 2017, 7: e017902. https://doi.org/10.1136/bmjopen-2017-017902. https://bmjopen.bmj.com/content/7/10/e017902 (2017)
28. Grover, A., Leskovec, J.: node2vec: scalable feature learning for networks. In: KDD '16: Proceedings of the 22nd ACM SIGKDD International Conference on Knowledge Discovery and Data Mining. August 2016, pp. 855–864. https://doi.org/10.1145/2939672.2939754 (2016)
29. HGNC: https://www.genenames.org/
30. Drugs@FDA: FDA-Approved Drugs. https://www.accessdata.fda.gov/scripts/cder/daf/index.cfm
31. FDA Adverse Event Reporting System (FAERS) Public Dashboard. https://www.fda.gov/drugs/questions-and-answers-fdas-adverse-event-reporting-system-faers/fda-adverse-event-reporting-system-faers-public-dashboard
32. Talukder, A.K., Haas, R E.: AIoT: AI meets IoT and Web in Smart Healthcare. (2021). https://dl.acm.org/doi/fullHtml/10.1145/3462741.3466650

33. Unified Medical Language System (UMLS): https://www.nlm.nih.gov/research/umls/knowle dge_sources/metathesaurus/index.html
34. Kim, S., et al.: Patient safety over power hierarchy: a scoping review of healthcare professionals' speaking-up skills training. https://pubmed.ncbi.nlm.nih.gov/32149868/ (2020)
35. 50 years of the inverse care law. The Lancet Editorial 2021. https://doi.org/10.1016/S0140-6736(21)00505-5. https://www.thelancet.com/journals/lancet/article/PIIS0140-6736(21)00505-5/fulltext

Combining Ontology and Natural Language Processing Methods for Prevention of Falls from Height

Sarra Ben Abbes[1(✉)], Lynda Temal[1(✉)], Guillaume Arbod[1],
Pierre-Luc Lanteri-Minet[2], and Philippe Calvez[1]

[1] CSAI Lab ENGIE, Paris, France
sarra.ben-abbes@external.engie.com, lynda.temal@engie.com
[2] GLOBAL CARE, ENGIE, Paris, France

Abstract. Safety on construction sites is the most important aspect that a company should guarantee to its employees. To reduce the risk, it is necessary to analyze HIgh POtential (HIPO) hazards. In this study, we focus on Fall From Height (FFH) risk which is one of the main causes of worker fatalities. In order to improve the prevention plan, artificial intelligence (AI) can help to determine the causes and the safety actions from FFH historical data. This paper aims to: (i) develop and populate an ontology for FFH with the help of domain experts, and (ii) analyze and extract key information from a HIPO database through Natural Language Processing (NLP). Experimental results are conducted in order to evaluate the proposed approach.

Keywords: High potential hazard · Fall from height · Ontology · Natural language processing · Risk · Prevention plan

1 Introduction

Safety is the most important aspect that a company should guarantee to its employees. To ensure this safety, companies provide employers and workers with safety training appropriate to the position they occupy in their work [12,14]. They also give them various safety guidelines and specifications to protect them from potential dangers on the construction site, such as the Unified Specification for Construction Safety Technology (GB50870-2013). Moreover, companies report all the situations related to occupational/near-miss accidents. Especially, companies in the construction sector are particularly affected by the Fall From Heights (FFH) hazard, which is considered as one of the leading causes of fatalities/injuries to employees. Eliminating or reducing drastically this hazard becomes the main priority of these companies. In addition to all the actions concerning safety such as safety training, and sensitization campaigns, some companies engage a lot of efforts to explore the historical data of previous accidents and hazardous situations. This information is generally exploited by classical statistics methods in order to generate some figures for the business reports. The information is currently not exploited efficiently. Artificial Intelligence (AI) methods can bring a significant enhancement in the safety management field, by turning the data on knowledge, allowing to exploit the data at different levels of abstraction. This topic motivated several researchers to eliminate hazard accidents by creating the missing link between safety management and ontology knowledge. Different domains on AI are explored: i) symbolic representation of domain knowledge is used. Thus we developed an ontology representing the knowledge that is implicitly hidden in the HIPO schema, and expressed by the domain expert, and ii) Symbolic and statistical Natural Language Processing (NLP) techniques are used to analyze and extract key information from free texts. A knowledge graph is created based on the HIPO data and ontology, and exploited with a powerful query language that allows extracting information at different levels of abstraction.

This paper is organized as follows: Sect. 2 gives the most common state-of-the-art approaches dedicated to improve the domain of safety management. Section 3 presents the methodology proposed and gives a detailed description of the different steps. Section 4 shows experimentation and highlights the results of our methodology. Finally, a conclusion and some future works are given in Sect. 5.

2   Related Work

Fall from height, the most important cause of mortality after road accidents, has prompted researchers to investigate ways to improve prevention and to reduce risks and hazardous situations.

State-of-the-art approaches have been proposed to demonstrate the importance of investigating around fall from height factors based on ontologies. Ontologies are complex artifacts (composed of concepts, hierarchical relations, and roles) that are built according to different points of view and purposes. The most widely cited definition of an ontology is a formal, explicit specification of a shared conceptualization, used to help humans and machines to share common knowledge [6,8]. This definition highlights the following characteristics of an ontology: (i) Conceptualization defines the objects, concepts, and other entities that are assumed to exist in some area of interest and the relationships that hold among them [5]. A conceptualization is an abstract, simplified view of the world that we wish to represent for some purpose [6], (ii) Explicit corresponds to the precise definition of the concepts and the constraints of their use, (iii) Formal refers to the fact that the expressions must be machine-readable, and (iv) Shared refers to a common understanding of domain knowledge among people or agents.

[16] proposed ontology-based semantic modeling to capture safety management knowledge. The ontology utilizes construction-based safety information (e.g. Occupational Safety and Health Administration (OSHA) regulation 1926, Occupational Injury and Illness Classification Manual) in order to support safety

hazard knowledge on construction sites. Based on this construction safety ontology, SWRL (Semantic Web Rule Language) rules are applied to represent OSHA regulations. [17] proposed in this work an evaluation of the structure and content of the construction safety ontology based on interviews and surveys with construction experts. This ontology is expected to provide benefits in safety planning tools (e.g. when the schedule of a project changes, the user can re-run the prototype and quickly receive updated results of the hazard analysis).

Similarly, [10] presented an ontology for construction safety checking named *CSCOntology* (Construction Safety Checking Ontology). This ontology is defined by five primary safety checking concepts, "Line of work", "Task", "Precursor", "Hazard" and "Solution". These concepts are linked by semantic relationships (e.g. the relation "hasTask" connecting the concept "Line of work" to the concept "Task"). This work also describes some knowledge rules such as the constraints which are involved in the safety checking process and it is represented by SWRL as a standard rule language.

[9] developed an ontology of control measures for fall from height, called *FFH-Onto*, in order to facilitate knowledge reuse and sharing in the construction industry. The proposed ontology consists of nine concepts i.e. actor, task, building element, hazard, construction method, constraint, safety resource, hazard control measure, and residual risk, and its relationships. These concepts are part of three ontological modules namely problem, context, and solution.

[18] applied an ontological model to represent hazards implied in construction images. A structure of HowNet [2] is used and is extended to cover the construction potential hazard analysis.

[3] combined an ontology of falls from height with computer vision in order to automate the identification of risks of hazards on a construction site. The authors used computer vision algorithms to extract knowledge from images and ontological knowledge to identify unsafe behaviors on several construction sites.

The main drawback of all these approaches is that the ontologies created are not always available. In addition, the state-of-the-art approaches used some standards of FFH to design their ontologies, but in the real world and with the real data, we have more and more information about, for example, the cause of the hazard, action to take, etc., which are not mentioned in public standards. It is important to start creating an ontology from standards and reusing existing ones, but real data can also widely enrich the designed ontology.

3 Proposed Methodology

As previously mentioned, the goal of this paper is to explore efficiently the information of FFH hazards stored in the HIPO database. For this purpose, we used two research fields in AI: on one side, ontology is used to represent the knowledge related to the FFH domain; on the other side, NLP is used to extract information about causes and prevention actions from the FFH description stored in free text format (see Fig. 1).

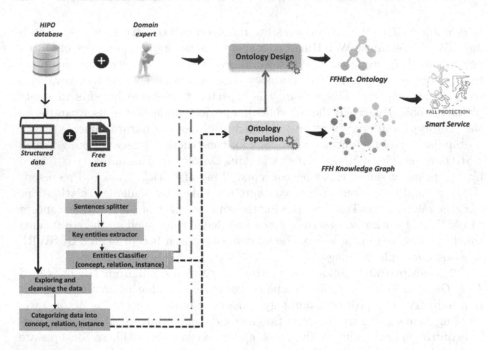

Fig. 1. Overview of our approach for creating the FFH knowledge graph from HIPO resources

3.1   Step 1: Extract of Key Entities and Relations

Our methodology takes as input the HIPO database that includes two data formats; structured and unstructured data (texts). In this section, we describe the process applied to explore/analyze, and classify this data.

The analyzing, the exploring, and cleaning phase, of the HIPO structured dataset leads to address a list of concepts and relations that are necessary to represent them. Among these concepts we can cite: the project, its location (Place and Country), the company responsible of the project, the agent involved in the project, and the company that employs him. This structured data is directly transformed into semantic data, thanks to SPARQL-Generate tool[1].

For the unstructured data, we describe the process that takes advantage of NLP methods for extracting entities and relationships from FFH free texts and integrates them in the ontology (Step 2) and in the knowledge graph (Step 3). We exploited the following tasks after splitting sentences:

- Key entities extractor: the aim of this task is to detect the most used words and verbs in order to select the important information from FFH textual descriptions. We exploited Stanford Core NLP[2] for analyzing entities and relations among them. Each sentence that contains at least a subject and a

---

[1] https://ci.mines-stetienne.fr/sparql-generate/.
[2] https://stanfordnlp.github.io/CoreNLP/.

verb, is analyzed. A parse tree of the input text is built based on a logistic regression classifier for words dependency (from governor to dependent nodes) and the semantic contexts. In our approach, we keep a set of entities and their relationships that are referred to the target domain. The sentences that are corresponding to these entities are also extracted.

- Entities Classifier: in this task, it might happen that different entities may assign the same concept with alternative forms. These entities can also represent specific or generic concepts that are less recognizable to the domain. The aim of this task is to process the extracted entities and their relationships by merging alternative labels, removing ambiguous entities, and detecting generic entities. Some punctuation (e.g. dots, apostrophes) and stop-words (e.g. pronouns) are deleted and replaced by some other entities if necessary (e.g. if "it" is referred to the FFH, so it will be replaced by this latter). Entities might be too generic to describe the FFH domain (e.g. environmental cause, behavior cause, preventive action). We also merged entities with the same meaning by using the lemmatizer form i.e. singular and plural forms. Synonym entities are also detected by using a lexical resource (e.g., WordNet[3]). All these entities that are described in the same concept will be specified as a set of alternatives of this concept. Exploring the dependency tree of the extracted entities, we found the important relation, represented as a predicate in the ontology, for each pair of entities. The knowledge expert might use this result to enrich the designed ontology and also the FFH knowledge graph (see example in Fig. 2).

3.2   Step 2: Ontology Design

Several methods exist to build an ontology: i) Top-down method, which is based mainly on the needs expressed by the experts, and description by possible scenarios and competency questions that an ontology should respond to. ii) Bottom-up method [15], which is based on the concrete data, from which an abstraction process is made, and iii) hybrid method, that mixes the top-down and the bottom-up in iterative manner [4]. In this work, we used the hybrid method to build the ontology. On one hand, the bottom-up method is used by exploiting the HIPO data stored on a Database (structured and free texts), and on another side, the top-down method, is used, by working together with the expert to identify the scope of the ontology and competency questions that are formulated in natural language.

To design the ontology, instead of creating a new ontology from scratch, we apply the best practice that advises reusing existing ontologies. Then, we reuse FFH-onto [9] that defines the core concepts of FFH domain, and extend it with new concepts that are not covered by FFH-onto. In the rest of this paper, we call the extended ontology *FFHExt*.[4]

---

[3] https://wordnet.princeton.edu/.

[4] https://sites.google.com/view/kgswc-2022-ia-ffh-resources/ontology.

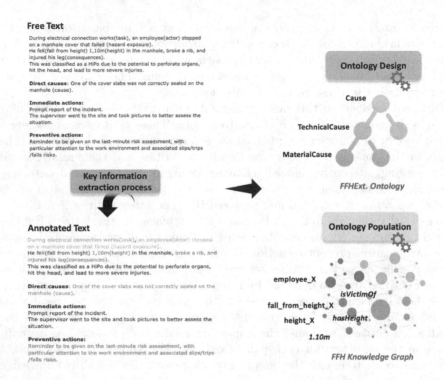

Fig. 2. Example of FFH key information extraction: From texts to Ontology and Knowledge Graph

In the following, we detail the elements to represent several notions extracted from the HIPO database in addition to entities (concepts/relations) and competency questions identified during the discussion with the domain expert. To design this missing entities, we reuse another well-known ontology called DOLCE+DnS Ultralite (DUL), which is a simplification and an improvement of some parts of foundational ontology DOLCE (Descriptive Ontology for Linguistic and Cognitive Engineering) Lite-Plus library[5] and Descriptions and Situations ontology[6].

Project Context. The project and its context enclose some information that is important to be represented such as i) the geographical location of the project, ii) the organization that is responsible for the project, and iii) the employees involved in the project.

As shown in the Fig. 3: *dul:Project dul:definesTask ffh:Task* allows to describe the link between the project and the different tasks defined in the project. A distinction is made between a task (*ffh:Task*) which is a description defined

---

[5] http://www.ontologydesignpatterns.org/ont/dul/DUL.owl.
[6] http://www.ontologydesignpatterns.org/ont/cdns/cDnS.owl.

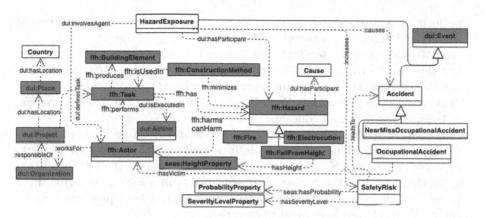

Fig. 3. Overview of the *FFHExt.* ontology

in a project, and the action (*dul:Action*) that represents the execution of the task (*ffh:Task dul:isExecutedIn dul:Action*). The action is executed by an agent (*dul:Action dul:involvesAgent ffh:Actor*). The agent works for an organization (*ffh:Actor worksFor dul:Organization*) which is responsible of the project (*responsibleOf dul:Project*). Both project and organization have location in some place (*dul:hasLocation dul:Place*).

Near-miss, Accident and Risk. It is important to report all hazardous situations, HIPO includes not only accidents that involve victims, but also near-misses that have no victims. The goal is to assess safety requirements and evaluate the level of safety risk increased by exposure to hazards. In order to represent this information, we introduce the following concepts and relations (see Fig. 4):

- *OccupationalAccident* is an *Accident* which refers to an unexpected event that causes injury or death to a person while in their place of work or while

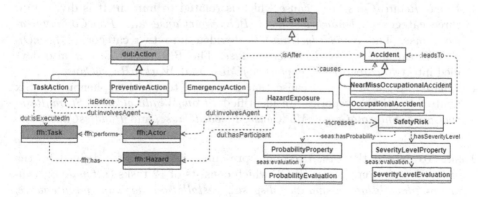

Fig. 4. Extract of the accident module

performing a work-related activity. Every occupational accident involves at least one victim (*OccupationalAccident hasVictim ffh:Actor*).

- *NearMissOccupationalAccident* is an *Accident* that has the potential to cause, but does not actually result in human injury. Every *NearMissOccupation-alAccident* does not involve any victim (*NearMissOccupationalAccident No (hasVictim some ffh:Actor)*) *NearMissOccupationalAccident* and *Occupation-alAccident* are disjoined concepts.
- *SafetyRisk* is a *Risk* which can result into (*leadsTo*) an accident (near-miss or occupational). This safety risk is increased by the exposition of work-ers to some hazards *HazardExposure*. Safety risk assesses the likelihood that given the Exposure, the projected consequences will occur. To describe this a probability property *ProbabilityProperty* and severity level properties *Sever-ityLevelProperty* are associated to each risk.
- *HazardExposure* is the event where some worker is exposed to some hazards that can cause (*Accident*). Every *HazardExposure* involves haz-ards (*dul:hasParticipant ffh:Hazard*) and involves agent (*dul:involvesAgent ffh:Actor*). A hazard (*ffh:Hazard*) is related to the actor (*ffh:Actor*) by two links:(i) harms (*ffh:harms*) that damages the health of the actor, and (ii) can harm (*:canHarm*) that relates the hazard to the potential actor.

Causes. In the real world, every hazardous situation is caused by various ele-ments. To manage these hazards, having better categorization of these causes is mandatory. Based on HIPO database, and the expert knowledge that defined some competency questions related to causes, the causes are divided in four disjoint categories (see Fig. 5):

1. *TechnicalCause* which is a cause that are related to technical issues such as material cause. *MaterialCause* is also divided in three categories : *Equip-mentCause, MachineCause* and *RevetementCause*.
2. *OrganizationalCause* is a cause which is related to organization issues. It is divided on threes categories : *InsufficientQualityOfRiskAnalysisCause, Poor-PreperationEquipmentCause* and *LackOfClearInstructionCause*
3. *HumanRelatedCause* is a cause which is related to human. It is divide into three categories: *KnowledgeCause, BehaviourCause* and *PoorCommunica-tionCause*. The *KnowledgeCause* is devided into two categories: *LackOf-SkillsCause* and *LackOfExperienceCause*. The *BehaviourCause* is also dev-ided into two categories *ComplacencyCause* and *WorkInHurryCause*
4. *EnvironmentalCause* is a cause that is related to weather phenomena. Six kind of weather categories are identified *StrongWindingCause, StrongRain-ingCause, SnowingCause, MudCause, FloodedCause,* and *PuddlesCause*.

Task. To extend the task taxonomy presented in FFH-onto, defined by the Industry Foundation Classes (IFC), which consists of 12 tasks (*attendance, con-struction, demolition, dismantle, disposal, installation, logistic, maintenance, move, operation, removal, and renovation*), we introduce -according to the

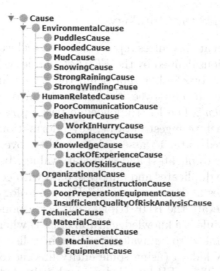

Fig. 5. Causes Taxonomy

domain expert needs- an intermediate level (see Fig. 6) to distinguish between *ProjectTask* that is defined in a very big project, such as *ConstructionTask* and *HeavyMaintenanceTask* and *OperatingTask* that is defined in less important project such as *InstallationTask*, *OperatingAndMaintenanceTask* and *Facility-ManagementTask*.

Fig. 6. Tasks taxonomy

Action. In the *FFHExt.* ontology, in order to represent information about the action that should be executed in different cases, we added a hierarchy of actions. Actually, three types of actions can be distinguished:

1. *TaskAction* is an action which executes a task (*dul:executesTask ffh:Task*) that is defined in a project (*ffh:Task dul:isDefinedIn dul:Project*)
2. *PreventiveAction* is an action which is executed before starting to execute the task action (*before TaskAction*).
3. *EmergencyAction* is an action that is executed immediately after an accident is happen (*isImmediatelyAfter some (OccupationalAccident or NearMissOccupationalAccident)*)

3.3   Step 3: Populate the Ontology

This module aims to create instances representing the reel world entities based on the concepts and relations defined by the *FFHExt.* ontology. The FFH knowledge graph[7] is a semantic network that represents the linked knowledge of our fall use case. For example, an employee *"Ali"* is a victim of a *"fall from a roof"* from a height of *"1.10m"*. The FFH knowledge base applies the semantic data transformation approach to ingest data i.e. key information from texts and key entities from structured data, to assess data quality, overcome quality issues (e.g., missing values or duplicates) and aggregate values. In a knowledge graph, we may need to reduce duplicates and annotate data with concepts and relations from existing ontologies and/or from our *FFHExt.* ontology. We use SPARQL-Generate[8] tool to generate the RDF triples that correspond to the input data. The output of this module is a knowledge graph of FFH where concepts and their relationships can be linked to equivalent entities in a list of knowledge graphs such as DBpedia or Wikidata (using some entity linking tools like Falcon[9] and DBpedia Spotlight[10]). This FFH knowledge graph is loaded and materialized in a RDF format and stored in a triplestore.

3.4   Step 4: Query the Ontology

Once the FFH knowledge graph is created, it can be explored and queried using the SPARQL query language posed over the query processing engine. Such an engine can be integrated into triplestores (e.g. GraphDB, Virtuoso, Stardog) or can be through federated query processing engines (e.g. FedX [13], Semagrow [11]). In a simple case, the answer for a simple query does not require any inference on the facts of the knowledge graph. However, we can receive a complex query that needs more inference using the *FFHExt.* ontology for getting answers. Figures 8 and 9 show examples of queries (simple and complex) and their answers. Query 1 is simple for selecting actors who are involved in a task. Query 2 is more complex to extract human causes that integrate several kind of causes such as *KnowledgeCause*, *BehaviourCause*, *PoorCommunicationCause*. These queries can be integrated into a smart service on preventing falls from heights to reduce risks and injuries ("Work smart, work safe when working at height"[11]).

4   Experimentation and Validation

This section is dedicated to present the use-case of FFH and highlight the impact of artificial intelligence in the prevention of falls from height process.

---

[7] https://sites.google.com/view/kgswc-2022-ia-ffh-resources/knowledge-graph.
[8] https://ci.mines-stetienne.fr/sparql-generate/.
[9] https://github.com/SDM-TIB/falcon2.0.
[10] https://github.com/dbpedia-spotlight/dbpedia-spotlight.
[11] https://medium.com/@iapsgroup/work-smart-work-safe-when-working-at-height-37e161b25296.

4.1   Fall from Height Use Case

To demonstrate the validity of our proposed approach, we can focus on identifying the causes of falls from height and also proposing the actions to take into account. We evaluate our approach using the data from high potential risks database of our company. The HIPO database represents information on all accidents, significant hazardous situations with high-potential severity reported by our company. It also provides safety and installs adequate fall protection systems in the organization. This data is represented by two formats; structured and unstructured information (texts). This does not facilitate the use of the database for major risk prevention purposes. Around 32500 records have been stored. In our experiment, we describe results on an extract of this given dataset. For each concerned record, we explore information from structured data and FFH event description (see Fig. 7).

Fig. 7. An extract of HIPO database

In the following, both kinds of data are exploited to design the ontology and to create the knowledge graph. The given knowledge graph will be queried by domain users.

4.2   Results of Our Approach and Discussion

Evaluating *FFHExt*. Ontology. A large effort has been devoted to propose the criteria of ontology evaluation in [1, 7]. We discuss three criteria that are matched to the objectives of the FFH use case and the *FFHExt*. ontology requirements such as coverage/completeness/competency, clarity, and adaptability, in order to measure the quality degree of the designed ontology.

- **Coverage/Completeness/Competency** presents how effectively the ontology reports concepts from the fall from heights domain. To ensure domain coverage, we reuse existing domain ontologies that contained industry standards and guidelines, integrate manually the knowledge of business experts and the information collected from the HIPO database. The ontology and the knowledge graph answer to the competency questions that are defined by the business expert (see next section). The *FFHExt*. ontology includes all relevant concepts and their lexical representations.

- **Clarity** defines how effectively the ontology communicates the intended meaning of the defined semantic entities (concepts and relations) without ambiguity [1]. The definitions of concepts and relations are ensured by the interaction with the business expert and the data sources of our company. These latter are generated by experts of industry sites, renewable energy, etc., who are responsible to provide concepts with explicit definitions. If two terms are synonyms in the text, it is represented by *skos:altLabel* in the ontology (e.g. *Operating&MaintenanceTask skos:altLabel "O&M Task"*) to get a machine-readable knowledge.
- **Adaptability** indicates the anticipation use of the ontology and its possibility to be extended [1]. Our proposed ontology is an effort to extract the most important concepts and relations of the falls from height domain. The *FFHExt.* ontology is modular and scalable to integrate new concepts and relations.

We have focused on the quality of our proposed ontology but more evaluations should be provided when the *FFHExt.* ontology will be updated and maintained.

Querying FFH Knowledge Graph. Once the knowledge graph creation process is established and facts are generated as RDF in form of triplets and hosted on an RDF triplestore like GraphDB or AllegroGraph, we ask some intuitive competency questions from real-world examples. As the knowledge base is defined through mapping to the *FFHExt.* ontology, the query processing engine can process queries posed using the SPARQL query language. In general, if knowledge is stored in a triplestore, then the knowledge base can be accessed using a visual SPARQL query over the query engine embedded in the triplestore. In our case, we use one such visual query triplestore given by GraphDB.

Table 1. Some competency questions

Query	Complexity	Competency questions
Q1	Simple	Who are victims of falls from height?
Q2	Complex	What are the human causes of falls from height?
Q3	Complex	What are the behavior causes of falls from height?
Q4	Complex	What are the organizational causes of falls from height?

Table 1 shows some competency questions that the business expert wants the ontology to answer. Two categories of queries are identified: (i) a simple query that explicitly selects the right triples from the knowledge graph, and (ii) a complex query that needs more inference rules from the domain ontology to answer it.

The performance of our research results is based on two aspects: (1) the ability to formalize these natural language questions in terms of SPARQL queries and (2) the capacity to answer to these questions using the *FFHExt.* ontology and the corresponding knowledge graph. Figures 8 and 9 show for each question posed in Table 1, we have specific and precise answers.

Fig. 8. SPARQL query corresponding to the simple natural language question and its answer

Fig. 9. SPARQL query corresponding to the Complex natural language question and its answer

In the case of queries 2, 3 and 4, we extract knowledge either from the most generic concept (e.g. to select all causes of FFHs, we get all sub-concepts (e.g. *BehaviourCause*, *OrganizationalCause*, etc.) of the concept *Cause*) to the most specific (e.g. to extract *LackOfSkillsCause*).

Based on these performance aspects, our proposed approach proves that using the ontology to capitalize information and knowledge, gives more accurate knowledge on the quality of results. That means the ontology guarantees interoperability by keeping the definitions of the correct concepts of the use case requirements and competency questions. To improve our identifying of unsafe behaviors, causes, and actions to take, we need more business questions and exploit more historical data for this domain.

5   Conclusion and Perspectives

This paper proposes and illustrates two Artificial Intelligence methods to facilitate information extraction from the database containing falls from height events. Our method is divided into several steps: (i) Extract key entities and relations from unstructured and structured data (ii) Model an ontology, and (iii) Populate the ontology and query the resulted knowledge graph of falls from height. Our future research will focus on (1) enhancing the *FFHExt.* ontology notably by integrating more prevention rules; (2) exploring additional experiments on the complete HIPO datasets; and (3) deploying a digital FFH prevention tool for workers based on the developed falls from height ontology.

References

1. Brank, J., Grobelnik, M., Mladenić, D.: A survey of ontology evaluation techniques. In: 8th International Multi-conference Information Society (2005)
2. Dong, Z., Dong, Q., Hao, C.: Hownet and its computation of meaning. In: The 23rd International Conference on Computational Linguistics: Demonstrations (2010)
3. Fang, W., Ma, L., Love, P.E., Luo, H., Ding, L., Zhou, A.: Knowledge graph for identifying hazards on construction sites: integrating computer vision with ontology. Autom. Constr. **119**, 103310 (2020)
4. Francesconi, E., Montemagni, S., Peters, W., Tiscornia, D.: Integrating a bottom-up and top-down methodology for building semantic resources for the multilingual legal domain. In: Semantic Processing of Legal Texts: Where the Language of Law Meets the Law of Language (2010)
5. Genesereth, M.R., Nilsson, N.J.: Logical foundations of artificial intelligence (1987)
6. Gruber, T.R.: A translation approach to portable ontology specifications. Knowl. Acquisition **5**(2), 199–220 (1993)
7. Gruber, T.R.: Toward principles for the design of ontologies used for knowledge sharing? Int. J. Hum. Comput. Stud. **43**(5–6), 907–928 (1995)
8. Guarino, N.: Formal ontology, conceptual analysis and knowledge representation. Int. J. Hum. Comput. Study **43**(5–6), 625–640 (1995)
9. Guo, B., Goh, Y., Scheepbouwer, E., Zou, Y.: An ontology of control measures for fall from height in the construction industry. In: the 35th International Symposium on Automation and Robotics in Construction (2018)
10. Lu, Y., Li, Q., Zhou, Z., Deng, Y.: Ontology-based knowledge modeling for automated construction safety checking. Saf. Sci. **79**, 11–18 (2015)
11. Mami, M.N., Scerri, S., Auer, S., Vidal, M.-E.: Towards semantification of big data technology. In: Madria, S., Hara, T. (eds.) DaWaK 2016. LNCS, vol. 9829, pp. 376–390. Springer, Cham (2016). https://doi.org/10.1007/978-3-319-43946-4_25

12. Safe work Australia: preventing falls in housing construction: code of practice. https://www.safeworkaustralia.gov.au/system/files/documents/1705/mcop-preventing-falls-in-housing-construction-v2.pdf (2016)
13. Schwarte, A., Haase, P., Hose, K., Schenkel, R., Schmidt, M.: Fedx: optimization techniques for federated query processing on linked data. In: The Semantic Web, ISWC 2011–10th International Semantic Web Conference, Proceedings (2011)
14. Song, Y., Wang, J., Liu, D., Guo, F.: Study of occupational safety risks in prefabricated building hoisting construction based on HFACS-PH and SEM. IJERPH 19(3), 1550 (2022)
15. van der Vet, P., Mars, N.: Bottom-up construction of ontologies. IEEE Trans. Knowl. Data Eng. 10(4), 513–526 (1998)
16. Zhang, S., Boukamp, F., Teizer, J.: Ontology-based semantic modeling of safety management knowledge. In: 2014 International Conference on Computing in Civil and Building Engineering (2014)
17. Zhang, S., Boukamp, F., Teizer, J., Zhang, S., Boukamp, F., Teizer, J.: Ontology-based semantic modeling of construction safety knowledge: towards automated safety planning for job hazard analysis (JHA). Autom. Constr. 52, 29–41 (2015)
18. Zhong, B., Li, H., Luo, H., Zhou, J., Fang, W., Xing, X.: Ontology-based semantic modeling of knowledge in construction: classification and identification of hazards implied in images. J. Constr. Eng. Manage. 146(4), 04020013 (2020)

Learning to Automatically Generating Genre-Specific Song Lyrics: A Comparative Study

Tze Huat Tee, Belicia Qiao Bei Yeap, Keng Hoon Gan[✉], and Tien Ping Tan

School of Computer Sciences, Universiti Sains Malaysia, USM, 11800 Pulau Pinang, Malaysia
{tee.tze.huat.ucom20,beliciayqb}@student.usm.my, {khgan, tienping}@usm.my

Abstract. The impact of music on the many dimensions of human life can be partly attributed to its linguistic component – the lyrics. In hopes of helping song-writers reach their full potential, researchers have implemented advanced artificial intelligence (AI) technology to automatically generate song lyrics. These efforts, however, were met with challenges that accompany the distinctive qualities of song lyrics, such as word repetition, structural pattens, and line breaks; all of which are dependent on the music genre. Seeing as most previous research either focuses on a given approach or genre, or performs the task without consideration of lyric variation among genres, this study attempts to address the gap by exploring and comparing the capabilities of three promising methods, specifically Markov chains, long short-term memory (LSTM), and gated recurrent units (GRU), in algorithmically generating lyrics for six selected music genres, namely rock, pop, country, hip-hop, electronic dance music (EDM), and rhythm and blues (R&B). Our findings show that LSTM scored better in the average readability index in overall, however, GRU produced the overall highest Rhyme Density score.

Keywords: Lyrics · Text generation · Natural Language Processing (NLP) · Markov chains · Long Short-Term Memory (LSTM) · Gated Recurrent Units (GRU)

1 Introduction

It is generally believed that music has been deeply ingrained in our societies since the dawn of humanity, with a significant amount of ancient musical instruments dating back as far as the Middle and Upper Palaeolithic [1]. Indeed, the tremendous influence music has on people of all ages from pre-schoolers [2], to adolescents [3], and to seniors [4]; is undeniable.

One of the fundamental elements of music is its linguistic content, i.e., the lyrics. In addition to intensifying emotions such as sadness, nostalgia, and astonishment, song lyrics have been observed to activate certain psychological mechanisms, including episodic memory, evaluative conditioning, contagion, and visual imagery [5].

B. Villazón-Terrazas et al. (Eds.): KGSWC 2022, CCIS 1686, pp. 62–75, 2022.
https://doi.org/10.1007/978-3-031-21422-6_5

Moreover, despite being initially centered on a limited number of themes, lyrics have, since the 1960s, evolved into a vessel for writers and performers to convey a broad spectrum of symbolic messages [6]. In particular, a considerable number of artists have leveraged the capacity of song lyrics to raise awareness in important issues, such as mental health, gender equality, and racial harmony [7].

2 Problem Statement

With the purpose of helping songwriters overcome the many challenges of lyric writing, notable efforts in the automatic generation of song lyrics have been made. Nevertheless, the application of artificial intelligence (AI) to the writing of lyrics has been proven to be no easy feat. Due to its unique features, an in-depth understanding of songwriting techniques, on top of sound knowledge in natural language processing (NLP), is crucial [8]. The necessity of modelling the line breaks, stylistic elements (e.g., flow, rhyming, and repetition), and structural layout and components (e.g., verse, refrain, chorus, and bridge) observed in lyrics adds another layer of complexity to the already difficult task [9]. Furthermore, these linguistic attributes may vary among different music genres. For instance, it has been demonstrated that rap songs incorporate significantly more word repetition as compared to country songs [9].

Regardless of the intricacies, several methods, including Markov chains [10], long short-term memory (LSTM) [9], and gated recurrent units (GRU) [11], have been shown to produce promising results on separate occasions. Therefore, it would be interesting to expand on previous research, such as that of Gill et al., and explore and compare the performance of these three approaches in the algorithmic generation of song lyrics. In this case, sub-genres are classified into their parent genres (e.g., categorizing metal as part of rock) due to computational and time constraints. This study thus focuses on six popular music genres of the English language, namely rock, pop, country, hip-hop, electronic dance music (EDM), and rhythm and blues (R&B) [12, 13].

3 Literature Review

3.1 Generating Non-genre-specific Lyrics

In 2010, Settles presented two interactive computational creativity tools designed to aid the song-writing process – Titular, a text synthesis algorithm capable of generating song titles semi-automatically, and LyriCloud, which displays a cloud of suggested lyrics based on a word input [14]. These intelligent tools were developed based on the criteria that their recommendations should be both unlikely and meaningful. Although the results were semantically satisfactory, they failed to exhibit any notion of stylistic qualities such as lyrical wordplay (e.g., rhyme) and other devices of creative writing (e.g., repetition).

On the other hand, Pudaruth et al. attempted to generate the lyrics of an entire song using context-free grammars (CFGs) [8]. By imposing grammatical rules and statistical constraints, they successfully produced lyrics that were grammatically correct and rather convincing, with more than half (52%) of their respondents evaluating one of their generated lyrics as an existing song. However, their output often lacked semantic meaning due to the impossibility of defining all grammatical rules which exist in the English language.

The studies above approached the task at hand without taking into account the influence of the genre on a song's lyrics, though Pudaruth et al. examined a few themes (i.e., love, pain, and cause) commonly found in popular songs [8]. Since writing is usually performed with an audience in mind [9], capturing the differences among genres, be it semantically or stylistically, could be an essential matter.

3.2 Generating Lyrics for a Specific Genre

An article published by Barbieri et al. in 2012 describes a framework of Constrained Markov Processes which generates lyrics in the style of a particular writer while maintaining the structural properties (in terms of rhyme and meter) of a provided text [10]. Apart from these features, their demonstration of mapping Bob Dylan's songwriting style onto the structure of the Beatles' "Yesterday" showed syntactic correctness and semantic relatedness. Nevertheless, additional cases should be investigated to ensure that this technique can be generalized to different writers and styles.

A more recent study by Fernandez et al. compared the performance of three character-level deep learning models, namely plain recurrent neural network (PRNN), long short-term memory (LSTM), and gated recurrent units (GRU), in the composition of rap lyrics [11]. The resulting lyrics achieved positive overall evaluation, convincing 67% of the participants who are familiar with rap lyrics in one of the instances, in spite of low rhyme density. Consequently, they suggested incorporating rhymes and intelligibility in the algorithm to improve rhythmic flow and coherency.

Despite promising results, these methods were formulated to address the issue for a specific genre (e.g., rap). In view of the broad spectrum of music preferences, it would perhaps be useful to explore the application of these approaches to other genres to appeal to a wider audience.

3.3 Generating Lyrics for Multiple Genres

In 2020, Gill et al. proposed a method which uses state-of-the-art long short-term memory (LSTM) to automatically generate lyrics for a specified music genre [9]. Upon evaluating their output using linguistic metrics, it was found that their model performed better in capturing the characteristics of pop and rap lyrics, in comparison to other genres such as rock, metal, country, and jazz. Seeing as only a single technique, i.e., LSTM, was employed, further research should be conducted to explore and compare the potential of other algorithms in computationally composing lyrics of various genres.

4 Methodology

The following section consists of descriptions of the dataset used in this study as well as details regarding data pre-processing, exploration, and cleaning.

4.1 Dataset Description

The dataset is self-collected by using Geniuslyrics API (Genius 2020) and Spotify Web API.

At the beginning, an account is required in Spotify to request access to Spotify Web API. After Spotify verified and approved the application, the client key and client secret are granted for access to Spotify Web API. By using Spotify Web API, the categories provided in Spotify playlist are retrieved and the genre of each playlist (rock, pop, country, hip-hop, EDM, or R&B) is identified. Following that, the track details are extracted from the identified playlist.

On the other hand, setting up an account in the Genius Lyrics Website authorized the access to apply for API Clients. A new API Client can be created with the application name and application website URL information. Upon confirmation of the API Client, the page generated a Client ID and Client Secret that authorize the usage of Geniuslyrics API.

Once the Client ID and Client Secret are provided, the lyricsgenius package in Python called the API and scraped the lyrics based on the track details retrieved from Spotify Web API. To avoid duplication of songs, a filter is added to skip live, demo, and remix versions in the scraping process. The relevant attributes of the collected dataset are as described in Table 1.

Table 1. Description of attributes.

Attribute	Description	Data type
id	The unique identifier for the track generated by Spotify	Nominal categorical data
name	The title of the track	Nominal categorical data
artist	The artist(s) who performed the track	Nominal categorical data
album	The album of the track	Nominal categorical data
lyric	The lyric of the track	Nominal categorical data
genre	The music genre of the artist(s)	Nominal categorical data
popularity	The popularity of the track based on the total number of plays and the recency of plays	Continuous numerical data

4.2 Data Pre-processing

As mentioned above, the song lyrics are collected by using web scraping API, Genius-lyrics. Within the scraped data, there are unwanted strings such as "EmbedShare", "URL-CopyEmbedCopy", and new line "\n" etc. All the unwanted string are replaced with a space. Other than that, the null data for the lyric column is removed and only the top 100 rows being selected as our dataset in this experiment.

Next, the lyrics strings are converted to lower case and punctuation is removed. Finally, tokenization breaks the lyrics strings into tokens.

4.3 Data Analysis

Text analysis of the song lyrics is carried out to further understand the six different music genres in terms of their linguistic content. The most common words in the song lyrics are identified and visualized in a word cloud for each genre. Apart from that, bar charts are also created to visualize the frequency distribution of the number of words in the song lyrics for each genre.

As shown in Table 2, the highest average word length in song lyrics can be seen in hip-hop. On top of that, hip-hop also has the highest average unique word counts. This indicates that hip-hop has highest complexity among all genres and could possibly impact the model performance.

Besides that, the genres having the highest and second highest noun term frequencies can be determined as hip-hop and EDM respectively. These two genres are also the highest and second highest in terms of verb term frequencies in song lyrics. Thus, it can be deduced that the noun term frequencies and verb terms frequencies in song lyrics are correlated to each other.

Since the usage of adverbs in song lyrics are relatively close for every genre, this characteristic plays an insignificant role in analytics.

Interestingly, EDM has the greatest maximum number of words (3980) as well as the lowest minimum number of words (37) in song lyrics. In contrast, pop and R&B seem to have rather short lyrics in general as shown by their maximum number of words.

Table 2. Text analysis of lyrics

Characteristic	Genres & counts
Average word length in song lyrics	Rock: 282.44 Pop: 368.69 Country: 323.69 Hip-Hop: 581.32 EDM: 332.37 R&B: 444.04

(continued)

Table 2. (*continued*)

Characteristic	Genres & counts
Average unique word counts in song lyrics	Rock: 110.06 Pop: 132.25 Country: 139.77 Hip-Hop: 254.13 EDM: 107.42 R&B: 159.61
Noun term frequencies in song lyrics	Rock: 1188 Pop: 1209 Country: 1350 Hip-Hop: 2423 EDM: 1388 R&B: 1066
Verb term frequencies in song lyrics	Rock: 630 Pop: 637 Country: 662 Hip-Hop: 1124 EDM: 775 R&B: 678
Adverb term frequencies in song lyrics	Rock: 214 Pop: 231 Country: 232 Hip-Hop: 271 EDM: 232 R&B: 233
Maximum number of words in song lyrics	Rock: 1390 Pop: 748 Country: 1019 Hip-Hop: 1125 EDM: 3980 R&B: 747
Minimum number of words in song lyrics	Rock: 136 Pop: 167 Country: 140 Hip-Hop: 281 EDM: 37 R&B: 196

Figure 1 illustrates the word cloud generated from the lyrics of the collected country songs. From the diagram, the outliers and most common terms, such as "got", "yeah", "oh", and "know", are identified; all of which will introduce bias to the model.

The bar chart in Fig. 2 depicts the frequency distribution of the number of words in lyrics of the selected country songs. Based on Fig. 2, most of the number of words are

Fig. 1. Word cloud of country song lyrics.

scattered between 200 to 400. An outlier where the number of words is more than 1000 can also observed but it only occurred once.

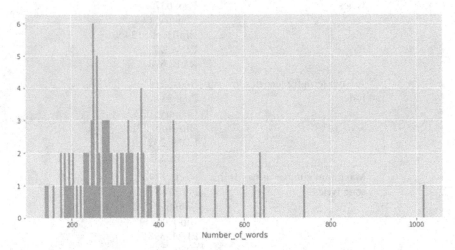

Fig. 2. Frequency distribution of the number of words in country song lyrics.

4.4 Markov Chains

A Markov chains model is a statistical tool that identifies the pattern dependencies in different kinds of systems, especially pattern recognition system [15]. As characters or words are normally characterized by dependencies between patterns, the Markov chain theory is suitable for implementation in the domain of natural language processing.

Markov chains is selected in our study as it is one of the basic methods for text generation. The core idea of Markov chains is a simple assumption that the next word is dependent on the previous word.

First, the song lyrics is tokenized into each token. Then, a dictionary is initialized to hold all the words and next words. After that, all the words will pair up with the next word and they will be stored in the previously created dictionary. Finally, a function can be created to generate consecutive words upon receiving an input text by referring to the dictionary iteratively. For Markov chain, the output will be measured based on the readability and density score.

4.5 Long Short-Term Memory (LSTM)

Long short-term memory (LSTM) is a type of recurrent neural network (RNN) that is able to learn the order dependencies that exists in sequence prediction problems [16]. These networks were introduced by Sepp Hochreiter et al. in 1997 [17]. A memory unit known as a "cell state" is introduced in the LSTM to address the existing failure of RNN in learning the presence of past observations that is greater than 5–10 discrete time steps between relevant inputs and their target signals [18]. The cell state acts as a carrier to transfer information or context over longer discrete steps, hence allowing adjustment of the network gradient descent in the information flow.

The layers of our trained LSTM model that output the best results with the limitation in hardware specification are as described in Table 3.

Table 3. Summary of LSTM model for pop genre.

Layer	Output	Param#
Embedding	(None, 756, 40)	117160
LSTM (bidirectional)	(None, 200)	112800
Dropout	(None, 200)	0
Dense	(None, 2929)	588729

The model trained with 30 epochs and achieved a range of accuracy from 0.6446 to 0.773 based on different kinds of genres. Then, the model is implemented to predict the class from the generated token list and the output will become the newly generated song lyrics.

4.6 Gated Recurrent Units (GRU)

In 2014, Kyunghyun Cho et al. introduced gated recurrent units (GRU), which is an improvement of the standard RNN [19].This is a relatively new method compared to RNN and LSTM as it is an improved version of them. GRU able to perform well in sequence learning tasks and handling the vanishing gradients problem seen in traditional RNN [20].

Compared to LSTM, GRU implements gates to control the flow of information and abandons the usage of cell states. GRU consists of only one hidden state and has a simpler architecture, thus it will shorten the training time of the model [21].

The layers of our trained GRU model that output the best results with the limitation in hardware specification are as detailed in Table 4.

Table 4. Summary of GRU model for pop genre.

Layer	Output	Param#
Embedding	(None, 719, 40)	102520
GRU (bidirectional)	(None, 200)	85200
Dropout	(None, 200)	0
Dense	(None, 2563)	515163

All models trained with 30 epochs with one exception of the EDM genre, which experienced early stopping in 25 epochs. They achieved a range of accuracy from 0.6993 to 0.8198 based on different types of genres. Then, the model is implemented to predict the class from the generated token list and the output will become the newly generated song lyrics.

5   Evaluation Criteria

Three evaluations were performed in this paper, namely model performance, average readability, and rhyme density score. In addition, we have also included sample of generated song lyrics from Markov chain, LTSM and GRU.

5.1   Model Performance

Based on Table 5, the GRU models for every genre slightly outperformed the LSTM models in terms of accuracy after 30 epochs. Due to hardware limitations (which will be further elaborated in the discussion section), the epoch is set to 30 as the maximum value. Therefore, it is believed that the LSTM models require more epochs to achieve higher accuracies based on the theory stated above.

Table 5. Comparison of model performance.

Genre	LSTM accuracy	GRU accuracy
Pop	0.7420	0.8068
Rock	0.7426	0.8198
Country	0.6945	0.7738
EDM	0.7773	0.8142
Hip-Hop	0.6517	0.6993
R&B	0.6698	0.7271

5.2 Average Readability

Readability is the ease with which a reader is able to understand a written text and is measured by the complexity of the text's vocabulary and syntax [22]. In this experiment, the average readability of the generated lyrics is obtained by using the textstat library in Python. The higher the average readability, the better the generated song lyrics.

Based on Table 6, the average readability of generated lyrics for Markov chains are the highest in every genre. As a result of the stored dictionary that is implemented in Markov chains, the fixed structural and grammatical rules in the Markov chains approach enable it to obtain high scores in average readability. In the meantime, LSTM model outputs managed to score better than GRU model outputs in 4 different genres such as pop, rock, country, and R&B. However, GRU model outputs score better for the EDM and Hip-Hop genres that have huge number of tokens. As the GRU model trains faster along the epochs, the model is determined capable to handle the higher complexity and huge dimension dataset.

Table 6. Average readability of generated lyrics.

Genre	Average readability (Markov chains)	Average readability (LSTM)	Average readability (GRU)
Pop	21.42	20.92	20.32
Rock	22.88	22.10	20.62
Country	21.84	21.62	20.04
EDM	22.60	18.9	23.12
Hip-Hop	22.52	19.78	20.60
R&B	22.30	20.52	19.62

5.3 Rhyme Density Score

Rhyme density score referred to the total number of rhymed syllables that divided by total number syllables in the corpus or song lyrics in our case [23]. It is part of evaluation criterion to determine whether which approach able to generate the best lyrics as the output. For this measurement, the higher the rhyme density represent the better the generated song lyrics.

Referring to Table 7, the GRU model has the highest score for Rhyme density score in overall. In the meantime, the Markov chains score the lowest due to the randomness retrieval from the stored dictionary and form the lyrics. Besides that, the pop genre songs more likely to score higher compared to the other genre. It could be due to the chorus and word repetition in the pop genre songs.

Table 7. Rhyme density scores of generated lyrics.

Genre	Rhyme density score (Markov chains)	Rhyme density score (LSTM)	Rhyme density score (GRU)
Pop	0.6250	0.7114	0.7482
Rock	0.5652	0.5833	0.6715
Country	0.4385	0.5431	0.6419
EDM	0.5349	0.7634	0.5082
Hip-Hop	0.5556	0.6207	0.6275
R&B	0.6667	0.4715	0.5912

5.4 Sample Output of Generated Song Lyrics

Markov Chains Sample Output (Pop Genre)
Breathing just rub it never wanna keep you first baby, let's get your bad 'cause I got it, got me be alone in my records on everything seems like you leave me, girl? not the things that I never does why? you are you so I see one is that.

LSTM Sample Output (Pop Genre)
Happy for me out of myself I am I think I'm gonna get so I've been thinking I know what you know that I was born to run I don't belong to everybody but you're not to me I don't deserve someone loyal to me I don't want to be a

GRU Sample Output (Pop Genre)
Sad so don't say oh woah oh but yeah I hate you I don't wanna be my spot I've been work out baby it's just like this might be so bitter ooh ooh ooh ooh ooh just sayin' this what you know that you're hiding something I know it's true it's

6   Discussion

6.1   Models

Throughout the processes, all the methods are compared to each other based on their differences, time required, and the output of the generated song lyrics.

First, LSTM model retains even more information further down the sequence when it compared to GRU model. Meanwhile, Markov chains approaches implemented a simple method to generate dictionary on top of the corpus to generate the song lyrics randomly based on the stored dictionary.

Besides, Markov chains took the shortest time to implement among all the approaches as it doesn't involve complex model training process. Then, GRU model is faster than

LSTM model due the number of gates in the neural network architectures. LSTM has three gates, but GRU only has two gates in the network.

Despite the Markov chains are fast, however the average readability of the generated song lyrics outputs is highest among all the methods but due to its randomness in generating the lyric. Thus, it is not suitable to select as the right approach for lyrics generation. In the meantime, when comparing the outputs of the GRU and LSTM models, LSTM scored better in the average readability index in overall. However, the GRU have the overall highest Rhyme Density score.

Overall, GRU is the most favorable approach for small datasets that was applied in our paper as it has fast computational speed and better output.

6.2 Limitations

In our experiments, LSTM is the model that required high computation power and long hours to train. In the first few trials in training the LSTM, the time taken to complete for a model took around 8 h. Due to that issue, different kinds of approaches being implemented to improve the overall training time or speed One of the approaches is instead of using CPU in the tensorflow library, the CUDA and GPU driver are installed to enable the tensorflow-gpu. The GPU that being applied in this experiment are NVIDIA GeForce GTX 1650. There is an obvious improvement in the training time which reduced to 3 to 4 h for training the LSTM model. It has been very challenging for us to train to the models for LSTM and GRU models for every genre in total 12 models as the training model are time intensive.

Other than that, the huge dataset also is one of the limitations for our experiments. Apparently, our hardware insufficient RAM to train huge dataset that exceeded around 2GB. Thus, the dataset required to limit down to 2GB so that it can fit into the model and carry out training process. For example, due to large dimension for EDM genre in our dataset, therefore it reduced to 80 song tracks in order to train.

7 Conclusion and Future Work

In this paper, three different algorithms, specifically Markov chains, long short-term memory (LSTM), and gated recurrent units (GRU) have been implemented to generate song lyrics. Our experimental results show that the GRU has the best output based on the song lyrics. Based on our trials in training all the stated model, a larger dataset is required to produce a better outcome. However, our hardware resources are limited, and the GPU memory is unable to support a bigger dataset. Therefore, our future work includes collecting more data, using upgraded hardware to train the models, and observing the outcome.

References

1. Montagu, J.: How music and instruments began: A brief overview of the origin and entire development of music, from its earliest stages. Front. Sociol. **2**, 8 (2017)

2. Levinowitz, L.M.: The importance of music in early childhood. General Music Today **12**(1), 4–7 (1998)
3. North, A.C., Hargreaves, D.J., O'Neill, S.A.: The importance of music to adolescents. Br. J. Psychol. **70**(2), 255–272 (2000)
4. Cohen, A., Bailey, B., Nilsson, T.: The importance of music to seniors. Psychomusicol. J. Res. Music Cogn. **18**(1–2), 89–102 (2002)
5. Barradas, G.T., Sakka, L.S.: When words matter: a cross-cultural perspective on lyrics and their relationship to musical emotions. Psychol. Music **50**(2), 650–669 (2022)
6. Astor, P.: The poetry of rock: song lyrics are not poems but the words still matter; another look at Richard Goldstein's collection of rock lyrics. Pop. Music **29**(1), 143–148 (2010)
7. Russell, E.: 25 pop songs with social messages. In: PopCrush (2015). https://popcrush.com/pop-songs-social-messages. Accessed 11 Aug 2022
8. Pudaruth, S., Amourdon, S., Anseline, J.: Automated generation of song lyrics using CFGs. In: 2014 Seventh International Conference on Contemporary Computing (IC3), pp. 613–616. IEEE, Noida, India (2014)
9. Gill, H., Lee, D.T., Marwell, N.: Deep learning in musical lyric generation: an LSTM- based approach. Yale Undergrad. Res. J. **1**(1), 1–7 (2020)
10. Barbieri, G., Pachet, F., Roy, P., Esposti, M. D.: Markov constraints for generating lyrics with style. In: Proceedings of the 20th European Conference on Artificial Intelligence (ECAI 2012), pp. 115–120. IOS Press, Amsterdam, Netherlands (2012)
11. Fernandez, A.C.T., Tarnate, K.J.M., Devaraj, M.: Deep rapping: character level neural models for automated rap lyrics composition. Int. J. Innov. Technol. Explor. Eng. (IJITEE) **8**(2S), 306–311 (2018)
12. Lowder, J.: The 10 most popular music genres. In: Audio Captain (2021). https://audiocaptain.com/most-popular-music-genres. Accessed 11 Aug 2022
13. Clark, B.: The top 10 genres in the music industry. In: Musician Wave (2021). https://www.musicianwave.com/top-music-genres. Accessed 11 Aug 2022
14. Settles, B.: Computational creativity tools for songwriters. In: Proceedings of the NAACL HLT 2010 Second Workshop on Computational Approaches to Linguistic Creativity, pp. 49–57. Association for Computational Linguistics, Los Angeles, California, USA (2010)
15. Al-Anzi, F.S., AbuZeina, D.: A survey of Markov chain models in linguistics applications. In: Computer Science and Information Technology (CS & IT) Conference Proceedings, vol. 6, no. 13, pp. 53–62. AIRCC Publishing Corporation, Chennai, Tamil Nadu, India (2016)
16. Brownlee, J.: A gentle introduction to long short-term memory networks by the experts. In: Machine Learning Mastery (2017). https://machinelearningmastery.com/gentleintroduction-long-short-term-memory-networks-experts. Accessed 11 Aug 2022
17. Hochreiter, S., Schmidhuber, J.: Long short-term memory. Neural Comput. **9**(8), 1735–1780 (1997)
18. Rathore, A.S.: LSTM — Introduction in simple words. In: Medium (2020). https://medium.com/nerd-for-tech/lstm-introduction-in-simple-words-fe544a45f1e7. Accessed 11 Aug 2022
19. Cho, K., et al.: Learning phrase representations using RNN encoder–decoder for statistical machine translation. In: Proceedings of the 2014 Conference on Empirical Methods in Natural Language Processing (EMNLP), pp. 1724–1734. Association for Computational Linguistics, Doha, Qatar (2014)
20. Shen, G., Tan, Q., Zhang, H., Zeng, P., Xu, J.: Deep learning with gated recurrent unit networks for financial sequence predictions. Proc. Comput. Sci. **131**, 895–903 (2018)
21. Saxena, S.: Introduction to gated recurrent unit (GRU). In: Analytics Vidhya (2021). https://www.analyticsvidhya.com/blog/2021/03/introduction-to-gated-recurrent-unit-gru. Accessed 11 Aug 2022

22. Banga, S.S.: Readability index in Python (NLP). In: GeeksforGeeks (2021). https://www.gee
ksforgeeks.org/readability-index-pythonnlp. Accessed 11 Aug 2022
23. Hirjee, H., Brown, D.G.: Using automated rhyme detection to characterize rhyming style in
rap music. Empir. Musicol. Rev. **5**(4), 121–145 (2010)

DLIME-Graphs: A DLIME Extension Based on Triple Embedding for Graphs

Yoan A. López[1]([✉]) [iD], Hector R. Gonzalez Diez[1] [iD],
Orlando Grabiel Toledano-López[1] [iD], Yusniel Hidalgo-Delgado[1] [iD], Erik Mannens[2] [iD],
and Thomas Demeester[3] [iD]

[1] Departamento de Informática, Universidad de Las Ciencias Informáticas, Havana, Cuba
{yalopez,hglez,ogtoledano,yhdelgado}@uci.cu
[2] Ghent University - iMinds, Sint-Pietersnieuwstraat, 9000 Gent, Belgium
erik.mannens@ugent.be
[3] Ghent University, Gent, Belgium
thomas.demeester@ugent.be

Abstract. In the last years, several research works have been proposed for the Knowledge Graph Completion task. However, like most Machine Learning models, most Knowledge Graph Completion models are opaque and lack interpretability. In order to achieve transparency, several interpretable and explainable models have been proposed. The Deterministic Local Interpretable Model-Agnostic Explanations (DLIME) was proposed to solve the lack of stability of the Local Interpretable Model-Agnostic Explanations (LIME), one of the most popular surrogate models. However, using DLIME to explain Machine Learning models in graphs becomes an issue due to its experiments being published only with tabular data. Therefore, this work aims to propose an interpretable method for graphs as an extension of DLIME named DLIME-Graphs. As a triple representation, DLIME-Graphs uses triple embeddings computed by SBERT which in turn, are reduced by the UMAP technique. Instead of using Hierarchical Clustering as DLIME, DLIME-Graphs uses HDB-SCAN to get clusters. To explain a test triple, DLIME-Graphs proposes to train two interpretable models: logistic regression and decision tree plus getting the most similar triples by a k-NN algorithm. The demonstration through a study case showed that DLIME-Graphs is able to give explanations for 100% of the triples in the test dataset through the former models offering transparency and interpretability.

Keywords: Triple embeddings · Knowledge graphs · Interpretable method · DLIME · SBERT · TSDAE

1 Introduction

Knowledge graphs (KGs) like Freebase [1], Yago [18] and WordNet [11] have numerous facts in the form of (h, r, t) where $h, t \in E$, the set of entities, and $r \in R$, the set of relations. They are useful resources for many Artificial Intelligence tasks such as web search, recommendations and question answering [14]. Since KGs often suffer from

© The Author(s), under exclusive license to Springer Nature Switzerland AG 2022
B. Villazón-Terrazas et al. (Eds.): KGSWC 2022, CCIS 1686, pp. 76–89, 2022.
https://doi.org/10.1007/978-3-031-21422-6_6

incompleteness and noise, a number of recent techniques have proposed models to predict facts within the task of knowledge graph completion (KGC) [14].

Much research work has been devoted to knowledge graph completion [8]. A common approach, knowledge graph embedding (KGE), also called knowledge representation learning (KRL), represents entities and relations in triples as real valued vectors and assesses the plausibility of triples with these vectors [24]. However, like most Machine Learning models, they are opaque and sometimes the binary "yes" or "no" answer is not sufficient and questions like "why" is more significant [17]. Triple Prediction models only evaluate the accuracy of KGEs at predicting missing triples, while in real applications, explanations for predictions are valuable, as they will likely improve the reliability and transparency of predicted results [25]. Understanding the reasons behind predictions is quite important in assessing trust, which is fundamental to take action based on a prediction [17, 24]. Thus, we stated the first research question as follows: a) How to explain the prediction of a missing triple based on observed triples in the knowledge graph?

To achieve transparency, several interpretable and explainable models have been proposed. These models have taken one of three main routes: i) approximating models with a simpler surrogate model [9, 15, 17], ii) highlighting relevant aspects of the computation within the provided model [6], and aiming at interpretable Machine Learning models in graphs, iii) relying the explanations on relevant graph components [14, 24, 25]. Existing interpretable models in graphs use paths [25], influence triples [14] and subgraphs [24] to explain the predictions, however, searching for these structures in the whole graph remains a challenge.

Surrogate models establish decision rules from features [9]. Local Interpretable Model-Agnostic Explanations (LIME) [17] is one of the most popular techniques in this route. It generates an explanation for a single prediction by any Machine Learning model by learning a simpler interpretable model around the prediction by random perturbations. However, the process of perturbing the points randomly makes LIME a non-deterministic approach, lacking "stability", a desirable property for an interpretable model [15]. To cope that limitation, a Deterministic Local Interpretable Model-Agnostic Explanations (DLIME) framework was proposed [15]. DLIME uses Hierarchical Clustering (HC) to partition the dataset into different groups instead of randomly perturbing the data points around the instance, and its explanations generated are stable on each iteration. In order to use DLIME to explain Machine Learning models in graphs, there is an issue due to its experiments were published with tabular data, nevertheless, recognizing the good of the transparency of DLIME, this work aims at proposing an interpretable method for graphs which is an extension of DLIME: DLIME-Graphs.

Since for a computer to understand human-readable text (triples in graphs), it is needed to convert it into a machine-readable format, in order to find a triple representation able to allow the search for similar triples in graphs, we stated the second research question as follows: What machine-readable representation for triple to use for calculating similar triples in graphs? Recent advances in knowledge-graph-based research include Transformer-based knowledge encoding [8]. Transformer-based models [19] have boosted text representation learning, and there is a presence of Transformers-based graph representation. CoKE [21] employs Transformers to encode edges and path sequences in graphs. Similarly, KG-BERT [8, 23] borrows the idea from language model pretraining and takes Bidirectional Encoder Representations from Transformer (BERT [5]) model as an encoder for entities and relations in graphs. However, most existing approaches (including models not based on transformers i.e. RDFsim [4]) have only focused on learning feature vectors for nodes and predicates. Triple2Vec [7] was the first approach, to learn triple embeddings in knowledge graphs, opening a novel class of downstream applications. However, Triple2Vec needs to compute the weights of edges in terms of relatedness between predicates beforehand.

Keeping on the same vein of above Transformers-based works, and taking advantages of the similarity between graph triples and text sentences, i.e.: " <zutsu_manemon nationality japan>", and "Zutsu Manemon has Japanese nationality", in this work, we propose a new way of generating triple embeddings in graphs through sentence-BERT (SBERT) [16]. SBERT is a modification of the pretrained BERT [5] to derive semantically meaningful sentence embeddings that can be compared using cosine-similarity. It has gained a new state-of-the-art performance on semantic textual similarity (STS) [16].

Overall, our interpretable method for graphs (DLIME-Graphs) is an extension of DLIME. It uses triple embeddings computed by SBERT as triple representation. We demonstrated our method through a study case where KG-BERT-made triple predictions are explained. The main contributions of this work are the following:

- An interpretable method for Machine Learning models in graphs which is an extension of DLIME. Unlike other interpretability methods in graphs, DLIME-Graphs is able to give explanations for 100% of the triples in the test dataset,
- DLIME-Graphs proposes triple embeddings based on SBERT. These novel triple embeddings can be used in other interpretable methods like CrossE [25], and CRIAGE [14] to look for similar triple components. Triple embeddings based on SBERT contribute to narrow the gap between graph triples and text sentences,
- Unlike DLIME, DLIME-Graphs proposes HDBSCAN as the clustering algorithm. HDBSCAN selects the best cut point to get clusters automatically,
- DLIME-Graphs continues boosting the insertion of Transformer architectures and Pre-Training models like BERT in the knowledge graph context,
- DLIME-Graphs method is in turn, a new approach of triple prediction in graphs.

The remaining of this paper is organized as follows: First, in Sect. 2, related work is described. Then, we present our interpretable method DLIME-Graphs in Sect. 3. Next, we introduce a study case addressing the application of DLIME-Graphs to explain KG-BERT-made triple predictions in Sect. 4, and finally, the conclusions and future work are presented in Sect. 5.

2 Related Work

Recent work aimed at explaining Machine Learning models has taken one of three main routes: i) approximating models with a simpler surrogate model [9, 15, 17], ii) highlighting relevant aspects of the computation within the provided model, for instance, inspecting feature gradients [6], and aiming at interpretable Machine Learning models in graphs iii) relying the explanations on relevant graph components, such as, paths, influence triples, subgraphs [14, 24, 25]. Despite, this third route rules in the context of interpretability methods in knowledge graphs, searching for these structures in the whole graph remains a challenge. In this work, we aim at developing a method based on a surrogate model. With emphasis on the routes i and iii, approaches related to ours are discussed below.

Surrogate models establish decision rules from features [9]. Local Interpretable Model-Agnostic Explanations (LIME) [17] is one of the most popular techniques in this route. It generates an explanation for a single prediction by any ML model by learning a simpler interpretable model around the prediction by random perturbations. However, the process of perturbing the points randomly makes LIME a non-deterministic app-roach, lacking "stability", a desirable property for an interpretable model. To cope that limitation, a Deterministic Local Interpretable Model-Agnostic Explanations (DLIME) [15] framework was proposed. DLIME uses Hierarchical Clustering (HC) to partition the dataset into different groups instead of randomly perturbing the data points around the instance, and its explanations generated are stable on each iteration. Aiming at applying DLIME and due to its experiments being published over opaque models with tabular data, in this work, we propose an extension of DLIME for graphs.

Looking at the third route, CrossE [25] proposed a kind of knowledge graph embed-dings named interaction embeddings. The process of generating explanations for one triple (h, r, t) is modeled as searching for reliable paths from h to t and similar structures to support explanations of predictions on graphs. Because of the volume of search space, path searching starts with selecting candidate entities and relations based on embedding similarity to reduce the search space. Explanations in CrossE depend on finding graph paths to explain the triples. Unlike CrossE, our method uses triple embeddings which contain the semantic of the whole fact and we are able to explain the 100% of triples in the graph. Moreover, triple embeddings facilitate the obtainment of similar triples included similar relations and entities, whether it is desired.

CRIAGE [14], another technique in this vein, is an investigation of robustness and interpretability of Link Prediction via Adversarial Modifications. It aims at finding min-imum changes in the graph structure such that the prediction of a target fact changes the most after the embeddings are relearned. Given the large of graphs, CRIAGE focuses on finding influence triples in the neighborhood of the target triple, while other facts in the graph could be influencing the prediction. Unlike CRIAGE, we aim at finding similar triples in the whole graph.

GNN Explainer [24] is an approach to explain graph neural networks (GNN) joining nodes information with the graph structure. It obtains node embeddings by recursively propagating information from its neighbors. The definition of a node neighborhood N is crucial, as it can affect the performance and scalability of the GNN model. Gnn Explainer provides a local interpretation by highlighting relevant features as well as an important subgraph structure with the edges that are most relevant to the prediction. However, GNN Explainer should decide the most convenient neighborhood for those computational graphs which is always a challenge.

Regarding DLIME-Graphs, it is an extension of DLIME for graphs. We propose to work with triple embeddings based on SBERT [16] as features. SBERT is the current-best models for producing information-rich representations of sentences and paragraphs.

In addition, we propose to use HDBSCAN [3] as cluster algorithm and due to the large of triple embeddings, we proposed UMAP function [10] to reduce the feature dimension. DLIME-Graphs relies on two interpretable models: logistic regression and decision tree, which along with similar triples obtained from a K nearest neighbor algorithm, offer explanations for the predictions outputted from uninterpretable models.

2.1 Triple Embeddings

Transformers [19] have wholly rebuilt the landscape of Natural Language Processing (NLP). With Transformer models, it is possible to use the same core of a model and simply swap the last few layers for different use cases (without retraining the core). This new property resulted in the rise of pretrained models for NLP. Pretrained Transformer models are trained on vast amounts of training data. One of the most widely used of these pretrained models is BERT (Bidirectional Encoder Representations from Transformers) [5]. BERT models can learn context-aware text embeddings with rich language information via pre-trained language models.

Looking at sentence similarity, BERT had one issue: it only produces vector embeddings for each word (or token) similar to word2vec, not sentence-level embeddings. Therefore, the solution was sentence-BERT (SBERT) [16] and the Sentence Transformers library[1]. SBERT is a modification of the pretrained BERT network that uses siamese and triplet network structures to derive semantically meaningful sentence embeddings that can be compared using cosine-similarity. SBERT is based on PyTorch and has gained a new state-of-the-art performance on Semantic Textual Similarity (STS) [16].

Emerging advances on knowledge-graph-based research include Transformer-based knowledge encoding [8]. CoKE [14] employs transformers to encode edges and path sequences in graphs. Similarly, KG-BERT [23] borrows the idea from language model pretraining and takes BERT model as an encoder for entities and relations. KG-BERT also incorporates textual information to enrich knowledge representation, outperforming most knowledge graph embeddings models which only use structure information in observed triple facts [23].

However, most existing knowledge representation learning approaches have only focused on learning feature vectors for nodes and predicates. Adding to interpretability reasons, triple embeddings can be useful for clustering, triple classification, user-item recommendations, triple similarity, fact-checking and reasoning [8]. In this regard, Triple2Vec [7] was the first approach to learn triple embeddings from knowledge graphs, which opened a novel class of downstream applications. Triple2Vec needs to compute the weights of edges in terms of relatedness between predicates beforehand.

Keeping the same vein of above Transformers-based works, in this work, we propose a new way of generating triple embeddings in graphs through SBERT, specifically, we borrow the unsupervised sentence embedding learning from TSDAE [20] which allows us to train the model on the target dataset itself. Thus, these triple embeddings are used in the explanations of the interpretable method.

---

[1] https://www.sbert.net/.

3   Interpretable Method: DLIME-Graphs

In this section, we present DLIME-Graphs, an extension of the DLIME method, for explaining the test triple prediction in graphs. Figure 1 shows the block diagram of DLIME-Graphs.

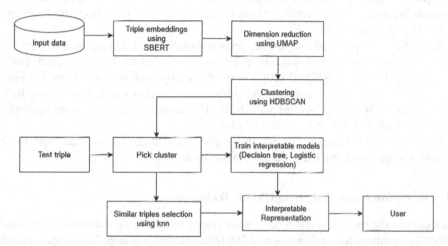

Fig. 1. A block diagram of DLIME-Graphs interpretable method.

DLIME-Graphs transforms graph triples into textual sentences, and those sentences are converted to sentence embeddings using SBERT and Sentence Transformers library. Due to those embeddings have a large dimension of 768 features, for the sake of interpretability and computation consume reduction, a dimension reduction is performed. By default, we do a dimension reduction to five features, and the user can change it according to the desired number of explanations. From here, the dataset is partitioned into clusters and the cluster that contains the test triple is selected. Then, all triples in the selected cluster are used to train two interpretable models. On the above, similar triples are shown to the user, complementing the explanations. The main components of the DLIME-Graphs method are explained further below.

3.1   Triple Embeddings Using SBERT

To obtain triple embeddings, we develop a SBERT model called sbert_test_mnr[2] from a pretrained BERT model (and tokenizer). The model was also fine-tuned with an unsupervised training by TSDAE approach [20][3]. As the original SBERT [16], we fine-tuned the model on the combination of the SNLI [2] and the multi-Genre NLI [22] datasets. We trained the model with 3 epochs, a batch-size of 16, Adam optimizer with learning rate 2e-5, a linear learning rate warm-up over 10% of the training data, and Multiple

---

[2] https://drive.google.com/file/d/1LMLGu7MIs2QslyUDBfpP7oBypq0APYMl/view?usp=sharing.

[3] https://github.com/yalopez84/Interpretable_method/.

Negatives Ranking (mnr) as loss function. Unlike the original SBERT model, we fine-tuned SBERT with mnr loss function because of according to the experiments, that is the best loss function in the Sentence Transformers library currently. Our default pooling strategy was MEAN.

Our model consists of two modules: the BERT transformer module followed by a mean pooling module. The transformer module was loaded from Hugging Face[4]. We evaluated the model on the semantic textual similarity dataset from Hugging Face, and we reached an accuracy of 0.84.

Regarding unsupervised fine-tuning, despite unsupervised training methods are not as effective as their supervised counterparts, it has proved its worth complementing the supervised training. TSDAE [20] is one of the approaches raised from Sentence Transformers library to get sentences embeddings in domains where there is very little labeled data, as normally happens with KGs. Therefore, we decided to implement a fine-tuning with TSDAE on the observed triples.

DLIME-Graphs proposes to get the triple embeddings only once, and these embeddings are saved in a file to be used in the future.

3.2 Dimension Reduction Using UMAP Technique

After getting the triple embeddings, it is necessary to reduce their dimension because SBERT embeddings have a dimension of 768 features, which is large for being clustered. By default, we do a reduction to five features, and the user can change it according to the desired number of explanations. Dimension reduction plays an important role in Data Science, being a fundamental technique of pre-processing in Machine Learning.

DLIME-Graphs proposes UMAP (Uniform Manifold Approximation and Projection) [10] as the dimension reduction technique. UMAP has no computational restrictions on embedding dimension, making it viable as a general-purpose dimension reduction technique. UMAP preserves more of the global structure than previous reduction dimension algorithms, with superior run time performance.

3.3 Clustering Using HDBSCAN

Just like DLIME [15], our method partitions the dataset into different groups, however, instead of Hierarchical Clustering (HC), we propose to use HDBSCAN [3]. HDBSCAN generates a complete density-based clustering hierarchy from which a simplified hierarchy composed only of the most significant clusters can be easily extracted. This allows HDBSCAN to find clusters of varying densities and be more robust to parameter selection. HDBSCAN detects the best cut point to select clusters automatically, which is a great advantage. The only parameter we need to set up is the minimum size of clusters. DLIME-Graphs establishes five as the minimum size of clusters, which is its default value.

---

[4] https://huggingface.co/.

3.4 Interpretable Models

Linear regression, logistic regression and decision tree are commonly used interpretable models [12]. DLIME [15] relies the explanations on a linear regression model, while DLIME-Graphs proposes to give explanations through: i) a logistic regression model, ii) a decision tree model, and iii) similar triples from a k nearest neighbor algorithm. In order to explain the prediction of a test triple, we selected the cluster that contains that triple and all triples belonging to that cluster are used to train the former interpretable models as we detail next.

3.5 Logistic Regression Model

Logistic regression [12] models the probabilities for classification problems with two possible outcomes. The logistic regression model, instead of fitting a straight line or hyperplane, uses the logistic function (depicted below) to squeeze the output of a linear equation between 0 and 1. We use 0.5 as the threshold for classification.

$$logistic(n) = \frac{1}{1 + exp(-n)} \tag{1}$$

The interpretation of the logistic regression model [12] tailored to DLIME-Graphs method is formalized as follows: let B_i the weight associated to the feature xi of the triple, and odds ratio (the probability that the prediction of the triple is true divided by the probability that the prediction of the triple is false), a change in the feature xi by one unit changes the odds ratio by a factor of exp (B_i). That is, a change in xi by one unit increases the log odds ratio by the value of the corresponding weight B_i. DLIME-Graphs shows the odds ratio for each feature.

3.6 Decision Tree Model

Decision tree models [12] split the data multiple times according to certain cutoff values in the features. Through splitting, different subsets of the dataset are created with each instance belonging to one subset. The final subsets are called terminal or leaf nodes, and the intermediate subsets are called internal nodes or split nodes. To predict the outcome in each leaf node, the average outcome of the training data in this node is used. The following formula describes the relationship between the outcome "y" and features "x".

$$\hat{y} = \hat{f}(x) = \sum_{m=1}^{m} c_m I\{x \epsilon R_m\} \tag{2}$$

Each instance falls into exactly one leaf node (=subset Rm). $Ix \epsilon Rm$ is the identity function that returns 1 if x is in the subset R_m and 0 otherwise. If an instance falls into a leaf node R_1, the predicted outcome is $\hat{y} = cl$, where cl is the average of all training instances in leaf node R_1. The interpretation in the decision tree model [12] tailored to DLIME-Graphs method is as follows: starting from the root node and following to the next nodes and the edges we get the dataset subsets. Once we reach the leaf node, the node states the predicted outcome. All the edges are connected by 'AND'. As the maximum depth of the decision tree, we propose to use five.

3.7 Similar Triples Selection Using KNN

Given N training vectors, the k-nearest neighbor algorithm identifies the k number of nearest neighbors. For classification, it assigns the most common class of the nearest neighbors of an instance. The tricky parts are finding the right k and deciding how to measure the distance between instances, which ultimately defines the neighborhood. DLIME-Graphs establishes cosine similarity as the distance measure and k equal to five.

knn differs from the former interpretable models because it is an instance-based learning algorithm. knn lacks an interpretability global model because the model is inherently local and there are no global weights or structures explicitly learned. Presenting the k nearest neighbors can give good explanations according to [12]. DLIME-Graphs proposes complementing the explanations of above interpretability models with the five most similar triples. DLIME-Graphs method is formally presented in the Algorithm 1.

Algorithm 1: DLIME-Graphs

Input: *Dtrain* Dataset for training, *X* Instance, *K* Lenght of explanations, X_p Uninterpretable model prediction

Output: E_x List of explanations

/* Data initializations   */

Data: $E \leftarrow$ [] triple embedings, $E_r \leftarrow$ [] reduced triple embedings, $C \leftarrow$ [] triple clusters, $C_x \leftarrow \emptyset$ selected triple clusters, $T_s \leftarrow$ [] target triple set, $W \leftarrow \emptyset$ weights of logistic regression, *log_reg_pred* $\leftarrow \emptyset$ prediction of logistic regression model, *dec_tree_pred* $\leftarrow \emptyset$ prediction of decision tree model, *rules* $\leftarrow \emptyset$ decision tree rules, *knn* $\leftarrow$ [] nearest neighbors, $E_x \leftarrow$ [] explanations

1 Read E;

2 **if** $E == \emptyset$ **then**

3      $E \leftarrow$ CALCULATE *SBERT_Triple_Embeddings(Dtrain, X)*

4 **end**

5 $E_r \leftarrow$ CALCULATE *UMAP_Reduced_Embeddings(E, K)*

6 $C \leftarrow$ CALCULATE *HDBSCAN_Clusters(E_r)*

7 $C_x \leftarrow$ GET *_cluster_of_X(C, X)*

8 **while** *no triples in Dtrain* **do**

9   **if** *cluster_of_triples* == C_x **then**

10   ADD triple to T_s

11   **end**

12 **end**

13 W, *log_reg_pred* $\leftarrow$ CALCULATE *Logistic_Regression_Prediction(T_s, X)*

14 *rules, dec_tree_pred* $\leftarrow$ CALCULATE *Decision_Tree_Prediction*

15 *knn* $\leftarrow$ CALCULATE *K_Nearest_Neighbors(T_s, X, K)*

16 $E_x \leftarrow$ ADD(W, *log_reg_pred, dec_tree_pred, rules, knn, X_p*)

17 **return** E_x;

---

4   Study Case

To show our method, we selected KG-BERT [23] as the uninterpretable model for knowledge graph completion. KG-BERT proposes to use pre-trained language models for knowledge graph completion. Indeed, it treats triples in knowledge graphs as textual sequences and proposes a novel framework named Knowledge Graph Bidirectional Encoder Representations from Transformer (KG-BERT) to model these triples. Experimental results on multiple benchmark knowledge graphs show that KG-BERT can achieve state-of-the-art performance in tasks of Knowledge Graph completion such as: triple prediction, link prediction and relation prediction. Hereafter, we focus on explaining KG-BERT-made triple predictions with DLIME-Graphs.

As datasets, we picked Freebase [1]. Freebase is a large knowledge graph of general world facts. The data in Freebase is collaboratively created, structured, and maintained. For the sake of reducing the consumption of resources in the study case, and following expert recommendations that state a training dataset with more than 20K of instances is good enough, we created two reduced subsets of Freebase. For training, we created a dataset of 22164 triples called FB13_train_reduced, and for test we created a dataset of 2253 triples called FB13_test_reduced.

The study case consisted of two parts: firstly, we reproduced the experiments of triple prediction with KG-Bert, where predictions on test dataset were saved in a file. Secondly, we gave explanations for these predictions supported by our interpretable method DLIME-Graphs.

4.1   Triple Prediction with KG-BERT

With KG-BERT, according to the original paper [23], we reproduced both: training and tests. Training started with a preprocessing of triples that included completing the dataset with negatives triples and extending triples with their entity and relation descriptions. Extended triples were tokenized and formatted according to the BERT sequence structure to feed the BertForSequenceClassification model used in KG-BERT. Once, we trained KG-BERT model on triples of FB13_train_reduced, it was saved to a file locally.

KG-BERT tests required to convert FB13_test_reduced triples into BERT sequences as well. Indeed, test triples were also extended with entity and relation descriptions. From the tests we got a loss of 0.39 and an accuracy of 0.87. This accuracy is below the original KG-BERT experiments, as expected due to the training dataset reduction. All experiment code to reproduce KG-BERT is available in github[5]. Prediction results of KG-BERT were saved in a file called "FB_test_results.txt" to be explained with our interpretable method as we detail below.

4.2   DLIME-Graphs Application

For explaining test triples, interpretability is aimed over triples of the training dataset. In order to demonstrate the DLIME-Graphs method, we picked one triple from the test dataset (any test triple can be picked): <john_glenn_beall_jr>   <nationality>

---

[5] https://github.com/yalopez84/Interpretable_method/.

<united_states> whose label is "1". Firstly, we calculated the training triple embeddings along with the test triple embedding based on the SBERT-TSDAE-based model[6] addressed earlier on. In order to take advantage of the contextual embeddings that SBERT offers, triples were extended with entity and relation descriptions before. We got an embedding tensor of dimension 22.165 ÃŮ 768 that includes training triples plus the test triple to be explained.

Afterwards, using UMAP technique, we reduced the dimension of these embeddings from 768 to 5 features. A dimension of 5 features is good enough to transparently approach the explanations to the user. Then, triple embeddings were clustered by HDBSCAN algorithm from which we got a total of 606 clusters. In turn, according to the method, we selected the cluster where the test triple was located to go to the explanations. All 1362 triples belonging to the selected cluster were used to train two interpretable models: logistic regression and decision tree. In addition, a k nearest neighbor algorithm was also trained for searching the five most similar triples to the test triple. The logistic regression model yielded a prediction of "1" to the test triple matching its label and the KG-BERT prediction as well. We offered a table (depicted in Fig. 2) with the odds ratios per feature, which tells the user how much each feature contributes to the obtained prediction.

On the other hand, the decision tree model yielded a prediction of "1" to the test triple matching its label as well. We used the sklearn.tree.export_text class [13] to show the decision tree (depicted in Fig. 3). With the decision tree and the test triple features, we could find the complete route that supports the given prediction, starting from the root node and following to the next nodes down to the leaves.

```
Logistic Regression prediction: 1
    Features    Weight  Odds Ratio
0  Feature_0  0.012896   1.012979
1  Feature_1  0.025218   1.025539
2  Feature_2  0.151296   1.163341
3  Feature_3  0.405281   1.499724
4  Feature_4  0.187955   1.206779
```

Fig. 2. Odds ratio per feature in the Logistic Regression model

```
|--- feature_2 <= -3.39
|   |--- class: 1
|--- feature_2 >  -3.39
|   |--- feature_1 <= 9.99
|   |   |--- class: 1
|   |--- feature_1 >  9.99
|   |   |--- class: 0
```

Fig. 3. Decision tree

In addition, the most five similar triples yielded by the knn algorithm allowed to show why former models are able to draw a prediction of "1" to the test triple. Given graph triples are seen like textual sentences in this work, similar triples can be seen like influence triples and so, they are interesting for the predictions. Figure 4 illustrates i) the triple target, ii) similar triples, iii) the prediction from KG-BERT and iv) the predictions from the former interpretable models. All code related to explanations through DLIME-Graphs is available in github[7].

---

[6] https://github.com/yalopez84/Interpretable.

[7] https://github.com/yalopez84/Interpretable_method/.

```
Triple target :  john_glenn_beall_jr   nationality   united_states   label:1
Predictions: KG BERT:1    Log Regression:1   Decision Tree:1   knn:1
```

Out[127...

Influential triples:

Id_triple_Training	Subjects	Predicates	Objects	Label
0	train-3652	rienzi_melville_johnston	nationality	united_states	1
1	train_corrupt-6635	arthur_vicars	nationality	united_states	0
2	train-7521	john_whiteaker	nationality	united_states	1
3	train-10240	green_b_raum	nationality	united_states	1
4	train-10492	lynn_frazier	nationality	united_states	1

Fig. 4. Explanations for the test triple.

5 Conclusions and Future Work

Most knowledge graph completion models are opaque and lack interpretability. To achieve transparency, several interpretable and explainable models have been proposed. Deterministic Local Interpretable Model-Agnostic Explanations (DLIME) was proposed to solve the lack of stability of the Local Interpretable Model-Agnostic Explanations (LIME), one of the most popular surrogate models. However, in order to use DLIME to explain Machine Learning models in graphs, there is an issue due to its experiments were published only with tabular data. Therefore, this work aimed at proposing an interpretable method for graphs which is an extension of DLIME: DLIME-Graphs. Indeed, a study case was presented to demonstrate the value of our method.

As a result, unlike other interpretable models in graphs, DLIME-Graphs is able to give explanations for 100% of the triples in the tests dataset. We showed that the SBERT-TSDAE-based model offers a good representation for graph triples that can be even used by other interpretable models. Clustering with HDBSCAN allowed to select the clusters automatically. Two interpretable models: logistic regression and decision tree yielded further explanations to the user. These explanations along with the most similar triples outputted from a K nearest neighbor algorithm, offered transparency and interpretability to uninterpretable triple prediction models like KG-BERT. DLIME-Graphs continues boosting the insertion of Transformer architectures and pre-training models in the knowledge graph context. In addition to the interpretability goal, DLIME-Graphs can be seen as a novel approach for knowledge graph completion. As ongoing work, we plan to continue with the evaluation of the method on the whole Freebase dataset and other general-purpose datasets like Yago and Word-Net. The main advantage of DLIME-Graphs is that local surrogate models are based on instance features. Instead of finding structures like paths, neighbor triples, etc. in the whole graph, once you have the triple representation you can work with it, which is simpler.

Acknowledgments. This research has been partially sponsored by VLIR-UOS Network University Cooperation Programme-Cuba.

References

1. Bollacker, K., Evans, C., Paritosh, P., Sturge, T., Taylor, J.: Freebase: A collaboratively created graph database for structuring human knowledge. In: Proceedings of the 2008 ACM SIGMOD International Conference on Management of Data, pp. 1247–1250. SIGMOD 2008, Association for Computing Machinery, New York, NY, USA (2008). https://doi.org/10.1145/137 6616.1376746. (event-place: Vancouver, Canada)
2. Bowman, S.R., Angeli, G., Potts, C., Manning, C.D.: A large annotated corpus for learning natural language inference. In: Proceedings of the 2015 Conference on Empirical Methods in Natural Language Processing, pp. 632–642. Association for Computational Linguistics, Lisbon, Portugal (2015). https://doi.org/10.18653/v1/D15-1075, https://aclanthology.org/D15-1075
3. Campello, R.J.G.B., Moulavi, D., Sander, J.: Density-based clustering based on hierarchical density estimates. In: Pei, J., Tseng, V.S., Cao, L., Motoda, H., Xu, G. (eds.) Advances in Knowledge Discovery and Data Mining, LNAI, vol. 7819, pp. 160–172. Springer, Berlin (2013). https://doi.org/10.1007/978-3-642-37456-2_14
4. Chatzakis, M., Mountantonakis, M., Tzitzikas, Y.: RDFSIM: similarity-based browsing over dbpedia using embeddings. Information 12(11), 440 (2021)
5. Devlin, J., Chang, M.W., Lee, K., Toutanova, K.: BERT: pre-training of deep bidirectional transformers for language understanding. In: Proceedings of the 2019 Conference of the North American Chapter of the Association for Computational Linguistics: Human Language Technologies, vol. 1 (Long and Short Papers), pp. 4171–4186. Association for Computational Linguistics, Minneapolis, Minnesota (2019). https://doi.org/10.18653/v1/N19-1423, https://aclanthology.org/N19-1423
6. Erhan, D., Bengio, Y., Courville, A., Vincent, P.: Visualizing higher-layer features of a deep network. Univ. Montreal 1341(3), 1 (2009)
7. Fionda, V., PirrÀš, G.: Learning triple embeddings from knowledge graphs. Proc. AAAI Conf. Artif. Intell. 34(04), 3874–3881 (2020). https://doi.org/10.1609/aaai.v34i04.5800, https://ojs. aaai.org/index.php/AAAI/article/view/5800
8. Ji, S., Pan, S., Cambria, E., Marttinen, P., Yu, P.S.: A survey on knowledge graphs: representation, acquisition, and applications. IEEE Trans. Neural Networks Learn. Syst. 33(2), 494–514 (2022). https://doi.org/10.1109/TNNLS.2021.3070843
9. Lakkaraju, H., Kamar, E., Caruana, R., Leskovec, J.: Interpretable & Explorable Approximations of Black Box Models. CoRR abs/1707.01154 (2017)
10. McInnes, L., Healy, J., Melville, J.: UMAP: Uniform Manifold Approximation and Projection for Dimension Reduction (2018). https://doi.org/10.48550/ARXIV.1802.03426
11. Miller, G.A.: WordNet: a lexical database for English. Commun. ACM 38(11), 39–41 (1995). https://doi.org/10.1145/219717.219748. (New York, NY, USA Publisher: Association for Computing Machinery)
12. Molnar, C.: Interpretable Machine Learning (2022)
13. Pedregosa, F., et al.: Scikit-learn: machine learning in python. J. Mach. Learn. Res. 12(85), 2825–2830 (2011). http://jmlr.org/papers/v12/pedregosa11a.html
14. Pezeshkpour, P., Tian, Y., Singh, S.: Investigating Robustness and Interpretability of Link Prediction via Adversarial Modifications. CoRR abs/1905.00563 (2019).
15. Rehman Zafar, M., Mefraz Khan, N.: Dlime: a deterministic local interpretable model-agnostic explanations approach for computer-aided diagnosis systems. arXiv e-prints arXiv-1906 (2019)

16. Reimers, N., Gurevych, I.: Sentence-BERT: sentence embeddings using siamese BERT-networks. In: Proceedings of the 2019 Conference on Empirical Methods in Natural Language Processing and the 9th International Joint Conference on Natural Language Processing (EMNLP-IJCNLP). pp. 3982–3992. Association for Computational Linguistics, Hong Kong, China (2019). https://doi.org/10.18653/v1/D19-1410, https://aclanthology.org/D19-1410
17. Ribeiro, M.T., Singh, S., Guestrin, C.: "Why Should I Trust You?": explaining the predictions of any classifier. In: Proceedings of the 22nd ACM SIGKDD International Conference on Knowledge Discovery and Data Mining. pp. 1135–1144. KDD 2016. Association for Computing Machinery, New York, NY, USA (2016). https://doi.org/10.1145/2939672.293 9778 (event-place: San Francisco, California, USA)
18. Suchanek, F.M., Kasneci, G., Weikum, G.: Yago: a core of semantic knowledge. In: Proceedings of the 16th International Conference on World Wide Web, pp. 697–706. WWW 2007. Association for Computing Machinery, New York, NY, USA (2007). https://doi.org/10.1145/1242572.1242667 (event-place: Banff, Alberta, Canada)
19. Vaswani, A., et al.: Attention is all you need. In: Proceedings of the 31st International Conference on Neural Information Processing Systems, pp. 6000–6010. NIPS 2017, Curran Associates Inc., Red Hook, NY, USA (2017) (event-place: Long Beach, California, USA)
20. Wang, K., Reimers, N., Gurevych, I.: TSDAE: using transformer-based sequential denoising auto-encoder for unsupervised sentence embedding learning. In: Findings of the Association for Computational Linguistics: EMNLP 2021, pp. 671–688. Association for Computational Linguistics, Punta Cana, Dominican Re-public (2021). https://doi.org/10.18653/v1/2021.fin dings-emnlp.59, https://aclanthology.org/2021.findings-emnlp.59
21. Wang, Q., et al.: CoKE: Contextualized Knowledge Graph Embedding. CoRR abs/1911.02168 (2019)
22. Williams, A., Nangia, N., Bowman, S.: A broad-coverage challenge corpus for sentence understanding through inference. In: Proceedings of the 2018 Conference of the North American Chapter of the Association for Computational Linguistics: Human Language Technologies, vol. 1 (Long Papers), pp. 1112–1122. Association for Computational Linguistics, New Orleans, Louisiana (2018). https://doi.org/10.18653/v1/N18-1101, https://aclanthology.org/N18-1101
23. Yao, L., Mao, C., Luo, Y.: KG-BERT: BERT for Knowledge Graph Completion. CoRR abs/1909.03193 (2019)
24. Ying, R., Bourgeois, D., You, J., Zitnik, M., Leskovec, J.: GNN Explainer: A Tool for Post-hoc Explanation of Graph Neural Networks. CoRR abs/1903.03894 (2019)
25. Zhang, W., Paudel, B., Zhang, W., Bernstein, A., Chen, H.: Interaction embeddings for prediction and explanation in knowledge graphs. In: Proceedings of the Twelfth ACM International Conference on Web Search and Data Mining, pp. 96–104. WSDM 2019. Association for Computing Machinery, New York, NY, USA (2019). https://doi.org/10.1145/3289600.329 1014, (event-place: Melbourne VIC, Australia)

Edge-Labelled Graphs and Property Graphs - To the User, More Similar Than Different

Paul Warren[(✉)] [iD] and Paul Mulholland [iD]

Knowledge Media Institute, The Open University, Milton Keynes MK7 6AA, U.K.
{paul.warren,paul.mulholland}@open.ac.uk

Abstract. The two dominant paradigms for graph databases, edge-labelled graphs and property graphs, may appear quite different. Yet, to the user, they have strong similarities. A usability study, comparing RDF-star/SPARQL-star and Cypher, found evidence for only limited differences in preferences between the modelling paradigms. This suggests the possibility of a convergence of the two paradigms; indeed in one case (Stardog) this is already happening. We also found little difference between the paradigms in users' ability to detect valid and non-valid queries. In one specific case, the use of a reverse arrow in Cypher was interpreted significantly more accurately than the ^ symbol in the SPARQL query language; this argues, where possible, for the more intuitive Cypher notation.

Keywords: Graph databases · Cypher · SPARQL-star · Reification · Empirical study

1 Introduction

Over the past several decades, two graph database formats have arisen: edge-labelled graphs and property graphs. The former has been standardized by the W3C as the Resource Description Framework (RDF), along with an accompanying query language (SPARQL). These have been the subject of considerable theoretical investigation. The latter has been the subject of proprietary standards and implemented and used widely.

Our study sought to:

- compare the ease of use of the two language paradigms;
- identify preferences for alternative semantically equivalent models, and determine whether these preferences differed between the two paradigms;
- determine where there were difficulties in querying models.

We found that the similarities between the paradigms' usage outweighed the differences.

For edge-labelled graphs, we have investigated an extension of RDF, known as RDF-star (RDF*), along with a parallel extension of SPARQL known as SPARQL-star (SPARQL*), using the Blazegraph (https://blazegraph.com/) implementation. These extensions have been proposed to enable reification, and are now the subject of a W3C

Community Group Draft Report) [1][1]. As an example of a property graph language, we took Cypher, developed by neo4j (https://neo4j.com/).

Section 2 describes related work and Sect. 3 provides an overview of the study. The study was divided into five sections, and Sects. 4–8 describe each of these. Sections 4 and 5 are not concerned with the extended features of RDF and SPARQL, but compare the more basic features of the two paradigms. The remaining sections are concerned with how RDF* and SPARQL* permit the creation and querying of metadata, and how this compares with the use of metadata in Cypher. Section 6 compares how the two paradigms deal with predicate metadata. Sections 7 and 8 are concerned with metadata about metadata; in the case of Sect. 7, metadata about node metadata, in the case of Sect. 8, metadata about predicate metadata. Finally, Sect. 9 draws some conclusions and makes some recommendations for further development. Warren and Mulholland [2] provide a more detailed discussion.

2 Related Work

There have been few studies looking specifically at the usability of graph database languages. Holzschuher and Peinl [3] have compared the readability of Cypher, Gremlin, SQL and a Java API. Based on a readability metric, they concluded that Cypher was the most readable of the four languages. Warren and Mulholland [4] have investigated how accurately and rapidly users can reason about the main SPARQL features. Of relevance to this current study, they found that questions with reverse predicates were answered less accurately and more slowly than analogous questions with forward predicates.

Whilst we are not aware of any usability comparison of the two graph database paradigms, Hartig [5] has produced algorithms to transform between the two paradigms. For edge-labelled graphs, he uses RDF*. For property graphs, he uses his own formalization of the property graph model. However, because of the greater power of RDF*, there are limitations on what can be transformed from RDF* to property graphs.

3 Overview of the Study[2]

The study was organized as a between-participants study, i.e. each participant answered questions relating to one of the two paradigms. Participants were contacted through three routes: specialist mailing lists; internal mailing lists within our own organization; and personal contacts. Participants were provided with a tutorial document which specified all they needed to know for the study, to refer to before and during the study. The study was implemented using an online survey tool. Participants were initially asked to

---

[1] At the time of undertaking the study, we only had available to us the echelon ($<< \ldots >>$) syntax for quoted triples. The assumption at the time was that embedded triples would be asserted and this was implicit in the study questions. The current assumption is that, when using the echelon syntax, embedded triples are not asserted. There is now an additional syntax, the annotation syntax, which permits embedded triples to be asserted concurrently with their being quoted.

[2] The study was approved by the Open University Human Research Ethics Committee (HREC/3568).

provide some basic demographic information; and also asked about their expertise in graph database languages.

There were then two practice questions; one concerned with data models, the other with queries. The results of these questions were not analyzed. The next five sections of the study were the questions proper. There were two types of questions: modelling questions and querying questions. To minimize participants' time, and for simplicity of analysis, we sought participants' responses to our prepared models and queries. For the modelling questions, participants were given an English-language description of the model, and one or more English-language queries to be answered by the model. Participants were shown several models, in either RDF* or Cypher; and asked which of these models were correct, in the sense of representing the English description and being able to answer the queries. They were also asked to rank the models, with the possibility of ranking models equally. In the following we only report the results of the model ranking. This is partly for reasons of space, but also because we found that, in some cases, participants were classifying a model as incorrect, but then making a comment that they realized the model was technically correct but that they thought it unsuitable to the task, e.g. because it makes difficult future extensions to the data model. In all but one case, a modelling question was followed by one or more querying questions. Each querying question made use of a correct model, in RDF* or in Cypher, from the previous modelling question. One of the English-language queries from the modelling question was repeated, along with a number of queries in the target query language. Participants were asked to indicate which of these queries corresponded to the English-language query; in some cases there were more than one correct query.

In the reporting of results, model rankings have been normalized to a scale of 0 (lowest) to 1 (highest). The accuracy of assessing queries is also shown on a scale of 0 to 1. In the figures in Sects. 4–8, we present a 95% confidence interval for each statistic. We also use the conventional 95% level when discussing statistical significance. The R language was used for all statistical calculations reported in this paper [6].

Table 1. Expertise of participants in the two conditions

Condition	Language	No knowledge at all	A little knowledge	Some knowledge	Expert knowledge
RDF*/ SPARQL*	SPARQL	0%	11.5%	19.2%	69.2%
	Cypher	76.9%	19.2%	3.8%	0%
Cypher	SPARQL	61.1%	11.1%	16.7%	11.1%
	Cypher	38.9%	11.1%	33.3%	16.7%

There were 26 participants for RDF*/SPARQL*, and 18 participants for Cypher; in both cases they were chiefly drawn from Europe and the Americas. Because participants were anonymous, it is not possible to be sure that there was no overlap between the two groups. However, the targeting of participants was designed to reduce the likelihood of this. Table 1 shows the level of expertise in SPARQL and Cypher for participants in the two conditions, indicating that the level of relevant expertise was greater in the RDF*/SPARQL* condition. Where appropriate we have controlled for this in our statistical analysis.

4 Nodes, Literals and Reverse Predicates

This section describes five questions. The first is a modelling question. The remaining four questions are querying questions which compare the ^ notation in SPARQL with the arrow notation in Cypher. The questions in this section do not make use of the extended features of RDF* or SPARQL*.

Figure 1 shows the modelling question. In model 1, the RDF represented a location as a literal, whilst Cypher used a property value. In model 2, the RDF represented the location with an IRI, whilst Cypher used a node.

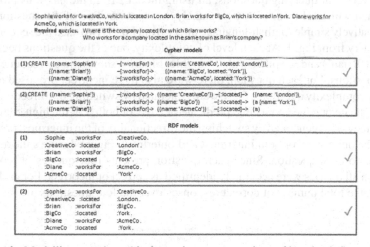

Fig. 1. Modelling question with alternative representations of location information

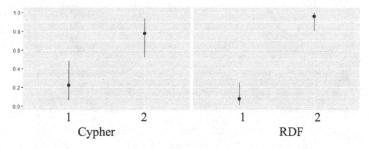

Fig. 2. Preference ratings for modelling question of Fig. 1

Figure 2 shows the participants' mean preference ratings with 95% confidence intervals. In both conditions there was a strong preference for model 2, in which the locations are represented by IRIs or nodes. A two factor analysis of deviance showed a significant difference in preference between the models ($\chi^2(1) = 55.741$, $p < 0.001$) and a significant interaction effect between models and languages ($\chi^2(1) = 5.418$, $p = 0.020$), indicating that the difference in rankings was significantly less for Cypher than for RDF. This is consistent with the greater use of strings to denote entities in property graphs, see a participant's comment below.

The participants' comments mostly confirmed the preference for model 2. A participant commented on RDF model 1 that it "hinders future extensions of the database, as no attributes or relations can be added to strings". Another participant commented "using strings to represent cities is such a bad modelling decision that this needs to be answered as an incorrect model". Similarly, in the Cypher condition a participant commented that "locations are best treated as entities - and thus, nodes - in their own right". On the other hand, in response to RDF model 1, a participant did observe "in Property Graphs it is not uncommon to denote entities by simple strings".

There were four querying questions, all using model 2 from the previous modelling question, and with analogous SPARQL and Cypher queries. Two of the questions used the first, relatively simple English-language query from Fig. 1; two used the second, more complex query from Fig. 1. At each level of complexity, one of the questions began with a constant ('Brian') and the other with a variable. Figure 3 shows the correct queries for the four questions. For each question, there were also three incorrect proposed queries. For the low complexity questions, these were generated by switching the directionality of the first predicate or edge, the second predicate or edge, or both. For the high complexity questions, they were generated by switching the directionality of the innermost predicates or edges, the outermost, or both innermost and outermost. Figure 4 shows the accuracy of response to each question. Since each question presents four queries, a score of 4 would mean all queries were accurately identified as correct or incorrect. For each of the conditions, the total number of correct responses over the four questions, i.e. out of 16,

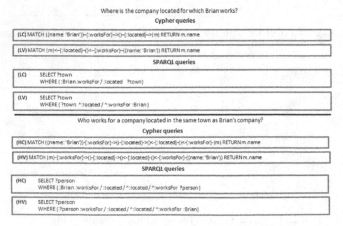

Fig. 3. Correct queries based on model 2 of Fig. 1

was computed for each participant. An ANOVA, after controlling for the participants' expertise in the language being studied, showed no significant difference between the languages $(F(1, 37) = 0.0095, p = 0.923)$.

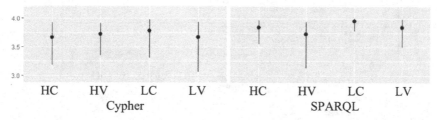

Fig. 4. Accuracy of responses to querying questions shown in Fig. 3 LC = low complexity, constant first; LV = low complexity, variable first; HC =high complexity, constant first; HV = high complexity, variable first.

5   Class Hierarchies

The questions in this section were concerned with modelling and querying class hierarchies. They were partially motivated by what seems to be a limitation in the property graph paradigm. As in the previous section, the questions did not make use of the extended features of RDF* and SPARQL*. For the modelling question, we presented two models. The natural way to model hierarchies in RDF is using *rdfs:subClassOf*. Cypher has node labels, which are generally used to indicate class membership. However, there is no natural way to represent hierarchies. One way to create hierarchies in Cypher is to use the *labels()* function to extract the strings representing the labels of a node. Another way is to emulate the use of *rdfs:subClassOf* in RDF. These are the approaches of Cypher models 1 and 2 respectively. For the RDF model 1, we created a model which mixed IRIs and strings, comparable to the use of the Cypher *labels()* function. The corresponding query uses the *STR()* function to extract the string representation of an IRI. For RDF model 2, we followed the *rdfs:subClassOf* approach. However, we did not use *rdfs:subClassOf* but rather a predicate *:subGroupOf*, because participants might not necessarily be familiar with RDFS. The models are shown in Fig. 5.

Figure 6 shows the participants' mean preference ratings. For both languages, the majority of participants preferred model 2. This was the case in Cypher, even though the use of labels is the natural way to indicate class membership. An analysis of deviance showed a significant difference between the models $(\chi^2(1) = 59.027, p < 0.001)$ and a significant interaction effect between models and language $(\chi^2(1) = 6.158, p = 0.013)$, indicating that the preference was significantly stronger for RDF than for Cypher. This is consistent with *subClassOf* being the accepted approach in RDF. A number of participants criticized RDF model 1 for mixing "resources and literal values for the same concept". For Cypher model 1, one participant noted "Neo4j generally pushes the use of labels for "types", but the path query for recursive subgroups is going to be awkward, requiring an equality condition on a node label and a node's property value."

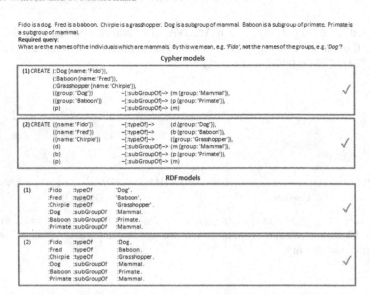

Fig. 5. Modelling question illustrating class hierarchies

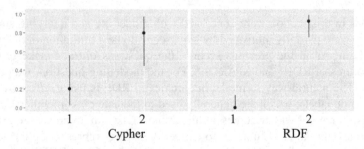

Fig. 6. Preference ratings for modelling question of Fig. 5

Fig. 7. Queries based on model 1 of Fig. 5

There were two querying questions, making use of each of the two models, and the English-language query from Fig. 5. Figure 7 shows the queries appropriate to model 1. For both conditions, only query 1 is correct.

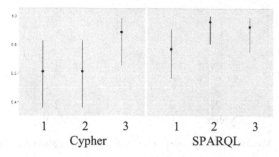

Fig. 8. Accuracy of responses to querying questions shown in Fig. 7

Figure 8 shows the accuracy of response to each question. For queries 1 and 2, accuracy of response for Cypher appears appreciably less than for SPARQL; for both these Cypher queries, accuracy was no greater than chance ($p = 0.240$). To compare the two languages, we calculated the number of accurate responses for each participant for each question. An ANOVA, after controlling for expertise in the language being studied, shown no significant difference in performance between the languages ($F(1,37) = 2.4872, p = 0.123$).

Figure 9 shows the queries appropriate to model 2. For both conditions, only query 3 is correct. Figure 10 shows the accuracy of response to each question. Based on the number of correct responses for each participant, an ANOVA, after controlling for expertise, showed no significant difference in performance between the languages ($F(1,37) = 0.2662, p = 0.609$). This question was answered much more accurately than the previous question: a mean of 2.89 correct compared with 2.43. A two-sided paired t-test indicated that this difference was significant ($t(43) = 3.3463, p = 0.002$).

Cypher queries	
(1) MATCH (x)−[:subGroupOf *]−>({group:'Mammal'}) RETURN x.name	X
(2) MATCH (x)−[:typeOf *]−>({group:'Mammal'}) RETURN x.name	X
(3) MATCH (x) −[:typeOf]−>()−[:subGroupOf *]−>({group: 'Mammal'}) RETURN x.name	✓

	SPARQL queries	
(1)	SELECT ?name WHERE { ?name :subGroupOf+ :Mammal }	X
(2)	SELECT ?name WHERE { ?name :typeOf+ :Mammal }	X
(3)	SELECT ?name WHERE { ?name :typeOf / :subGroupOf+ :Mammal }	✓

Fig. 9. Queries based on model 2 of Fig. 5

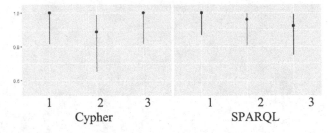

Fig. 10. Accuracy of responses to querying questions shown in Fig. 9

6 Metadata About Predicates

There were two questions in this section; a modelling question and a querying question. Figure 11 shows the modelling question, with the three alternative models used in the Cypher and the RDF*/SPARQL* conditions. For both conditions, model 1 associates the individuals' roles with the edge or predicate; model 2 associates roles with the individual; model 3 associates roles with the company. Cypher model 2 is incorrect because the model associates multiple roles with the node representing 'Adrian'. Cypher model 3 is similarly incorrect because it associates multiple roles with the node 'TransportCo'.

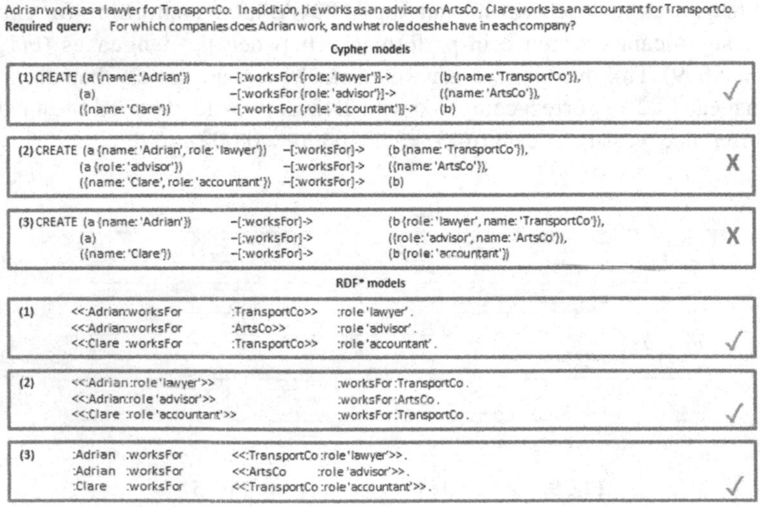

Adrian works as a lawyer for TransportCo. In addition, he works as an advisor for ArtsCo. Clare works as an accountant for TransportCo.
Required query: For which companies does Adrian work, and what role does he have in each company?

Cypher models

(1) CREATE (a {name: 'Adrian'})   –[:worksFor {role: 'lawyer'}]->   (b {name: 'TransportCo'}),
         (a)   –[:worksFor {role: 'advisor'}]->   ({name: 'ArtsCo'}),   ✓
         ({name: 'Clare'})   –[:worksFor {role: 'accountant'}]-> (b)

(2) CREATE (a {name: 'Adrian', role: 'lawyer'})   –[:worksFor]->   (b {name: 'TransportCo'}),
         (a {role: 'advisor'})   –[:worksFor]->   ({name: 'ArtsCo'}),   X
         ({name: 'Clare', role: 'accountant'})   –[:worksFor]->   (b)

(3) CREATE (a {name: 'Adrian'})   –[:worksFor]->   (b {role: 'lawyer', name: 'TransportCo'}),
         (a)   –[:worksFor]->   ({role: 'advisor', name: 'ArtsCo'}),   X
         ({name: 'Clare'})   –[:worksFor]->   (b {role: 'accountant'})

RDF* models

(1)   <<:Adrian:worksFor   :TransportCo>>   :role 'lawyer' .
       <<:Adrian:worksFor   :ArtsCo>>   :role 'advisor' .   ✓
       <<:Clare :worksFor   :TransportCo>>   :role 'accountant' .

(2)   <<:Adrian:role 'lawyer'>>   :worksFor :TransportCo .
       <<:Adrian:role 'advisor'>>   :worksFor :ArtsCo .   ✓
       <<:Clare :role 'accountant'>>   :worksFor :TransportCo .

(3)   :Adrian :worksFor   <<:TransportCo :role 'lawyer'>> .
       :Adrian :worksFor   <<:ArtsCo   :role 'advisor'>> .   ✓
       :Clare   :worksFor   <<:TransportCo :role 'accountant'>> .

Fig. 11. Modelling question using metadata about predicates

Figure 12 shows the participants' mean preference ratings. Participants' comments regarding the Cypher models generally confirmed their awareness that models 2 and 3 were incorrect. For the RDF* models, the responses confirmed the preference for model 1; one participant noting that "this one seems most intuitive". Regarding model 3, a participant commented "the model fulfils the query but the model is anyways not a good representation of the reality, therefore I say it is incorrect". Three participants

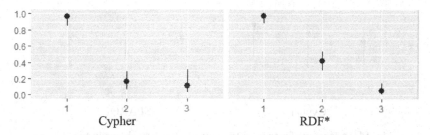

Fig. 12. Preference ratings for modelling question of Fig. 11

were unhappy about the use of string literals for roles. A few commented on the relative closeness of each of the models to the original English. In response to model 3, one participant noted "if you were to change the name of :worksFor to :worksAs, this would match the English almost perfectly".

For the querying question, we presented participants with the English-language query from the previous question, along with model 1 from that question and three proposed formulations of the query. These are illustrated in Fig. 13. For both conditions, query 1 is incorrect, whilst query 3 has been created by reversing query 2.

Figure 14 shows the accuracy with which the participants responded to the proposed queries. The only appreciable difference between the languages occurs with query 3, where directionality was reversed. For Cypher, 94% of the participants realised this was a correct query, against 85% for SPARQL*. An analysis of deviance, after controlling for expertise in the language being studied, showed a difference in performance between the two languages for this query ($\chi^2(1) = 5.2024$, p = 0.023).

Cypher queries

(1) MATCH (m {name: 'Adrian'})–[:worksFor]–>(n)	RETURN n.name, m.role	X
(2) MATCH ({name: 'Adrian'})–[e:worksFor]–>(n)	RETURN n.name, e.role	✓
(3) MATCH (n)<–[e:worksFor]–({name: 'Adrian'})	RETURN n.name, e.role	✓

SPARQL* queries

(1) SELECT ?company ?role
WHERE {
   <<:Adrian :role ?role>> :worksFor ?company .
}
	X

(2) SELECT ?company ?role
WHERE {
   <<:Adrian :worksFor ?company>> :role ?role
}
	✓

(3) SELECT ?company ?role
WHERE {
   ?role ^:role <<:Adrian :worksFor ?company>>
}
	✓

Fig. 13. Proposed queries based on model 1 of Fig. 11

Warren and Mulholland have previously noted that questions with reverse predicates in SPARQL were answered less accurately and more slowly than analogous questions [4]. We suggest that the difference between SPARQL and Cypher may be due to the more intuitive method of representing predicate directionality in Cypher.

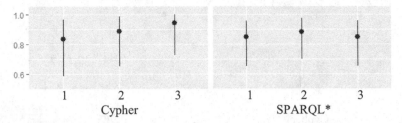

Fig. 14. Accuracy of responses to querying question shown in Fig. 13

7 Metadata About Node Metadata

This section contained only one question, a modelling question concerned with metadata about node metadata, specifically attaching date-stamped population values to a node representing a city. The models are shown in Fig. 15. For the RDF* model 1, the population is associated with the triple, and analogously for Cypher model 1, the population is associated with the edge. For model 2, it is the date which is associated with the RDF* triple and the Cypher edge. For both RDF* models 1 and 2, the population and date are represented as literals. For RDF* models 3 and 4, the date is an IRI; the difference between the two being that model 3 prepends 'year' before the numeric date. Cypher model 3 illustrates how an unlimited number of population-date pairs can be associated with a node, without creating any additional nodes.

Cypher models

```
(1) CREATE (l {name: 'London'}) –[:populationAt {size: 1011157}]–> ((date: 1801)),
        (l)                      –[:populationAt {size: 6226494}]–> ((date: 1901)),      ✓
        (l)                      –[:populationAt {size: 7172036}]–> ((date: 2001))
```

```
(2) CREATE (l {name: 'London'}) –[:hasPopulation {date: 1801}]–>   ((size: 1011157)),
        (l)                      –[:hasPopulation {date: 1901}]–>   ((size: 6226494)),   ✓
        (l)                      –[:hasPopulation {date: 2001}]–>   ((size: 7172036))
```

```
(3) CREATE (l {name: 'London'}) –[:hasPopulation {date: 1801, size: 1011157}]–>  (l),
        (l)                      –[:hasPopulation {date: 1901, size: 6226494}]–>  (l),    ✓
        (l)                      –[:hasPopulation {date: 2001, size: 7172036}]–>  (l)
```

RDF* models

```
(1)   <<:London    :populationAt 1801>> :size 1011157 .
      <<:London    :populationAt 1901>> :size 6226494 .        ✓
      <<:London    :populationAt 2001>> :size 7172036 .
```

```
(2)   <<:London    :hasPopulation 1011157>> :date 1801 .
      <<:London    :hasPopulation 6226494>> :date 1901 .       ✓
      <<:London    :hasPopulation 7172036>> :date 2001 .
```

```
(3)   <<:London    :hasPopulation 1011157>> :date :year1801 .
      <<:London    :hasPopulation 6226494>> :date :year1901 .  ✓
      <<:London    :hasPopulation 7172036>> :date :year2001 .
```

```
(4)   <<:London    :hasPopulation 1011157>> :date :1801 .
      <<:London    :hasPopulation 6226494>> :date :1901 .      ✓
      <<:London    :hasPopulation 7172036>> :date :2001 .
```

Fig. 15. Modelling question illustrating metadata about node metadata

Figure 16 shows the participants' mean preference ratings. For the RDF* models an ANOVA indicated a significant difference between the rankings ($F(3,100) = 14.527$, $p < 0.001$). A subsequent Tukey HSD analysis indicated a significant pairwise difference between all models, except 2 and 3, and 3 and 4.

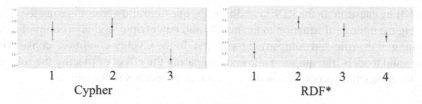

Fig. 16. Preference ratings for modelling question of Fig. 15

Comments about the RDF* models varied. Consistent with the low ranking for model 1, one participant said of this model "A date is more logical to be a qualifying statement than the population. I would not use this." On the other hand, two other participants viewed it favorably, commenting: "this seems the natural order" and "This model allows for the simplest query." Two participants explicitly complained about the use of a literal for a date. On the other hand, one commented on model 4: "I think this should be just an integer". The particular choice of predicate names may influence responses. Commenting on model 1, one of the participants said: "but I would not mind experimenting with something like <<London :existsAt :year1801>> :hasPopulation XXXX".

For Cypher, an ANOVA showed a significant difference between the rankings ($F(2,51) = 15.224$, $p < 0.001$). A subsequent Tukey HSD analysis indicated a significant difference between model 3 and the other two, but not between models 1 and 2. One respondent, in criticizing Cypher model 1, commented: "It seems more intuitive that the value of population should be the size, not the year". On the other hand, another participant said of Cypher model 2: "But horrible... Don't want a separate node for each number." There is a tension here. It seems more natural to make the object node the population, and to make the year a property. Yet, from a practical viewpoint, making the object node the year avoids arbitrarily creating nodes for particular numbers, and instead creates nodes for years which might be reused. A resolution of this tension was suggested by one participant, who commented "Allowing for composite properties - the design of which is under way - would be the best option." The third Cypher model was clearly less popular. One participant looked favourably on it, but made a suggestion which had some similarity to the comment above about composite properties: "The most correct and the best model, but really hard to read. I'd be arguing for nodes representing the measurement itself (e.g. node of type 'census report' or something similar, containing the date, size and information source.)".

The most striking result from this question is the significant preference amongst the RDF* models for the three which placed the date in the outer triple; compared with the lack of a significant preference between Cypher models 1 and 2. For the RDF* models, it seems that the participants saw the date as "more logical to be a qualifying statement than the population". This did not seem such a concern for the Cypher models.

8 Metadata About Predicate Metadata

This section contained a modelling question and a querying question. The models made use of doubly-nested RDF* statements, and their analogues in Cypher. Figure 17 shows the modelling question. In the RDF* condition, the question allowed comparing the effect of placing the timing information in the middle and outer triple, and also comparing the positioning of the role and company information. In the Cypher condition, considering the two valid models, the question allowed comparing the effect of placing the role and timing information as properties in the edge and as properties in the object.

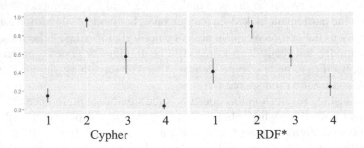

Cypher models

(1) CREATE ({name: 'Adrian'})–[:worksFor {role: 'engineer', from: 2000, until: 2010, role: 'manager', from: 2010}]–>({name: 'BuildCo'})	✗
(2) CREATE (a {name: 'Adrian'})–[:worksFor {role: 'engineer', from: 2000, until: 2010}]–>(b {name: 'BuildCo'}), (a) –[:worksFor {role: 'manager', from: 2010}]–>(b)	✓
(3) CREATE (a {name: 'Adrian'})–[:worksAs {company: 'BuildCo'}]–>({role: 'engineer', from: 2000, until: 2010}), (a) –[:worksAs {company: 'BuildCo'}]–>({role: 'manager', from: 2010})	✓
(4) CREATE ({name: 'Adrian'})–[:worksAs {company: 'BuildCo'}]–>({role: 'engineer', from: 2000, until: 2010, role: 'manager', from: 2010})	✗

RDF* models

(1) <<<<:Adrian :worksFor :BuildCo>> :from 2000>> :role 'engineer'. <<<<:Adrian :worksFor :BuildCo>> :until 2010>> :role 'engineer'. <<<<:Adrian :worksFor :BuildCo>> :from 2010>> :role 'manager'.	✓
(2) <<<<:Adrian :worksFor :BuildCo>> :role 'engineer'>> :from 2000. <<<<:Adrian :worksFor :BuildCo>> :role 'engineer'>> :until 2010. <<<<:Adrian :worksFor :BuildCo>> :role 'manager'>> :from 2010.	✓
(3) <<<<:Adrian :worksAs 'engineer'>> :for :BuildCo>> :from 2000. <<<<:Adrian :worksAs 'engineer'>> :for :BuildCo>> :until 2010. <<<<:Adrian :worksAs 'manager'>> :for :BuildCo>> :from 2010.	✓
(4) <<<<:Adrian :worksAs 'engineer'>> :from 2000>> :for :BuildCo. <<<<:Adrian :worksAs 'engineer'>> :until 2010>> :for :BuildCo. <<<<:Adrian :worksAs 'manager'>> :from 2010>> :for :BuildCo.	✓

Fig. 17. Modelling question illustrating metadata about predicate metadata

Figure 18 shows the participants' mean preference ratings. For RDF*, the two most preferred models had the timing information in the outer triple. An ANOVA of the rankings for the RDF* models shows a significant difference between the models ($F(3,100) = 21.59, p < 0.001$). A subsequent Tukey HSD indicated significant pairwise differences between all models, except models 1 and 3, and 1 and 4.

Fig. 18. Preference ratings for modelling question in Fig. 17

For Cypher, models 1 and 4 are ranked very low, reflecting the fact that these models are incapable of answering the query. Of the two correct Cypher models, there is a clear preference for model 2, i.e. in which the role and timing metadata are included as edge properties. An ANOVA of rankings for the Cypher models shows a significant difference between the models ($F(3, 68) = 66.8$, $p < 0.001$). Moreover, a subsequent Tukey HSD analysis shows a significant difference in ranking between all the models, except models 1 and 4, i.e. the two incorrect models. The significant difference between the two correct models indicates a firm preference for placing the metadata as properties in the predicate.

The querying question used model 2 from the previous question, along with the English-language query shown in Fig. 17. The proposed queries are shown in Fig. 19. For both conditions, question 3 was correct and questions 1 and 2 are distractors. In this case, it was not possible to create analogous distractors for the two conditions.

Cypher queries

(1) MATCH (m {name: 'Adrian'})–[:worksFor]–>({name:'BuildCo'}) RETURN m.role, m.from	X
(2) MATCH ({name: 'Adrian'})–[:worksFor]–>(n {name:'BuildCo'}) RETURN n.role, n.from	X
(3) MATCH ({name: 'Adrian'})–[e:worksFor]–>({name:'BuildCo'}) RETURN e.role, e.from	✓

SPARQL* queries

(1) SELECT ?role ?date WHERE { <<:Adrian :worksFor :BuildCo>> :role ?role . <<:Adrian :worksFor :BuildCo>> :from ?date}	X
(2) SELECT ?role ?date WHERE { <<<<:Adrian :worksFor :BuildCo>> :from ?date>> :role ?role}	X
(3) SELECT ?role ?date WHERE { <<<<:Adrian :worksFor :BuildCo>> :role ?role>> :from ?date}	✓

Fig. 19. Queries based on model 2 of Fig. 17

Figure 20 shows the accuracy of response to each question. The lack of equivalence between the distractors makes an overall comparison between the languages difficult. It is safer to compare the two correct queries, query 3. A logistic analysis of deviance, after controlling for the participants' expertise, showed no significant difference in performance between the languages ($\chi^2(1) = 1.2953$, $p = 0.255$).

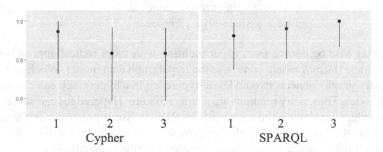

Fig. 20. Accuracy of responses to querying questions shown in Fig. 19

9   Conclusions and Future Directions

In Sect. 1 we set out three objectives: compare ease of use of the two paradigms; identify modelling preferences; and identify any particular difficulties with queries.

With one exception, described in Sect. 6, there was no evidence that one of the language paradigms was significantly easier than the other. The one exception suggests that, for the inversion of a predicate, the approach of Cypher may be preferable to the approach of SPARQL. However, we found no such effect for the questions in Sect. 4. Possibly, the benefits of Cypher's approach become apparent in complex situations. Generally, where models were analogous between the two paradigms, there was an agreement in ranking, underlining our thesis that the differences between the paradigms are not so great. In Sect. 4, there was agreement that cities should preferably betreated as nodes, rather than literals; perhaps to allow the possibility of making further statements about these cities, or representing an aversion to string comparisons in the required queries. A desire to avoid string manipulations showed itself in Sect. 5, where both sets of participants showed the same preference for constructing class hierarchies, despite the label in Cypher being a natural way to represent classes. This argues for support for basic query-time reasoning, e.g. with class hierarchies. In Sect. 8, there was agreement about the way the metadata should be ordered, perhaps corresponding to a natural way of thinking. The one exception to this agreement was in Sect. 7, where there was a significant difference between two RDF* models, but not between the analogous Cypher models.

When we consider the identification of valid and non-valid queries, we find that both sets of participants generally responded with a high degree of accuracy. The exception to this was in Sect. 5, where the query required string manipulation.

Turning to consider future directions, we suggest that, whilst the worlds of open Web and closed commercial applications may continue to have different needs, there is scope for bringing the paradigms closer together. Stardog (https://www.stardog.com/) has illustrated this, by implementing RDF* and SPARQL* with the syntax used in this paper, but also with a semantically equivalent syntax which is similar to Cypher. Thus, the following two lines are equivalent:

```
<<:Pcte a :Engineer>>    :since 2010 .
:Pete a { :since 2010 }:Engineer .
```

There may also be benefit from experimenting with more radical approaches. For example, Vaticle (https://vaticle.com/) use the hypergraph data model. Whereas an edge in an ordinary graph connects two nodes, a hyperedge in a hypergraph can connect any number of nodes. Thus, n-ary relations are a native feature. Hyperedges can also connect other hyperedges, extending the capability in RDF to use a predicate as the subject or object of another predicate. Vaticle's approach also supports query-time reasoning, through the ability to create hierarchies of entities, relations and attributes.

References

1. Hartig, O., Champin, P., Kellogg, G., Seaborne, A.: RDF-star and SPARQL-star Draft Community Group Report'. https://w3c.github.io/rdf-star/cg-spec/editors_draft.html. Accessed 07 Aug 2022
2. Warren, P., Mulholland, P.: Edge Labelled Graphs and Property Graphs; a comparison from the user perspective. ArXiv Prepr. arXiv:220406277 (2022)

3. Holzschuher, F., Peinl, R.: Querying a graph database–language selection and performance considerations. J. Comput. Syst. Sci. **82**(1), 45–68 (2016)
4. Warren, P., Mulholland, P.: A comparison of the cognitive difficulties posed by SPARQL query constructs. In: Keet, C.M., Dumontier, M. (eds.) EKAW 2020. LNCS (LNAI), vol. 12387, pp. 3–19. Springer, Cham (2020). https://doi.org/10.1007/978-3-030-61244-3_1
5. Hartig, O.: Reconciliation of RDF* and property graphs. ArXiv Prepr. arXiv:14093288 (2014)
6. R Core Team, R: A Language and Environment for Statistical Computing. R Foundation for Statistical Computing, Vienna, Austria (2013)

Knowledge Graph Supported Machine Parameterization for the Injection Moulding Industry

Stefan Bachhofner[1]([envelope]) [iD], Kabul Kurniawan[1,2] [iD], Elmar Kiesling[1] [iD],
Kate Revoredo[1] [iD], and Dina Bayomie[1] [iD]

[1] Institute for Data, Process and Knowledge Management,
Vienna University of Economics and Business, Vienna, Austria
{stefan.bachhofner,kabul.kurniawan,elmar.kiesling,kate.revoredo,
dina.bayomie}@wu.ac.at
[2] Austrian Center for Digital Production, Vienna, Austria

Abstract. Plastic injection moulding requires careful management of machine parameters to achieve consistently high product quality. To avoid quality issues and minimize productivity losses, initial setup as well as continuous adjustment of these parameters during production are critical. Stakeholders involved in the parameterization rely on experience, extensive documentation in guidelines and Failure Mode and Effects Analysis (FMEA) documents, as well as a wealth of sensor data to inform their decisions. This disparate, heterogeneous, and largely unstructured collection of information sources is difficult to manage across systems and stakeholders, and results in tedious processes. This limits the potential for knowledge transfer, reuse, and automated learning. To address this challenge, we introduce a knowledge graph that supports injection technicians in complex setup and adjustment tasks. We motivate and validate our approach with a machine parameter recommendation use case provided by a leading supplier in the automotive industry. To support this use case, we created ontologies for the representation of parameter adjustment protocols and FMEAs, and developed extraction components using these ontologies to populate the knowledge graph from documents. The artifacts created are part of a process-aware information system that will be deployed within a European project at multiple use case partners. Our ontologies are available at https://short.wu.ac.at/FMEA-AP, and the software at https://short.wu.ac.at/KGSWC2022.

Keywords: Semantic web · Knowledge graphs · Manufacturing process · Automotive industry · Failure mode and error analysis · Industry 4.0

This research has received funding from the Teaming.AI project, which is part of the European Union's Horizon 2020 research and innovation program under grant agreement No. 957402.

1   Introduction

In the manufacturing industry, the rise of the I4.0 paradigm facilitates complex data-driven use cases [12,13]. This has, however, only increased the need for deep domain knowledge to make sense of increasingly abundant data. In this context, Knowledge Graphs (KGs) have emerged as an important tool that is increasingly being adopted in manufacturing applications [5]. KGs can integrate data from a variety of sources and evolve their schema to accommodate growing requirements. Due to these properties, they have been used in industrial settings as a backbone for a variety of downstream tasks such as building digital twins, risk management, process monitoring, machine service operations, and factory monitoring [9]. Furthermore, they increasingly provide a foundation for machine learning and AI-driven applications in enterprise settings in general [2] and manufacturing in particular [27].

This opens up interesting opportunities in quality management. In this paper, we focus on the automotive industry and its supplier networks, in which the management of product quality along the production chain is crucial and the subject of various standards such as ISO/TS 16949:2009 [10]. In the injection moulding industry in particular – which supplies plastic parts to automotive manufacturers – the parameterization of production machines to achieve consistent output is a complex and delicate process that requires substantial domain knowledge [4]. Part of this domain-knowledge is codified in *injection process adjustment protocols*, which our industrial partner has integrated into their quality management processes (cf. Fig. 1). The main objective of such protocols is to document knowledge gained from (often long-term) experience and make it available in digestible form. In addition, Failure Mode and Effects Analysis (FMEA) documents are extensively used to describe those failure modes together with their potential causes and effects.

These documents are important tools to reduce the time machines spend in an unproductive state. In addition, they are used extensively as training materials for employees. However, both the adjustment protocol and FMEA are currently typically maintained in numerous spreadsheets and accompanied by *parameter sheets* in which injection technicians record parameter changes. This document-centric workflow makes it difficult for injection technicians to identify the root cause of product defects as well as to compare and link deviations in sensor measurements to parameter changes made by the injection technician according to the adjustment protocol. In addition, due to misspellings, lack of time, or an incentive mismatch, the recorded changes in the protocol are often plagued by data quality issues. Furthermore, the document-centric approach makes it difficult to operationalize unstructured FMEAs knowledge, for example relating it to issues on the shop floor to derive insights on how to resolve them.

In this paper, we address this challenge and contribute towards the vision of KG-based shop-floor support; more specifically, we propose an approach to enhance quality management on the shop floor that will serve as the backbone for various Artificial Intelligence (AI) applications within a larger process-aware Information System (IS). We integrate *injection process adjustment protocols* and

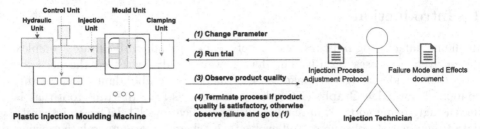

Fig. 1. Sketch of the current parameter adjustment process (plastic injection moulding machine adapted from [4]). Figure 2 shows the process formalized with Business Process Model and Notation (BPMN).

FMEAs within a KG and provide pipelines to iteratively update this knowledge graph from the respective documents. Furthermore, we replace the parameter spreadsheets with a KG that directly receives the changes from the injection moulding machine. To this end, we introduce an *FMEA-injection process adjustment protocol* ontology that extends an existing FMEA ontology [20]. In addition, we provide a software library that transforms spreadsheets into a KG representation. This lowers the entry barrier towards integrating the KG-based approach into the current workflow. We motivate and validate the approach with an application in injection moulding, but expect that our artifacts - the ontologies and the tool - are applicable more generally in other production settings with similar requirements.

The remainder of this paper is structured as follows. Section 2 describes the problem based on the current workflow; Sect. 3 introduces the adjustment protocol ontology, and the adapted FMEA ontology, which we develop based on the documents provided by our industrial partner. Furthermore, this section describes the accompanying software and provides summary statistics on the constructed KG. In Sect. 4, we describe the application of the constructed KG as the backbone for a parameter recommendation system. Section 5 discusses the broader context of the process-aware IS that the KG and components introduced in this paper are part of, outlines empirical evaluation strategies, future work, and limitations. In Sect. 6, we review related work before concluding the paper in Sect. 7.

2   Problem Statement

In this section, we describe the current workflow of the injection moulding processes and outline the research problem. This includes a description of the production process and how adjustments to the injection moulding machine are managed.

Our use case stems from a large European automotive supplier that produces plastic car parts through injection moulding. To develop the use case, we conducted a domain analysis and modeled the current processes in Business Process Model and Notation (BPMN) [19]. The injection moulding process itself is complex – due to space constraints, we describe only the high-level stages, which

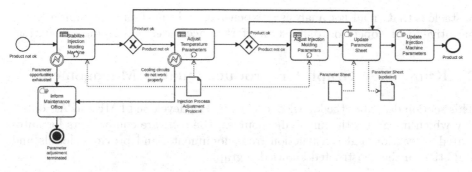

Fig. 2. Injection process adjustment protocol process in a BPMN model. The tasks of the process rely on documents, paper or digital, as an input or output.

are (cf. [4]): *(i)* *Filling*, i.e., feeding and melting the materials and injecting them into the mold; *(ii)* *Packing*, i.e., the pressing process to ensure the densely textured product is produced; *(iii)* *Cooling* until the molten material is fully solidified; and *(iv)* *Ejection*, i.e., the mold releasing process after the work piece has cooled to a given temperature. After these steps, the operator performs a quality inspection of the product and gives feedback on whether the product is *OK* or *Not OK (NOK)*. If the product is *NOK*, then the operator documents among other data points the type of defects and the condition under which the defect appeared. Based on these observations, the operator can adjust the parameters of the injection moulding machine according to specified procedures to address the issue.

The company manages these procedure during production with an *injection process adjustment protocol* (see Fig. 2). The protocol has three phases with *(i)* machine stabilization, *(ii)* temperature adjustment, and *(iii)* injection parameter adjustment. Within these phases, the protocol defines a sequence of actions that need to be taken to adjust a set of parameters; each action is associated with three aspects, which are *(i)* a parameter priority list, *(ii)* the direction in which the parameter should be changed, *(iii)* and the rate of change. Machine operators use this as a guiding tool, and it is supposed to shorten the time a machine spends in an unproductive state. The protocol requires the operator to manually readjust parameter values on the machine and update the parameter changes into a parameter sheet, which can be a source of error. Furthermore, the adjustment activities are carried out manually and are only supported by the guideline, which is provided in a text processing or spreadsheet format, without an automated feedback loop to the guideline. What is more, an FMEA document already exists that is supposed to be used to map defects to their specific causes and effects. This document is also in spreadsheet format, which makes it difficult to detect deviations from the guideline, identify the root cause of a given defect, and assess to what extent they are helpful, incorrect, and complete. Data quality is another key issue in this context, as it requires the shop floor employee to manually justify the reason for the deviation, which is error-prone and time-intensive. Another obstacle is that an ad-hoc analysis is expensive, and the absence of an immediate incentive for the operator to invest an effort in accurately describing a deviation.

3   Knowledge Graph for Product Quality Management

This section describes the knowledge sources, the developed FMEA-IPAP ontology which integrates the knowledge sources, the software components for automated extraction and construction from documents, and provides details and statistics on the constructed knowledge graph.

3.1   Knowledge Sources

The parameter adjustment knowledge graph integrates two major sources of knowledge with *(i)* the Failure Mode and Effects Analysis (FMEA) documents, and *(ii)* the Injection Process Adjustment Protocol (IPAP). We describe them next in greater detail.

Failure Mode and Effects Analysis (FMEA) is an engineering technique to define, review and identify potential failures and their effects and causes for systems, designs, processes or services [23]. The technique aims to describe, model, and analyze potential failures in order to ultimately improve product quality, increase productivity and reduce waste. This quality management tool has been widely used in many industrial applications and engineering domains. However, it is mostly conceived as a *"boring and complicated human activity"*, given its perception as complying with engineering regulations rather than improving product quality [24].

Injection Process Adjustment Protocols (IPAPs) contain a set of standard procedure definitions for injection parameter adjustment. Injection technicians use them, for instance, to adjust machine stabilization and mass temperature verification as well as parameter modification, where, depending on the type of failure mode (e.g., gloss or marbling) different sets of parameters such as compaction time or pressure, cooling time, and injection speed or pressure are changed individually and iteratively at a fixed rate until the production defect has been resolved. This protocol also includes information on the order in which parameters should be adjusted - the action priority. Action priority is a good example for knowledge derived over a long time period.

3.2   FMEA-IPAP Ontology

To represent concepts from both sources - described above - in an integrated *FMEA-IPAP ontology*, we developed *(i)* a parameter adjustment protocol ontology based on guideline documents used by the industrial partner, and *(ii)* an FMEA ontology based on an existing ontology and the FMEA documents also used by our industrial partner.

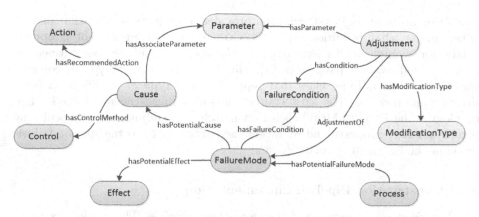

Fig. 3. FMEA Ontology (light blue) and IPAP Ontology (light green). Our FMEA ontology is adapted from [20]. (Color figure online)

To model the domain of interest, we started with a domain analysis that involved workshops and a review of the existing resources for FMEA and IPAP documents obtained from our industrial partner. Next, we conducted a survey of FMEA ontologies and identified the FMEA ontology proposed in [20] as a good candidate for reuse, due to the straightforward conceptual alignment with existing FMEA documents. For the representation of the knowledge from the IPAP documents, we did not find any suitable formalization in the literature and consequently used a bottom-up ontology construction approach [18] to create it.

As shown in Fig. 3, we combine the FMEA and IPAP concepts into a single integrated FMEA-IPAP OWL ontology[1]. The FMEA concepts (light blue) are partially adapted from an existing FMEA ontology proposed in [20]. The FMEA ontology consists of six main classes with seven object properties to link them. The FAILUREMODE class defines the individual failure modes that may happen, such as *Marbling, Gloss, and Burn*. Each failure mode may be associated with the property `hasPotentialCause` to a number of potential causes (defined as CAUSE). Each CAUSE can have a recommended ACTION (linked via `hasRecommendedAction`). The IPAP ontology (light green) consists of four main classes, and four object properties to relate the instances of the classes. We reused several data properties from the Dublin Core Ontology[2] – e.g., `title` and `description` – to describe class details. ADJUSTMENT is the main class of the ontology that represents the individual adjustment protocol. It has data properties such as `actionPriority` that define the priority level of the adjustment (e.g. first, second, and third), and `adjustmentRate` to define the rate of adjustment. Furthermore, the ADJUSTMENT class has object properties to the classes *(i)* `hasParameter` links to PARAMETER, a class that describes specific parameters that need to be adjusted, such as *injection speed, mass temperature,*

---

[1] https://w3id.org/teamingai/resources/ont/FMEA.
[2] https://www.dublincore.org/specifications/dublin-core/dcmi-terms/.

and decompression, (ii) `hasCondition` provides a temporal and spatial context – i.e., under what conditions (FAILURECONDITION) an adjustment is appropriate, for example *in the beginning, middle, last injection* (temporal aspects), *partially,* and *whole* (spatial aspects). *(iii)* `hasModificationType` to connect the MODIFICATIONTYPE class that represents the type of modification (i.e., *decrease* and *increase*) The ADJUSTMENT class also links to classes in the FMEA ontology – the FAILUREMODE class that matches the adjustment protocol (via `hasAdjustment` property and its inverse `adjustmentOf`) to the specific failure mode in the FMEA (e.g., marbling and gloss).

3.3   Construction Pipeline Implementation

We developed an extraction and transformation pipeline. The pipeline extracts the parameter adjustment rules and FMEA statements from spreadsheets, and then transforms them into a KG using the ontologies introduced in the previous section. We found that this automation dramatically reduces the entry barrier for the various stakeholders at our industrial partner. Moreover, it enables users unfamiliar with knowledge graphs to profit from our approach while they can use their familiar tool chain. To manage the creation of and interaction with the KG, and integrate it into existing workflows, we developed an Application Programming Interface (API) that provides three main functions – one function each for the transformations, and one function to recommend parameters. We designed the transformation functions to have the same function signature as they have the same responsibility, but for different ontologies. Function *recommend* is responsible for recommending a parameter adjustment action given a failure that arose under a condition. We refer the interested reader to the repository for further details.

3.4   Knowledge Graph Instance and Statistics

Figure 4 shows an excerpt of the knowledge graph output generated from both FMEA and Injection Process Adjustment Protocol (IPAP) documents. The two sources of knowledge are now integrated and linked. For example, a failure mode MARBLING links to potential cause EXCESSIVE INJECTION SPEED that has an associated parameter INJECTION SPEED; this parameter information has not been provided previously in the FMEA documents, but it is now being linked to the IPAP knowledge. The case is similar for the IPAP document, previously it had no information about the failure cause, but after integrating them into FMEA knowledge, we are able to directly trace the root cause of the failure – EXCESSIVE INJECTION SPEED. The failure mode is also linked to the condition DURING INJECTION PROCESS, which is the same entity as defined in the IPAP. We discuss the benefits of this integration and linking further in Sect. 4. Table 1 shows statistics of the developed ontology and the generated knowledge graph from both FMEA and IPAP. While they might evolve, we consider them as a static part, as the ontologies would change less compared to their instance data.

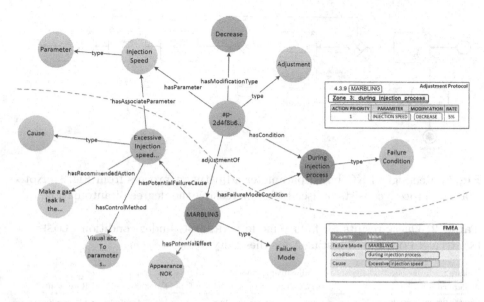

Fig. 4. Knowledge-Graph Output (excerpt) generated from FMEA and IPAP documents.

Table 1. Knowledge graph statistics[a]

	Type	FMEA	IPAP
Axioms	Static	38	73
Class count	Static	4	6
Object property count	Static	4	7
Data property count	Static	3	6
Individual (instance) count	Dynamic	112	222

[a]As per March 2, 2021.

4   Use Case: Shop Floor Parameter Adjustment Recommendation

In this section, we present our shop floor parameter adjustment application enabled by the transformation of heterogeneous semi-structured and non-structured sources to a homogeneous knowledge graph. We also build the foundation to address further issues we raised in Sect. 2. In particular, we migrate from a document-centric to a KG-centric paradigm with explicit semantic relations between the sources, which enables the company to shift tasks in the process towards computer-aided decisions. Figure 5 illustrates the redesigned process, in which the manual tasks from the original process (cf. Fig. 2), which were not computer-supported, have been replaced with computer-supported user tasks. Next, we provide a general description of the use case, followed by an example.

The parameter adjustment procedure makes heavy use of parameter adjustment recommendations, which are retrieved as follows. First, the failure mode

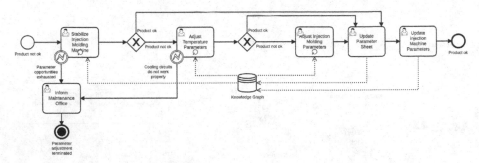

Fig. 5. Redesigned KG-based parameter adjustment process from Fig. 2 - Note that the process consists of user tasks, increasing the degree of automation.

Table 2. Query results for failure mode MARBLING under condition CLOSE TO INJECTION POINT - see Listing 1 for the query.

Failure mode and effects analysis	Adjustment protocol				
Failure cause	Action priority	Modification type	Parameter	Rate	Unit
Excess decompression	1	Decrease	Decompression	1.0	Millimeter
Lack of plasticizing back pressure	2	Increase	Back pressure	1.0	Bar
Excess temperature in hot chamber ...	3	Decrease	Mass temperature	10.0	Celsius

is determined based on automated or manual visual inspection on the shop floor. Next, we obtain a list of prioritized adjustments for the failure using the observed condition. Finally, we retrieve the failure mode, action priority, modification type, parameter, parameter rate, and the unit of the parameter – and sort the results ascending by action priority. The machine operator can then use this information to execute the parameter adjustment.

We illustrate the procedure for the shop floor parameter adjustment recommendation with an example. A machine operator observes a piece with a *Marbling* defect (failure mode) close to the injection point (condition). The operator enters this information on a Human–machine interface (HMI) in order to parameterize the query in Listing 1. The system assists the operator with a list of parameter adjustment recommendations ordered by their action priority - which is the order in which the parameters need to be changed (Table 2). Based on the query result, the operator learns, for instance, that the recommended first adjustment is to decrease the decompression parameter by 1 mm, and the last recommended adjustment is to decrease the mass temperature by 10 Celsius.

Table 2 lists the results for the query in Listing 1, i.e., all *fmea:FailureModes* and their respective failure conditions and failure causes. The failure cause is crucial, as it connects the FMEA with the IPAP via the parameter that is associated with a failure cause. This parameter is in turn connected to an adjustment, which has a modification type (indicating whether to increase or decrease the parameter), and an adjustment rate. The parameter is also connected to a unit. Finally, the query filters for the observed failure and condition.

```
PREFIX dcterm:<http://purl.org/dc/terms/>
PREFIX ap:
↪   <http://www.w3id.org/teamingai/resources/ont/adjustmentProtocol#>
PREFIX fmea: <http://www.w3id.org/teamingai/resources/ont/FMEA#>

SELECT DISTINCT ?failureCause ?actionPriority ?modifType ?param ?rate
↪   ?unit
WHERE {  ?failure a fmea:FailureMode;
                fmea:hasFailureCondition ?con;
                fmea:hasPotentialFailureCause ?failureCause.
         ?failureCause fmea:hasAssociateParameter ?param.
         ?adj ap:hasParameter ?param;
              ap:hasModificationType ?modifType;
              ap:actionPriority ?actionPriority;
              ap:adjustmentRate ?rate.
         ?param a ap:Parameter; ap:unit ?unit.
         FILTER regex(str(?failure),"MARBLING")
         FILTER regex(str(?con),"close.+to.+injection.+point")
         ... }
    ORDER BY ASC(?actionPriority)
    LIMIT 3
```

Listing 1. The parameter suggestion query for failure *marbling* under the condition *close to injection point* - see Table 2 for the query results.

5   Discussion

In this section, we discuss the wider context of the parameter recommendation system introduced in this paper. We start by discussing how the knowledge graph and transformation components are part of a larger envisioned software system that aims to translate human teamwork into the digital age. Next, we present our evaluation strategy. Finally, we finish the section discussing current limitations.

The Teaming.AI platform and KGs. The FMEA and IPAP are production system resources for quality management that we lift from a semi-structured to a semantically explicit structured form. This facilitates an increased degree of automation of the parameter adjustment process, and hence also of the production process. With this increase in automation, however, new challenges arise in all software development phases. It is, for instance, unclear how the system should act if it encounters an uncertain situation, and how such situations should be modelled in the first place.

The Teaming.AI platform – of which the artifacts of this paper are part of – addresses these challenges systematically. It is a software system built with human-computer interaction as the guiding principle in all development phases. This is driven by the growing number of tasks software systems can either take over partially or fully from humans - with the main objective to increase productivity and effectiveness. In this context, it is unclear whom of the two should be the performer, and whom the supporter of a task. To address this problem, the human-computer guiding principle is structured with the big five of teamwork [22] and the 4S interdependence framework [11]. These two frameworks are used to analyze and model tasks that require a form of interaction between humans and the software system, in other words who should be the performer and who the supporter. In these tasks, it is important for the human to have confidence in the decisions made by the software system. Knowledge modeling is a major pillar for the Teaming.AI platform, as it needs to support these requirements, and also application requirements. Because of the crucial role KGs have in this software system, it has two components. First, the dynamic KG is responsible for storing high level events, which are aggregated run time events needed for the process-aware IS. And second, the background KG, which is responsible for storing information on *(i)* organizational roles and responsibilities, *(ii)* products, *(iii)* production system resources, and *(iv)* production processes. In this paper, we describe an important building block for the *production system resources* in the background KG.

Evaluation. Overall, we found that the knowledge graph approach offers flexibility and reduces the cost of integrating additional data sources as well as interoperability within the organizations (and potentially beyond) enabled by Semantic Web standards. An evaluation of our approach beyond a qualitative validation through domain experts is out of the scope of the present paper, but we discuss potential quantitative evaluation approaches we plan in future work. To evaluate the system against the currently used approach, we will compare product quality metrics with and without our approach in place. Although quality can be measured in a variety of ways, we can define it pragmatically for our purposes as a fraction of parts meeting the quality standards relative to the total number of produced parts [15]

$$\text{quality} = \frac{\text{saleable parts}}{\text{produced parts}} \tag{1}$$

Another important variable to observe is the time a machine is in production mode, and the average time it takes to resolve an issue. We hypothesize that quality increases as a result of an increase in a productive state, which itself is a result of an increased speed in resolving issues.

Limitations. In this paper, we use a KG as a backbone for a quality management use case within the automotive industry. We validated the two ontologies and the software qualitatively in the context of the current use case of our industrial partner, but leave an investigation into the generalizability to other organizations and use cases for future work. We found that KGs are a useful approach in this context and support a broader vision of KG-enabled teaming, yet they are not the only candidate technology to address the use case requirements. For instance, a relational, key-value, or key-document data base could be used instead of a KG.

They are less flexible and less suitable for the broader vision of enabling Human-AI teaming, but the cost of integration may be similar or lower compared to using a KG. Furthermore, the pool of developers familiar with such technologies is also still larger, although this is not an inherent limitation of the proposed approach. Finally, an evaluation of our approach with the above mentioned evaluation strategies will only provide insights on whether computer aided support had an effect, but not that this effect is specific to the use of a KG.

6 Related Work

This paper contributes to the literature on Semantic Web (SW) technologies in manufacturing, specifically in the context of a real-world quality management use case from the automotive industry. In the following, we review related work on FMEA ontologies.

FMEA Ontologies. One of the first contributions focusing on structured FMEAs are Ref. [6, 14, 16, 21]. Lee et al. [16] introduces the DAEDALUS knowledge engineering framework that integrates product *"design and diagnosis"* with the purpose of *"exchanging and integrating design FMEA and diagnosis models"*. They therefore focus on connecting product design and diagnosis. Along similar lines, [6] discusses two management problems in relation to FMEA, and in knowledge management more generally. They find that *"relevant knowledge may often not be found in an explicit form like databases"*, and that *"the access to knowledge is encumbered with the problem that different actors use different terms to talk about the same topic"*. The authors conclude that *"It [ontologies] can solve the main shortcomings and the resulting problems as mentioned"*. Ref. [14] focus on the second problem raised in [6] and use ontologies to map different functional models to FMEA sheets. A number of approaches also have been developed that aim to make unstructured FMEAs information more structured and semantically explicit. For instance, [26] introduce an FMEA knowledge graph with ontologies in manufacturing processes; [17] describes an FMEA process and software tools for a lead-free soldering process; finally, [25] uses ontologies to ease sharing, reuse, and maintenance of FMEAs in manufacturing processes.

Similar to the approach in this paper, [20] address the problem of using natural language text for FMEAs, which makes it difficult to reuse this knowledge. We refer in this paper to documents in natural language text as non-structured documents. The organization as a result wasted resources on the FMEA document. Our work complements this and is also motivated by waste. In addition, we show how a structured and semantically explicit representation can help to increase the degree of automation. Another similarity to our work is the proposal of an FMEA ontology. Ref. [7] also state that the problem of FMEAs is *"in the form of textual natural language descriptions that limit computer-based extraction of knowledge for the reuse of the FMEA analyses in other designs or during plant operation."*. They also stress the need to move from non-structured to structured FMEAs and base their ontology on ISO-15926 to define general terms as a foundation so *"engineers can build new concepts from the basic set of concepts"*. Finally, Ref. [8] point out the importance of connecting a systems functional dependency description with the FMEA – using the heating, ventilation and cooling system example from the ISO 60812:2006 standard. Including functional dependencies enables *"automated reasoning to test and infer dependencies"* – which we have not considered so far, but will cover in future work.

7   Conclusion and Future Work

The main contribution of this paper is the integration of Injection Process Adjustment Protocol (IPAP) and Failure Mode and Effects Analysis (FMEA) knowledge; and the parameter recommendation applications this enables. We motivated the importance of this problem economically. First, the objective is to reduce waste caused by incorrect parameters. Second, the training of injection technicians is time-intensive and therefore costly and may take up to one year. Third, the long term objective is to fully automate the parameter adjustment process. We show how the integrated knowledge can be used by a parameter adjustment recommendation engine within a process-aware IS, and leave the application for conformance checking as future work. In particular, we show how the parameter adjustment recommendation engine changes the injection parameter adjustment process by switching from manual tasks to user tasks. This, combined with the conformance checking and other systems, may increase the automation even further, for example towards a fully automated process where humans have the supporter role and the software system the performer role.

To this end, we introduced a parameter adjustment protocol ontology and integrated it with a FMEA ontology - which we adapted to our purposes and is introduced in Ref. [20]. Both ontologies are based on a set of documents we received from our industrial partner. We accompany these ontologies with a software library for transforming spreadsheets. We argue that this is important, as it lowers the entry costs to these ontologies for third parties. These artifacts are the foundation for a KG that will be used as the backbone for AI applications for quality management and beyond. The KG can, for example, integrate sensor data from the injection moulding machine, and can itself be an input for machine learning models.

For future work, we plan to integrate data from the injection moulding machine via the parameter class. Specifically, we plan to integrate two sources: *(i)* the parameter settings from the machine, and *(ii)* the actual parameter values observed by sensors within the machine. Integrating the disparate, heterogeneous, and largely unstructured collection of information sources relating to FMEA and IPAP enables applications beyond the parameter adjustment recommendation engine we present here. Conformance checking, for instance, can leverage the KG to compare actual versus recommended parameter changes, which may reveal an incomplete and partially incorrect adjustment protocol. Furthermore, we used BPMN so far as a process modeling language to document and redesign the parameter adjustment processes, but not as part of the developed system. In future work, we aim to add process context to the KG, for which we aim to adopt the concept of a modular KG to organize different contexts [1]. This is linked to the idea of a layered KG, where a KG has different meanings to different stakeholders - which we plan to explore [3]. Finally, adding the just mentioned process context and using it as an active system component is an important enabler towards higher degrees of automation using service or script tasks. Using the process model actively is also important as it makes the performer and supporter roles mentioned in the previous paragraph usable in production. Maybe even more important, it makes these roles explicit for all stakeholders.

Acknowledgements. This work received funding from the Teaming.AI project in the European Union's Horizon 2020 research and innovation program under grant agreement No. 95740.

References

1. Bachhofner, S., Kiesling, E., Kabul, K., Sallinger, E., Waibel, P.: Knowledge graph modularization for cyber-physical production systems. In: International Semantic Web Conference (Poster), Virtual Conference, October 2021
2. Bellomarini, L., Fakhoury, D., Gottlob, G., Sallinger, E.: Knowledge graphs and enterprise AI: the promise of an enabling technology. In: 2019 IEEE 35th International Conference on Data Engineering (ICDE), Macau SAR, China, pp. 26–37. IEEE, April 2019
3. Bellomarini, L., Sallinger, E., Vahdati, S.: Knowledge graphs: the layered perspective. In: Janev, V., Graux, D., Jabeen, H., Sallinger, E. (eds.) Knowledge Graphs and Big Data Processing. LNCS, vol. 12072, pp. 20–34. Springer, Cham (2020). https://doi.org/10.1007/978-3-030-53199-7_2
4. Bozdana, A., Eyercioglu, Ö.: Development of an expert system for the determination of injection moulding parameters of thermoplastic materials: EX-PIMM. J. Mater. Process. Technol. **128**(1), 113–122 (2002)
5. Buchgeher, G., Gabauer, D., Martinez-Gil, J., Ehrlinger, L.: Knowledge graphs in manufacturing and production: a systematic literature review. IEEE Access **9**, 55537–55554 (2021)
6. Dittmann, L., Rademacher, T., Zelewski, S.: Performing FMEA using ontologies. In: 18th International Workshop on Qualitative Reasoning, Evanston, IL, USA, pp. 209–216, August 2004
7. Ebrahimipour, V., Rezaie, K., Shokravi, S.: An ontology approach to support FMEA studies. Expert Syst. Appl. **37**(1), 671–677 (2010)
8. Hodkiewicz, M., Klüwer, J.W., Woods, C., Smoker, T., Low, E.: An ontology for reasoning over engineering textual data stored in FMEA spreadsheet tables. Comput. Ind. **131**, 103496 (2021)
9. Hubauer, T., Lamparter, S., Haase, P., Herzig, D.M.: Use cases of the industrial knowledge graph at siemens. In: International Semantic Web Conference (P&D/Industry/BlueSky), Monterey, CA, USA, October 2018
10. ISO Central Secretary: Quality management systems - particular requirements for the application of ISO 9001:2000 for automotive production and relevant service part organizations. Standard ISO/TS 16949:2009 (2009)

11. Johnson, M., Vera, A.: No AI is an island: the case for teaming intelligence. AI Mag. **40**(1), 16–28 (2019)
12. Kagermann, H., Wahlster, W., Helbig, J., et al.: Recommendations for implementing the strategic initiative Industrie 4.0: final report of the Industrie 4.0 working group. Technical report, Berlin, Germany (2013)
13. Klingenberg, C.O., Borges, M.A.V., Antunes, J.A.V., Jr.: Industry 4.0 as a data-driven paradigm: a systematic literature review on technologies. J. Manuf. Technol. Manag. (2019)
14. Koji, Y., Kitamura, Y., Mizoguchi, R., et al.: Ontology-based transformation from an extended functional model to FMEA. In: 15th International Conference on Engineering Design, pp. 323–324. Melbourne, Australia, August 2005
15. Kronos Incorporated: Overall Labor Effectiveness (OLE): Achieving a highly effective workforce (2007). https://workforceinstitute.org/wp-content/uploads/2008/01/ole-achieving-highly-effective-workforce.pdf. Accessed: 02 June 2022
16. Lee, B.H.: Using FMEA models and ontologies to build diagnostic models. AI EDAM **15**(4), 281–293 (2001)
17. Molhanec, M., Zhuravskaya, O., Povolotskaya, E., Tarba, L.: The ontology based FMEA of lead free soldering process. In: 34th International Spring Seminar on Electronics Technology (ISSE), Tratanska, Lomnica, Slovakia, pp. 267–273. IEEE, May 2011
18. Noy, N.F., McGuinness, D.L., et al.: Ontology development 101: a guide to creating your first ontology (2001)
19. OMG: Business Process Model and Notation (BPMN), Version 2.0.2, December 2013. http://www.omg.org/spec/BPMN/2.0.2
20. Rehman, Z., Kifor, C.V.: An ontology to support semantic management of FMEA knowledge. Int. J. Comput. Commun. Control **11**(4), 507–521 (2016)
21. Russomanno, D.J., Bonnell, R.D., Bowles, J.B.: Functional reasoning in a failure modes and effects analysis (FMEA) expert system. In: Annual Reliability and Maintainability Symposium 1993 Proceedings, Atlanta, GA, USA, pp. 339–347. IEEE, January 1993
22. Salas, E., Sims, D.E., Burke, C.S.: Is there a "big five" in teamwork? Small Group Res. **36**(5), 555–599 (2005)
23. Stamatis, D.H.: Failure Mode and Effect Analysis: FMEA from Theory to Execution. Quality Press, Milwaukee (2003)
24. Wu, Z., Liu, W., Nie, W.: Literature review and prospect of the development and application of FMEA in manufacturing industry. Int. J. Adv. Manuf. Technol. **112**(5–6), 1409–1436 (2021)
25. Xiuxu, Z., Yuming, Z.: Application research of ontology-enabled process FMEA knowledge management method. Int. J. Intell. Syst. Appl. 34–40 (2012)
26. Zhao, X., Zhu, Y.: Research of FMEA knowledge sharing method based on ontology and the application in manufacturing process. In: 2nd International Workshop on Database Technology and Applications, Wuhan, China, pp. 1–4. IEEE, November 2010
27. Zhou, B., Svetashova, Y., Pychynski, T., Kharlamov, E.: Semantic ML for manufacturing monitoring at Bosch. In: International Semantic Web Conference (Demos/Industry), Virtual Conference, November 2020

IPR: Integrative Policy Recommendation Framework Based on Hybrid Semantics

Divyanshu Singh[1] and Gerard Deepak[2(✉)]

[1] Department of Mathematics, Birla Institute of Science and Technology, Pilani, India
[2] Department of Computer Science and Engineering, Manipal Institute of Technology Bengaluru, Manipal Academy of Higher Education, Manipal, India
gerard.deepak.christuni@gmail.com

Abstract. Policy recommendations aim to inform people who are faced with policy decisions on specific issues about how research and evidence can assist them in making the best decisions possible. This paper proposes an ontology-focused semantical driven integrative system for policy recommendation. The recommendation is user query-centric and uses Structural Topic Modelling to find topics that can be correlated. The semantic similarities are computed using Resnik and concept similarity methods to achieve ontology alignment, and for the alignment of principle classes, three models, normalized compression distance, Twitter semantic similarity, and Hiep's Evenness Index, are used. The IPR achieves the best-in-class accuracy of 94.72% and precision of 93.14% for a wide range of recommendations over the other baseline models, making it an efficient and semantically compliant system for the policies recommendation.

Keywords: Twitter semantic similarity · Resnik and concept similarity · Hiep's Evenness Index · RDF · NCD · Structural topic modelling

1 Introduction

Policies are guidelines or principles that govern decisions and lead to positive outcomes that benefit the community or unit. That leads to the development of procedures and protocols to ensure that policies are executed appropriately. Public policies are based on balancing individual and social values. Policy analysis helps us find the solutions to practical problems brought to the government's plan. Understanding how public policy works can help us better resolve problems. Since the amount of content available on the internet has increased dramatically after the internet's inception. This overabundance has resulted in a spot of bother for the end-user, who generally ends up with a handful of data

B. Villazón-Terrazas et al. (Eds.): KGSWC 2022, CCIS 1686, pp. 121–132, 2022.
https://doi.org/10.1007/978-3-031-21422-6_9

that may or may not suit his needs. Thus arises the need to recommend accurate and relevant data to the user's needs. A recommender system is a filtration solution that provides relevant and necessary data to a user based on the user's input queries and data.

The lack of knowledge about these policies can create multiple losses or problems. One spends an intemperate amount of time finding the proper documents for the government policy. The system proposed in this paper enables a semantic approach to recommend such government policies.

The policy recommendation is quite crucial in the present-day time because policies are required to implement irrespective of the domain. Policies serve as standard guidelines to ensure that specific rules and regulations are followed. To implement rules and regulations, policies have to be met, and to do this the existing policies have to be recommended for which recommendation systems are required sometimes it is very difficult to find the policy document for specialized specific instances because of the large amount of policies documents but lack of relevant policy documents.

There is always a need for a semantically inclined approach to policy recommendation which would not only try to identify the documents containing policies but also be able to relate existing policies through the query and moreover give a crisp answer. Structuring documents for recommending policies is the need of the hour such that the external existing intelligent information system is able to relate the query with existing policies and rephrase and reframe the policies in accordance with the query that has been input. To know that a semantically inclined strategic approach for knowledge-centric policy recommendation is the need of the hour and semantically inclined models based on semantic rules-based inference with preferential learning and differential hybrid semantics ensure an integrative collective intelligence driven model for recommending the policies.

Motivation: With the web's current structure, namely the Web 3.0 or the semantic web, there emerges a need for Semantically Infused recommendation strategies. Recommendations involving highly specialized domains like the judicial domain are challenging using traditional means, mainly because a lot of domain knowledge is needed. Moreover, the judicial aspects differ from country to country in the socio-legal judicial domain. Hence, it is generally difficult to generalize such systems; specific domain knowledge from different countries is essential for the system's support. This bolsters the need for the semantically inclined recommendation of socio-legal judicial documents.

Contribution: A semantically inclined recommender system IPR is proposed for the recommendation of government policies. The model incorporates generated Resource Description Framework (RDF). Structural Topic Modelling (STM) is used for topic modelling and is subjected to RNN and deep learning classifiers. The auxiliary knowledge is provided through government policy portals, Wikidata, and government policy blogs. The semantic similarities are computed using Resnik and concept similarity methods to achieve ontology alignment, and for the alignment of principle classes, three models, normalized compression distance, Twitter semantic similarity, and Hiep's Evenness Index, are used.

Organization: The remaining paper is organized as follows: Section 2 depicts related works, Sect. 3 depicts the proposed system architecture, Sect. 4 depicts results and the whole paper is concluded in Sect. 5.

2   Related Works

Tong et al. [1] present a model based on text mining which consists of Policy structure division; Attribute extraction; Matching, and recommendation. Their approach suggestion results cover policy text pieces, related legislative organizations, quality substances, legitimate relations, and connections to the whole policy text. Alessandra et al. [2] put forward two approaches firstly, an ontology-driven approach that heavily relies on the expressive features of Description Logic (DL) languages. The second approach is a rule-based system that encodes policies as Logic Programming (LP) rules. They also describe a hybrid approach that exploits the expressive capabilities of both DL and LP approaches. Olga et al. [3] have put forth a recommendation system that can add adaptive navigation support to existing learning management systems to overcome the current limitations of organizing training systems in terms of personalization and accessibility. Hyunsook et al. [4] have discussed In-depth curriculum and syllabus ontologies that were created. They also propose a method for syllabus integration and classification based on the definition of the syllabus's semantic model, claiming that this approach aids adaptive concept sequencing and syllabus sharing. Joshi et al. [5] proposed a semantic-based machine-processable structure to observe digital security strategy and populate an information diagram that effectively captures various consideration and prohibition terms and rules inserted in the approach They depict this system using Natural Language Processing, Modal/Deontic Logic, and Semantic Web as well as AI innovations. Ge et al. [6] have proposed a model for instance matching using concept similarity and semantic distance in a highly cohesive environment like the Web 3.0. Lu et al. [7] have put forth a hybrid model which is semantically inclined for recommendations personalization in support of e-government businesses. The personalization was with respect to e-services rendered based on business to business model for a government linked scheme in support of semantic techniques. Deepak et al. [8] proposed an intelligent system for webpage recommendation encompassing semantics using ontologies. Ontologies served as indicators and provisioned Knowledge Map in support of recommendations for a Web 3.0 environment. Adithya [9] an ontology focused collective knowledge approach for requirement traceability modelling. In European, Asian, Middle Eastern, North African Conference on Management Information Systems. Fernando et al. [10]

have proposed a framework for improving electronic communication for government agencies targeting several ethic groups using ontologies in multi-faceted dialect encompassing Language Processing Techniques. Deepak et al. [11] have put forth a differential semantic algorithm for recommending web pages where differential vibrational thresholds on Adaptive Pointwise Mutual Information measure was applied. Fernando et al. [12] put forth an approach for sharing, retrieval and exchange of Legal Documents in support of e-governmental policies using Ontology Driven Methods. Panchal et al. [13] have proposed a framework which is an Ontology Driven Semantic Model in support of higher education in public universities. Sanju et al. [14] have formulated methodologies for representing domain knowledge focusing on Web of Things. In [15] Tiwari et al., have formalized the study of knowledge graph construction and put forth their opinion in building knowledge graphs across several interrelated domains in the real-world Web. Yethindra et al. [16] proposed a fashion recommendation model in support of Web 3.0 and its dense Open Linked Format by integrating auxiliary knowledge, domain based fashion experts opinion, personalized information from Browsing History and Machine Intelligence Incorporation through Logistic Regression, Variational Inferencing and Ontologies.

3   Proposed System Architecture

The proposed system architecture is the government policy recommendation model, which is ontology-focused semantical driven, in correlation with standards of Web 3.0. Figure 1 illustrates the architecture diagram for the proposed government policies recommendation system from web data and input queries provided by the user. Initially, the input user query is analyzed, and information for the policies recommendation is taken and sent for pre-processing; the input query obtained is too subject to pre-processing. The pre-processing of the query data entails Lemmatization, Tokenization, named entity recognition and stop word removal.

These processes are performed using the python natural language toolkit. To yield the individual query words, many people seek different ways. The user query results in the individual informative query words, which are visualized as a set; these query words are subject to ontology alignment. Ontology alignment is the mapping of concepts and the sub concept as well as individuals with each other, which is achieved by computing similarity between the concepts and sub-concepts; however, the semantic similarity between individuals has not been computed to avoid ambiguity and to ensure that the proposed model is computationally less complex.

The ontology used here is the enhanced government policy ontology used for matching, which is obtained by modeling government policy ontology based on human cognition. Manually modeled ontology based on human cognition is initially used as a seed ontology, the ontology of government policy. This is further subjected to the enrichment of term aggregation and ontology enhancement by gathering information specifically from several blogs, government policy portals,

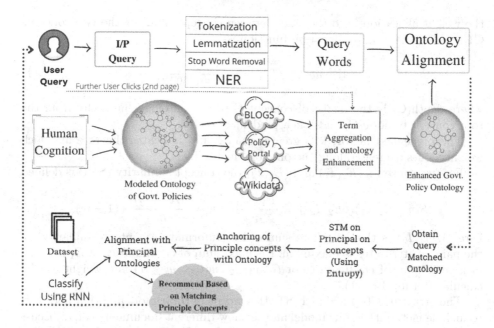

Fig. 1. Proposed system architecture design for the IPR model

and the scraped relevant data from the World Wide Web. We are also using wiki-data as a knowledge store in these cases. The wikidata knowledge store was also used before term aggregation and ontology enhancement which further yields the enhanced government policy ontology, which is used for ontology alignment based on the concept and sub concept alignment with that of the preprocessed query terms.

Ontology alignment is achieved by computing the semantic similarity using two models. The first model is Resnik similarity, and the second similarity is concept similarity. These two similarities are used here to work with the large volume and size of enhanced government policy ontology. Both Resnik and concept similarities are used at the threshold of 0.75, which is called ontology alignment. Aligned ontology with that of query words further used for Structural Topic Modeling (STM).

Resnik presents a new measure of semantic similarity based initially on an English lexical database of concepts and relations, WordNet. The measures include three augmenting path-based measures and two path-based measures with corpora information content statistics.

The Concept Similarity matching method based on semantic distance comprehensively takes into account the inheritance and semantic distance relationships among concepts, and uses semantic similarity to determine the degree of matching between concepts. The algorithm calculates semantic similarity between concepts using various macro steps and gains human intuition similarity. The whole concept similarity matching method is proposed in [6]. Here, the weight allocation is to the edge between concepts and for the two concepts C1 and C2, the weight allocation function is:

$$W[sub(C_1, C_2)] = 1 + \frac{1}{k^{depth(C_2)}} \tag{1}$$

where depth(C) is the depth of concept C in the ontology hierarchy from the root concept to node C, and k is a predefined factor.

The predefined factor, k is greater than 1 and indicates the rate at which weight values decrease along the ontology hierarchy.

For two concepts (E_1, I_1) and (E_2, I_2) the concept similarity (Sim) is defined as:

$$Sim((E_1, I_1), (E_2, I_2)) = \frac{|E_1 \cap I_1|}{r} * w + \frac{M(I_1, I_2)}{m} * (1 - w) \tag{2}$$

here, $M(I_1, I_2)$ is the set where sum of ics (Information content similarity) of the pairs of attributes is maximum, r is maximum of cardinalities of E_1 and E_2, m is maximum of cardinalities of I_1 and I_2, $w(0 \leq w \leq 1)$ is a weight can be calculated using Eq. (1).

The Structural Topic Model (STM) is a type of topic modeling that allows us to include metadata in our model and see how different documents could discuss the same underlying topic using different word choices. It is part of the Bayesian generation topic model, which assumes that each topic is a set of words and that each document is a combination of topics within the corpus. The STM allows for quick, transparent, and repeatable analyses with few a priori assumptions about the texts being studied. Researchers can use the STM algorithm to find topics that can be correlated and estimate their relationships to document metadata using the STM algorithm.

Here, STM is identified as the principal concept. This principal concept is determined by computing the entropy randomly on the obtained query-matched ontology. However, the entropy between the query word and the query matched ontology subset is computed and noted with the highest entropy considered the principal concept and structure topic modeling. This principal concept is subjected to STM and then encored with other concepts in the enhanced government policy ontology.

The pre-processed categorical government policy dataset is classified using the RNN. Artificial recurrent neural networks (RNNs) are a broad and diverse class of computational models inspired by biological brain modules in some way. RNN is a neural network with hidden states that allows previous outputs to be used as inputs. RNN is a deep learning algorithm that works with time series or sequential data. In an RNN, the information cycles through a loop. It considers the current input and what it has learned from the previous inputs and makes its decision. The reason for using RNN is that it is a deep learning model. Second, the recurrent neural network incorporates automatic feature selection so that a large volume dataset can easily be classified using RNN (Fig. 2).

$$h_{(t)} = \sigma_{(h)}(U_{(h)}.x_{(t)} + W_{(h)}.h_{(t-1)} + b_{(y)}) \tag{3}$$

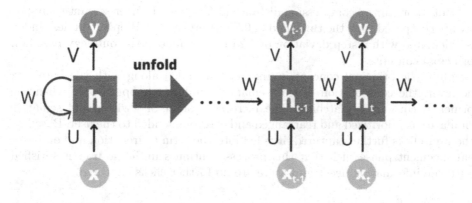

Fig. 2. Structure of Recurrent Neural Network (RNN)

$$y_{(t)} = \sigma_{(y)}(V_{(h)} \cdot h_{(t)} + b_{(y)}) \tag{4}$$

$x_{(t)}$: Input Vector, $y_{(t)}$: Output Vector, $\sigma_{(h)}$ and $\sigma_{(y)}$: Activation Functions, W and U : Parameter matrices, $b_{(h)}$ and $b_{(y)}$: Bias Vector

The principal classes are aligned again with the anchored principal concepts in the ontology cluster. For this alignment, we use three models: normalized compression distance (NCD), Twitter semantic similarity, and Hiep's Evenness Index.

The normalized compression distance (NCD) is a method of determining object similarity. The NCD is a set of distances that have been parametrized using the compressor Z. The higher Z is, the nearer the NCD approaches the NID, and also the better the results are. With Kolmogorov complexity function C(x), normalized compression distance NCD (x, y) is defined as:

$$NCD(x, y) = \frac{C(xy) - min(C(x), C(y))}{max(C(x), C(y))} \tag{5}$$

Twitter Semantic Similarity (TSS) is a semantic similarity measure supported social network Twitter that is time-dependent with static similarity measure. But it's a high temporal resolution for detecting real-world events and induced changes within the distributed structure of semantic relationships across the whole lexicon.

The Hiep's Evenness Index outperforms other evenness indices statistically in a low-diversity neighborhood of copepods with in benthos of shallow saline water habitat; it is the only index that shows no significant deviation from normality. Hiep's Evenness Index is defined as:

$$HE(Hiep's Evenness Index) = 1 + \frac{e^H - 1}{S - 1} \tag{6}$$

here H represents the Shannon's index of diversity and S represents species richness. The index ranges from 0 to 1 and measures how similarly the species extravagance adds to the all-out overflow or biomass of the local area.

Both normalized compression distance (NCD) and Twitter semantic similarity are computed with the threshold of 0.75. However, the Hiep's Evenness Index is calculated with a step deviation of 0.25 to recommend the ontology based on principal concepts.

Only the principal concept is first recommended along with the ontology; however, the computation of this semantic similarity happens only when the principal concept is matching. The increasing order of Twitter and semantic similarity is prioritized and rearranged and is recommended to the user. Based on the user clicks further captured, it is fed into the term aggregation and ontology enhancement phase such that this process continues until the user is satisfied with the information needs and there are no further clicks captured.

4 Results

The proposed IPR (integrative Policy Recommendation) framework is in comparison with two baseline modes, namely BizSeeker [7] and TMPR [1]. Since there is a lacune of models which recommend government policies. Experiments have been conducted by combining SVM, KNN with cosine similarity, and Adaboost with K-means clustering. Precision-recall accuracy, F-measure percentages, and the false discovery rate (FDR) are used as potential matrix precision-recall accuracy. F measure computes the relevance of the result. In contrast, FDR quantifies for several false positives discovered by the model. It is indicative in Table 1 that the proposed IPR yields the highest precision, recall, and accuracy of 93.14%, 96.29%, 94.72%, respectively, with the average F-measure of 94.69 with a very low FDR of 0.07.

The BizSeeker results average precision, recall, and accuracy of 78.68%, 80.44%, 79.56%, respectively, with average F-measure and FDR of 79.55, 0.22 respectively. TMPR yields average precision, recall, and accuracy of 88.24%, 90.18%, 89.21%, respectively, with an average F-measure of 89.20 and FDR of 0.12. SVM along with KNN and Cosine similarity results average precision, recall, and accuracy of 82.12%, 86.77%, 88.45%, respectively, with average F-measure and FDR of 84.38, 0.18 respectively, whereas Adaboost along with K-means clustering yields average precision, recall, and accuracy of 84.45%, 86.33%, 85.39%, respectively, with average F-measure of 85.38 and FDR of 0.16 (Table 2).

Table 1. Performance comparison of the proposed IPR with other baseline methods.

Models	Precision %	Recall %	Accuracy %	F-measure
BizSeeker [7]	78.68	80.44	79.56	79.55
TMPR [1]	88.24	90.18	89.21	89.20
SVM + KNN	82.12	86.77	88.45	84.38
Adaboost + K-means	84.45	86.33	85.39	85.38
Proposed IPR	93.14	96.29	94.72	94.69

Table 2. FDR values of the proposed IPR with other baseline methods.

Models	FDR
BizSeeker [7]	0.22
TMPR [1]	0.12
SVM + KNN	0.18
Adaboost + K-means	0.16
Proposed IPR	0.07

The proposed IPR yields the highest accuracy, average precision, average recall, F measure, and the lowest FDR mainly because it is semantically enriched and empowered. It is the hybrid model which uses strong semantics. Ontology alignment is the core of the model, along which structural topic modeling is selected using entropy. RNN and deep learning classifiers are used for the automatic classification of the dataset. Auxiliary knowledge is provided through government policy portals, Wikidata, and government policy blogs. Concept similarity, Resnik similarity, and Twitter semantic similarity are used for computing the semantic similarity at several stages, sometimes standing alone for ontology alignment with using concept similarity or computing the relevance they use hybridized with various differential thresholds. Combining three different semantic similarity models ensures that the relevance computation is much more at a higher rate. As a result, the proposed IPR yields the highest F measure, precision-recall accuracy with the lowest FDR rate.

The Bizseeker model is a very traditional renowned model that uses collaborating filtering. Collaborating filtering is lacking because item-based similarity item-item similarity has to be computed wherein the rating plays a vital role. Every government policy need not be read, and who raises government policies first of all the policy seekers are a very few and it is not very acting in seeking rating for government policy. It is not an eCommerce book store wherein every item might be rated. Government policy getting the rating for the policy itself is a very controversial phenomenon. As a result, collaborating filtering does not work practically; it is not feasible. Moreover, it does not yield any relevant results.

TMPR uses the text mining natural language processing methodologies for policy structure segment fragmentation, policy structure division, attribute extraction, attribution matching, elementary discourse, units' analysis, matching, and NER and other NLP methods constitute the text mining methods TMPR. Though the TMPR attempts to integrate text mining with NLP, it yields above-average results; however, it lacks auxiliary knowledge. As a result, the model does not perform up to the mark. It has to perform in a completely cohesive environment like Web 3.0.

SVM and KNN, and Cosine similarity hybridization use ensure that two naive classifiers and the conventional similarity measures have been used. It yields results, but the results are inappropriate because of the outdated classifier, and auxiliary knowledge is not included.

Although the classifier is robust in Adaboost and K-means clustering, ambiguity with K-means clustering improves the results. However, the classifier's lack of auxiliary knowledge and the lack of power ensure the model does not perform up to the mark.

The dataset used for experimentation is collected from multiple sources. It is a customized and curated dataset. The first dataset in this Combined Government Policy Dataset (CGPD) is the Australian government indigenous program and policy location, AGIL dataset. The AGIL metadata and the AGIL.csv file are taken; it is used as it is. However, the AGIL dataset is also annotated by using the RDF distiller. Specific crawlers are used to crawl online web documents relevant to the annotation, which are generated based on the metadata of the AGIL dataset.

The second curated dataset is based on documents which are based on several themes, namely the Aayushman Bharat Yojna, Skill India Scheme, Smart Cities Mission, Amrut, Pradhan Mantri Avas Yojana, Heritage city development and augmentation Yojna, Deen Dayal Upadhyay Gramin Kaushal Yojna, Gramin Bandara Yojna, JNNURM, Nipun Bharat Mission. These policies are based on the categories like agriculture, education, health. The policies like Digital India for urbanization, Atal Pension Yojna for pension, Bharat Maa Ki Suraksha Beema Yojna for insurance are considered. All the policy documents on the world wide web are crawled, directly or indirectly, and further indexed using an RDF distiller. This is an indigenous dataset specific to India.

The third one is from national portal of India. Policy for .in internet domain registration, auto policy, Kerla state women policy, web policy for Haryana, Industrial promotion policy, schemes - micro, small and medium enterprise government institute Goa, Information on Jammu and Kashmir Industrial Policy, Information on State Home for Women, Himachal Pradesh, Information on Mother Teresa Ashaya Matri Sambal Yojna, Himachal Pradesh, Logistics, Warehousing & Retail Policy, Haryana Textile Policy, Haryana Pharmaceutical Policy 2019, Information on Electric Vehicles Policy 2021, Assam, Information on Handloom Policy 2017–18, Assam, Information on Tourism Policy 2017, Assam, Bihar Industrial Investment Promotion Policy, 2016, Information on Industrial Policy 2019–2024, Chhattisgarh all these policies and related websites are crawled. Apart from the data crawled, 7864 documents comprising these policies were curated, out of which 44892 tags were generated for annotation.

The model is executed for 4481 queries, out of which 2681 queries are single-word queries, and the remaining queries are multiword. The ground truth has been validated by voting from 611 users based on the queries. Each user gave around 50 to 60 queries; however, only 30–40% of the total number of queries were collected for the remaining queries, the ground truth was assumed based on the dataset.

Implementation was conducted using the latest python version with Google Collaboratory as an online development platform.

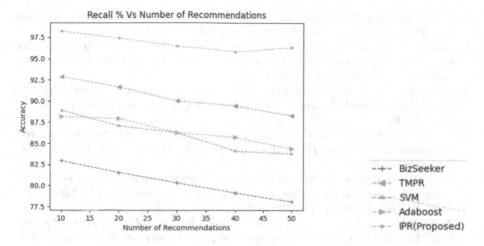

Fig. 3. Recall % vs Number of recommendations of the proposed IPR and other baseline models.

The accuracy at different number of recommendations of each baseline model is shown in Fig. 2. It is evident from the figure that the proposed Integrative Policy Recommendation (IPR) Framework has higher accuracy for the number of recommendations when compared to other baseline models.

5 Conclusion

A model with semantic infused artificial intelligence-driven skills, IPR, is proposed to recommend policies. The model is based on RDF and incorporates user queries web utilization statistics for policy recommendation. The Combined Government Policy Dataset (CGPD), different semantic similarity methods are used at several stages is used, along with it, using ontology alignment and STM results are obtained which are validated by ground truth too. As the proposed IPR is semantically enriched and empowered, it results average accuracy of 94.72% and yields much better results than the other baseline models and makes it an efficient and semantically compliant system for the policies recommendation.

References

1. Zhang, T., Liu, M., Ma, C., Tu, Z., Wang, Z.: A text mining based method for policy recommendation. In: 2021 IEEE International Conference on Services Computing (SCC), pp. 233–240. IEEE (2021)
2. Toninelli, A., Bradshaw, J., Kagal, L., Montanari, R.: Rule-based and ontology-based policies: toward a hybrid approach to control agents in pervasive environments. In: Proceedings of the Semantic Web and Policy Workshop (2005)
3. Santos, O.C., Boticario, J.G.: Requirements for semantic educational recommender systems in formal e-learning scenarios. Algorithms 4(2), 131–154 (2011)

4. Chung, H., Kim, J.: An ontological approach for semantic modeling of curriculum and syllabus in higher education. Int. J. Inf. Educ. Technol. **6**(5), 365 (2016)
5. Joshi, K., Joshi, K.P., Mittal, S.: A semantic approach for automating knowledge in policies of cyber insurance services. In: 2019 IEEE International Conference on Web Services (ICWS), pp. 33–40. IEEE (2019)
6. Ge, J., Qiu, Y.: Concept similarity matching based on semantic distance. In: 2008 Fourth International Conference on Semantics, Knowledge and Grid, pp. 380–383. IEEE (2008)
7. Lu, J., Shambour, Q., Xu, Y., Lin, Q., Zhang, G.: BizSeeker: a hybrid semantic recommendation system for personalized government-to-business e-services. Internet Research (2010)
8. Deepak, G., Ahmed, A., Skanda, B.: An intelligent inventive system for personalised webpage recommendation based on ontology semantics. Int. J. Intell. Syst. Technol. Appl. **18**(1–2), 115–132 (2019)
9. Adithya, V., Deepak, G.: OntoReq: an ontology focused collective knowledge approach for requirement traceability modelling. In: Musleh, A.-S., Abdalmuttaleb, M.A., Razzaque, A., Kamal, M.M. (eds.) EAMMIS 2021. LNNS, vol. 239, pp. 358–370. Springer, Cham (2021). https://doi.org/10.1007/978-3-030-77246-8_34
10. Ortiz-Rodriguez, F., Tiwari, S., Panchal, R., Medina-Quintero, J. M., Barrera, R.: MEXIN: multidialectal ontology supporting NLP approach to improve government electronic communication with the Mexican Ethnic Groups. In: DG. O 2022: The 23rd Annual International Conference on Digital Government Research, pp. 461–463 (2022)
11. Deepak, G., Priyadarshini, J.S., Babu, M.H.: A differential semantic algorithm for query relevant web page recommendation. In: 2016 IEEE International Conference on Advances in Computer Applications (ICACA), pp. 44–49. IEEE (2016)
12. Ortiz-Rodriguez, F., Medina-Quintero, J. M., Tiwari, S., Villanueva, V.: EGODO ontology: sharing, retrieving, and exchanging legal documentation across e-government. In: Futuristic Trends for Sustainable Development and Sustainable Ecosystems, pp. 261–276. IGI Global (2022)
13. Panchal, R., Swaminarayan, P., Tiwari, S., Ortiz-Rodriguez, F.: AISHE-Onto: a semantic model for public higher education universities. In: DG. O2021: The 22nd Annual International Conference on Digital Government Research, pp. 545–547 (2021)
14. Tiwari S., Garcia-Castro R.: A Systematic Review of Ontologies for the Water Domain. In: ISTE Book, ISBN 9781786307644, Wiley (2022)
15. Tiwari, S., Gaurav, D., Srivastava, A., Rai, C., Abhishek, K.: A preliminary study of knowledge graphs and their construction. In: Tavares, J.R.S., Chakrabarti, S., Bhattacharya, A., Ghatak, S. (eds.) Emerging Technologies in Data Mining and Information Security. LNNS, vol. 164, pp. 11–20. Springer, Singapore (2021). https://doi.org/10.1007/978-981-15-9774-9_2
16. Yethindra, D.N., Deepak, G.: A semantic approach for fashion recommendation using logistic regression and ontologies. In: 2021 International Conference on Innovative Computing, Intelligent Communication and Smart Electrical Systems (ICSES), pp. 1–6. IEEE (2021)

Convolutional Neural Networks Applied to Emotion Analysis in Texts: Experimentation from the Mexican Context

Juan-Carlos Garduño-Miralrio[✉], David Valle-Cruz, Asdrúbal López-Chau, and Rafael Rojas-Hernández

Universidad Autonoma del Estado de México, Toluca, Mexico
jgardunom010@alumno.uaemex.mx, {davacr,alchau,rrojashe}@uaemex.mx

Abstract. This work is twofold. First, it presents the state of the art of deep learning applied emotion analysis and sentiment analysis, highlighting the convolutional neural networks behavior over other techniques. Second, it presents experimentation on a convolutional neural network performance in the emotion analysis for the Mexican context, considering different architectures (with different number of neurons and different optimizers). The accuracy achieved in the proposed computational models is 0.9828 and 0.8943 with loss values of 0.1268 and 0.2387 respectively; however, the confusion matrices support the option of improving these models, giving the possibility of improving the values obtained and achieving greater accuracy.

Keywords: Convolutional neural network · Emotion analysis · Sentiment analysis · Accuracy · Social networks · Optimizers

1 Introduction

The human desire to recognize and reproduce emotions is not uncommon. Mankind has been doing it for thousands of years, especially in the field of artificial intelligence, using various computing techniques. Sentiment analysis within affective computing is a branch dedicated to detecting, interpreting, processing and/or simulating human emotions in different domains, but the performance of these tasks is often not precise enough [5].

In addition to this, the variety of expressions that people present in emotions and affections represents a decisive factor in the lack of 100% functional techniques to achieve the objectives of emotional computing, especially emotional analysis. However, thanks to several studies it has been possible to identify a particular pattern that is commonly presented when emotions are externalized,

B. Villazón-Terrazas et al. (Eds.): KGSWC 2022, CCIS 1686, pp. 133–148, 2022.
https://doi.org/10.1007/978-3-031-21422-6_10

on the one hand, there is the research in the psychological field that has led to the consolidation of the emotional frameworks of various characters such as Parrott, Tomkins, Plutchik, and Ekman [2,7]; in the technological environment, research is presented that addresses the improvement of tools and even experimentation for the comparison of the behavior of different technologies in the performance of this task [10,12,18,21,25], which has served as a basis for the development and improvement of precision computing techniques.

The emotion analysis by means of computer technology has been a work in progress for several years, however, it is difficult to merge it with a technique that allows to identify, analyze, process and/or simulate these emotions with total precision. According to Zatarain and colleagues "emotion detection is a complex task even for humans. Using only text or dialogue to detect emotions is a difficult task" [25], because each person has a different perspective on emotions, and in particular, each brain can react differently to a sensation, complicating identifying them.

Affective computing, is an artificial intelligence technique, that aims to research and develop systems and devices capable of recognizing, interpreting, processing, and stimulating human emotions. Besides, it is an interdisciplinary field that includes computer science, neuroscience, and cognitive science [1].

Emotion and sentiment analysis is a way of discovering feelings, opinions, and emotions that people express in their texts, body expressions, and voices, using natural language processing (NLP) by implementing algorithms to determine whether it is positive, negative or neutral in the case of feelings or otherwise the presence of emotions. As this is a complex task, an instrument capable of guaranteeing total accuracy has not been designed; however, over the years the techniques used have improved to the point of obtaining results with more than 90% accuracy, as is the case of the work of Valle-Cruz et al. [21].

This paper aims to present the state of the art of deep learning applied to affective computing, emotion analysis, and sentiment analysis, highlighting the behavior of convolutional neural networks over other techniques. And presents experimentation on a convolutional neural network (CNN) performance in the emotion analysis in the Mexican context.

The paper is divided into six sections, including the foregoing introduction. The second section shows the potential of CNN to emotion analysis and senti- ment analysis. The third section presents the state of the art of deep learning applied to emotion analysis and sentiment analysis. The fourth section pro- poses the construction of the computational model. The fifth section presents the implementation. The final section shows conclusions, limitations, and future work.

2   CNN Applied to Emotion Analysis and Sentiment Analysis

One of the main applications that have been proposed for the emotion analysis and/or feelings is the timely detection of emotions classified as negative (sadness, anger, hatred, resentment, revenge), impatience, jealousy, envy, among others) in order to take preventive or corrective actions, or simply to identify opportunities for improvement or to generate recommendation engines [19] in social networks or streaming platforms such as the Spotify or Netflix recommendation algorithms [16]. Emotion analysis is performed as a classification task with three levels of classification: document-level emotion analysis, sentence-level emotion analysis, and target or appearance-level emotion analysis [10].

Convolutional Neural Networks (CNN) are a deep learning technique that (although it is not a technique of recent innovation) has adopted an important relevance due to the performance it exhibits in its implementation in different nature tasks for example: visual search, recommendation engines, predictive analytics in medicine and more [17, 19].

Convolutional Neural Networks are models that have many applications in artificial intelligence. CNN can be classified into two major fields, the first is the application of CNNs in image processing (as facial recognition in social networks) [4, 24] and the second is the application of CNNs in word processing (such as document recognition and validation or grammar and spelling analysis) [11, 15], each category in turn being applied to different tasks.

CNN applied to text processing has the potential to detect emotions and feelings boosting sentiment analysis. In this regard, sentiment analysis is a very common job in natural language processing, its purpose is to determine the underlying emotion of a text, whether praise or criticism, support or objection. For example, sentiment analysis can be used to analyze social media comments and thus gather people's opinion about something and with further analysis it is possible to obtain trends and audience orientation [15].

3   The State of the Art of Deep Learning Applied to Emotion Analysis and Sentiment Analysis

This section describes relevant research on the application of machine learning and deep learning techniques in the detection of emotions and psychological aspects that encompass the understanding and classification of emotions.

3.1   Important Works of Machine Learning and Deep Learning Techniques

Some important works on the application of machine learning and deep learning techniques in emotion and sentiment analysis are described below.

Casas García, Alejandro, and Villena Román [3] succeeded in designing an algorithm with an efficiency of 42% for a 5-step polarity system and 64% for a 3-level system. Valle-Cruz et al. [21] describes the process in which an algorithm for the analysis of impressions was designed, which had an accuracy of over 90%, thus identifying impressions on trending topics on Twitter.

Experimental results by Konate and Du [10] show that single-layer CNNs and CNN Long Short-Term Memory Network (CNN-LSTM) achieve higher accuracy than LSTM/Long Short Term Memory (BSLTM) based models and Stochastic Gradient Descent (SGD) classifier. As with the classic models, Naïve Bayes (NB) trains slightly better on trigrams than from term frequency-inverse document frequency (TF-IDF). However, similar to deep learning models, support-vector machines (SVM) achieves a lower level of accuracy on trigrams than TF-IDF in comparison with the k-nearest neighbors (KNN) model.

In the study by Zatarain Cabada et al. [25] emotion recognition was performed with different techniques, obtaining the following results according to the implemented techniques shown in Table 1.

Table 1. Accuracy in data processing techniques with deep and machine learning according to the study by Zatarain Cabada et al. [25]

Implemented techniques	Accuracy
Bernoulli NB	76.77%
Multinomial NB	75.31%
SVC	75.79%
Linear SVC	74.69%
SGDC Classifier	76.69%
KNN	68.46%
CNN + LSTM	88.26%

A similar experience is posited in the work of SHI et al. [18], in which three artificial intelligence techniques are tested, an NB classifier, an SVM and an accumulation neural network whose works can highlight the advantage of accuracy in various feature analyzes was performed.

From the obtained results, CNN's results in the analysis of BF and CF features stand out with 90.36% while the NB classifier obtains 78.64% and the SVM is presented with 83.56%. In contrast, the study of Lee et al. [12] propose a short-term emotion recognition method based on single Photoplethysmography (PPG) pulses whose aim is to achieve high accuracy in a short period of time, thus dividing the raw PPG signal into a single pulse. fed into a one-dimensional convolutional neural network (1D CNN) to extract features and classify emotions. The output of the 1D CNN is a binary classifier indicating whether the excitation and valence are high or low. Tripto and Ali [20] proposed a deep learning model to detect sentiments and emotions from YouTube comments in Bengali. After obtaining their data, they preprocessed the text to remove stop words. They then obtained the embedded representation of the words using both Skip-Gram and Continuous Bag of Words (CBOW) in Word2Vec. The output is then introduced as an input to a LSTM architecture defined in the first phase of the model, followed by a second phase consisting of the CNN architecture. Their research results show that they applied deep learning methods; LSTM and CNN are significantly superior to traditional ML methods such as SVM and NB with emotion classification accuracy of 59.2% and multiclass (3 and 5) sentiment labels accuracy of 65.97% and 54.24%, respectively.

3.2   Emotional Frameworks

In the field of neuroscience, various opinions have been given on the classification of emotions, mainly to determine which are the most basic emotions, highlighting 4 works.

Tomkins. According to Tomkins (1984), emotions are grouped into two dimensions: positive (amused, surprised, delighted) and negative (fear, anxiety, anger, shame, disgust). All have innate response patterns that are activated in response to certain stimuli, and facial expression is the primary form of facial expression as the primary vehicle for their expression. One of Tomkins' main contributions was to provide a theoretical framework for the study of facial expression [2].

Plutchik. Plutchik (1958) presented a major and minor emotion model, based on Darwin's evolutionary tradition. Plutchik's model is based on two principles: a) emotions are the body's response to problems in life to better adapt; b) Emotions are structured in pairs of opposites suitability; b) Emotions are structured in pairs of opposites [7]. Accordingly, emotions have value. On the other hand, having emotions represents a step forward in adapting to a more positive situation (positive emotions). On the other hand, there are dangerous situations, difficulties, obstacles that must be overcome in order to adapt (negative emotions). For Plutchik, there are four fundamental issues: identity, timing (reproduction), hierarchy, and territoriality. These problems are shared by individuals and animals [2,14].

OCC Model. The OCC model [7], presents 22 types of emotions, grouped into pairs of opposites: pride-shame, admiration-reproach, happiness-resentment, gloating-compassion, hope-fear, joy-stress, satisfaction-fear-confirmation, relief-disappointment, gratification-memory, gratitude-anger, and love-hate.

Ekman. Paul Ekman is one of the main authors of this method. He contributed to the study of facial expression. Ekman and Friesen (1978) developed a procedure for analyzing facial muscles by analyzing facial muscle movements, called FACS (Facial Action Coding System) [2].

4   Construction of the Computational Model Applied to Emotion Analysis and Sentiment Analysis

In this section, the authors describe technical and scientific specifications taken into account for the design of the computational models of the CNNs used in the experimental section of emotion analysis in texts for the Mexican context.

4.1 Optimizers

The goal of neural network training is to minimize the cost function by finding appropriate weights for the edges of the network (thus ensuring good generalization). These weights are determined by a numerical algorithm called backpropagation. The optimizer is responsible for generating better and better weights: its importance is decisive. Its main function is based on calculating the slope of the cost function (partial derivative) for each weight (parameter/dimension) of the network [22]. The most important optimizers are presented below.

AdaGrad. The AdaGrad algorithm introduces a very interesting variation of the training coefficient concept: instead of considering a uniform value for all weights, a specific training coefficient is maintained for each weight. It is impractical to specifically compute this value, so from the initial training coefficient, AdaGrad scales and adjusts it for each dimension relative to the cumulative gradient at each iteration [8, 23].

RMSProp. It is based on using a moving average of the squares of the gradient and normalizing this value using recent amplitudes of previous gradients. The normalization is done, because otherwise the fact of squaring means that in the case of denormalization we get an excessively high gradient value (namely squaring). RMSProp is a similar algorithm. It also maintains a different training factor for each dimension, but in this case the training factor is scaled by dividing it by the mean of the exponential decrease of the square of the gradient [9, 23].

Adam. Adam's algorithm combines the advantages of AdaGrad and RMSProp. A training coefficient for each parameter is maintained, and in addition to calculating the RMSProp, each training coefficient is also affected by the average gradient pulse [22].

AdaDelta. AdaDelta is a variant of AdaGrad, where instead of calculating the scale of the training factor of each dimension taking into account the gradient accumulated since the start of the implementation, we limit ourselves to a size window fixation of the last n gradients [8].

SGD. It is not possible to compute the partial derivative of the cost function for each weight of the network for each observation, given the number of different weights and observations, it is not possible. Therefore, the first optimization involves the introduction of a random (random) behavior [8]. SGD does something as simple as limiting the derivative to one observation (per batch). For example, there is some variation based on the selection of several observations instead of one (SGD minibatch) [23].

5   Implementation

The following section details the way in which the authors implemented deep learning in a series of experiments with the purpose of testing convolutional neural networks in emotion analysis in texts for the Mexican context, this section is divided into three subsections, which are:

1. Data collection and preprocessing
2. Text polarity classification (positive and negative) with multiple models of different numbers of neurons and different optimizers
3. Emotion analysis in texts for the Mexican context

5.1   Data Collection and Preprocessing

For the experimental phase with the convolutional neural network models, the data collection was performed from the social network Twitter implementing the IFTTT server or in Spanish SIEEE, which is a type of web service that allows to create and schedule actions to automate various tasks and actions on the Internet, from its website but also from its app [6].

Once the data were collected, they were processed, a process of data cleaning by removing unwanted information that interferes with sentiment analysis. The data collection and cleaning procedure based on the Valle-Cruz, et al. [21] process is described below.

- Data collection: Data analysis was performed on the 5000 texts collected by using the IFTTT server.
- Generation of the corpus: A corpus was created, selecting 2500 texts at random from the data obtained. Each one of them was read independently by three people to assign a label to each one, with the values of positive and negative for the case of experiment 1 and of the emotions identified in them for the case of experiment 2, in those cases in which the labels of two of the people do not coincide in a text the label of the third person is implemented with the purpose of having an agreement in the same ones.
- Preprocessing: Preprocessing was a phase prior to feature extraction for emotion analysis, but of utmost importance that involved the following steps:
  - Eliminate auxiliary words or stopwords.
  - Replace accented vowels with their unaccented equivalents.
  - Eliminate junk data from texts such as URLs, usernames, emoticons, numbers, spaces, and punctuation marks.
  - Apply stemming or lemmatization: which is to relate an inflected or derived word to its canonical form or lemma (Urdaneta-Fernández, 2019). All the above preprocessing steps are performed automatically using the Python programming language.
  - Feature extraction: Once the texts were labeled and preprocessed, it was necessary to extract useful features to apply deep learning methods for emotion analysis.
  - Training of convolutional neural networks for emotion analysis.

Example of Text Labeling Procedure. For experiment 1. Considering the text "*I like animals, especially dogs*", there are no negative words or phrases in the text, so in the field In this case, the text can be classified as POSITIVE by all three people. Now, if one considers the text "*I'm still not ready for such a task*", despite the fact that it appears to be a common sentence at first glance, it is classed as a NEGATIVE sentence because it expresses the feeling of words. refuse' to complete a task. In case person 1 marks the sentence as POSITIVE, then the opinion of the third person is taken into account, i.e. if person 1 marks as POSITIVE, person 2 as NEGATIVE and person 3 as NEGATIVE, then the text version will be labeled for the exercise as NEGATIVE.

For experiment 2. Considering the text "*I love animals, especially dogs*" and the primary emotions defined by Parrot (love, joy, surprise, anger, sadness and fear) we can identify in the text the emotions: love and joy which would be the value of the labels. In the event that any of the three people consider that there is some other emotion, the equivalent labels are maintained, for example, if person 1 identifies the emotions LOVE and JOY, person 2 identifies the emotions LOVE and JOY, then these two labels remain as there is a concordance between person 1 and 2. Now, if one considers the text "*I am not yet ready for such a task*", one can identify the emotions SURPRISE and FEAR. For this particular case it is likely that the presence of the emotion of sadness is considered so that, if person 1 identifies the emotions SURPRISE, FEAR and SADNESS, person 2 identifies the emotions SURPRISE and FEAR and person 3 identifies the emotions SURPRISE and FEAR then the latter remain as labels as there is a concordance between persons 2 and 3.

Example of Data Preprocessing. Considering the following text in the Spanish language: "¡¡ Estoy feliz!!!! ☺ for the gift I received from Franck. Thank you so much pequeñín!!!" the preprocessing steps would be as follows. Elimination of the auxiliary words: the words "I am", "by", "the", "that" and "of" belong to the set of auxiliary words, so the resulting text will be: "happy!!!". ☺ gift I received Franck. Thank you so much pequeñín!!!".

Replace accented vowels and vowels with special characters with their non-special equivalents. The words "I received" and "pequeñín" are the only words in the sentence that contain accents so the text would look like, "happy!!!". ☺ gift received Franck. Thank you so much pequeñin!!!".

Removal of unnecessary data such as names, emoticons, numbers, and punctuation marks. In this section the emoticons and punctuation marks present in the text are removed. The result of the action is: "happy gift received Franck Thank you very much little one". Conversion of words to their canonical form or lemma. The word "pequeñin" is considered a diminutive or adaptation of the word "pequeño". The result of the lemmatization would be: "happy gift received Franck Thank you very much pequeño".

5.2   Text Polarity Classification (Positive and Negative) with Multiple Models of Different Numbers of Neurons and Different Optimizers

The analysis of emotions and feelings in texts using computational techniques can range from what may be a simple task for humans, such as classifying text according to its content as positive or negative, to more complex ones such as impression detection. In this experimental phase, we set the goal of implementing a convolutional neural network to classify the polarities of the data set resulting from the above preprocessing. The model that has been configured for this test is described below, starting from a basic model or architecture for a CNN.

– Sequential input with dimensions adjusted to the maximum number of words reached by the tweet.
– Embedding matrix representing the class of the same name. A first step that must be taken before provisioning a neural network is to convert the text into a digital format suitable for the network. For this task, the Keras Tokenizer class is used, which has several methods for converting text into a digital format.
– Bigram layer with convolution layer and corresponding pooling layer, the activation function is relu.
  The Trigrams layer retains the properties of the previous class.
– A quadrilaterals layer with the same characteristics. These three previous layers are implemented based on n-grams, a technique that according to those who study the statistical properties of language it has been found that studying linear sequences of linguistic units can provide many things about a text, in this case about emotions. These linear sequences are known as bigrams+ (2 units), *trigrams (3 units) or more generally as n-grams.
– Hidden layer with 256 neurons activated by relu function. This hidden layer is a function of the weighted sum of the inputs. The function is the activation function and the values of the weights are determined by the estimation algorithm.
– Output layer with a neuron activated by a sigmoid function. This layer contains the target (dependent) variables.

Taking as reference the previous model, 6 variants of the same model were proposed considering the AdaGrad optimizer that according to PlatoAiStream (2021) [13] is more reliable than the gradient descent algorithms and its variants, and achieves a faster convergence, this is due to the fact that different learning rates are used for each iteration. Changes in the learning rate depend on different parameters during training. The more the parameters are changed, the less the learning rate changes. This change is very beneficial because the real data set contains sparse and dense features. Therefore, it is unfair to have the same learning rate value for all features. The advantage of using Adagrad is that you do not have to manually change the learning rate. More reliable than the steepest descent algorithm and its variants, reaching convergence faster [23].

AdaGrad is the second optimizer involved in this experiment. This optimization algorithm further extends the stochastic gradient descent method to update the network weights during training. Unlike maintaining a single learning rate throughout SGD training, the Adam optimizer updates the learning rate for the weights of each network individually [8,23]. The developers of the Adam optimization algorithm know the benefits of the AdaGrad and RMSProp algorithms, which are also extensions of the stochastic gradient descent algorithm. Therefore, the Adam optimizer inherits the functionality of the AdaGrad and RMSprop algorithms. Instead of adjusting the learning rate based on the first (average) momentum as in RMSProp, Adam also uses the second momentum in the gradient [13].

One more aspect to consider in the variation of the proposed models is the number of neurons in the hidden layer of the neural network. The original model assumes a hidden layer of 256 neurons, in addition to this, two more dimensions are added for the variation being 128 and 512 the number of neurons proposed.

Then, the computational models are conformed as follows: an input of adjustable size, an integration shade, a bigram layer with its respective convolution and integration layer activated by a relu function, a trigram layer with the same characteristics of the previous layer, a class of quadrigrams with the same properties of the previous layers, a hidden layer of variable size between 128, 256 and 512 neurons and an output layer with a neuron activated by the sigmoid function; all this supported by a variable optimizer between Adam and AdaGrad. The resulting models are shown in Table 2.

Table 2. Proposed models with variant architectures

Model	Number of neurons in the hidden layer	Optimizer
1	128	Adam
2	256	Adam
3	512	Adam
4	128	AdaGrad
5	256	AdaGrad
6	512	AdaGrad

For each of the proposed models, training and testing is performed with the same data set on 15 occasions to identify the aspects that each of them have. The results of accuracy obtained are shown in Table 3.

With the results obtained in Table 3, basic statistical tests are carried out to determine which is the optimal model to continue with the experimentation. The following is obtained from these analyses. Application of the Shapiro test using Python to identify whether the results follow a normal distribution or not, as shown in Table 4.

Table 3. Accuracy of the proposed models with various architectures

Exercise	Model 1	Model 2	Model 3	Model 4	Model 5	Model 6
1	85.78	89.70	91.06	87.94	90.48	90.42
2	86.41	90.88	91.42	87.21	90.55	88.40
3	86.34	88.31	91.33	88.75	90.70	88.78
4	87.26	88.20	91.30	90.72	89.54	89.72
5	86.13	87.31	90.18	88.79	89.79	90.33
6	87.51	90.77	91.25	91.85	90.43	89.74
7	87.59	90.75	89.86	86.02	90.62	88.27
8	84.14	88.59	91.58	90.59	89.03	89.92
9	89.69	88.12	90.79	89.48	91.45	89.44
10	87.91	88.05	89.59	88.36	90.21	89.13
11	87.86	87.06	90.03	91.38	91.66	89.08
12	83.39	88.09	89.64	88.90	91.00	90.67
13	89.44	90.85	91.82	91.37	89.62	89.50
14	89.53	90.55	90.84	86.91	90.71	90.44
15	89.12	90.77	90.76	88.41	89.35	88.20

Based on the data shown in Table 4, it can be deduced that *model* 1 and *model* 5 are the most reliable architectures to continue with the experimentation process in the detection of emotions in texts.

The confusion matrix for models 1 and 5 is shown in Fig. 1 below.

If we look at the confusion matrix of *model* 1 for the test data, we can deduce that in most cases the classes are not confused with each other, in the case of the confusion matrix of *model* 5 we can see that in this case there are more errors with the negative class (True Negatives increase and False Positives decrease), however, there are more errors with the positive class (False Negatives increase and True Positives decrease).

Table 4. Results Shapiro's test, detection of normal distribution in the results of the different models.

Model	Result Shapiro test (p)	Distribution of results
1	0.09439	Normal
2	0.00034	Non-normal
3	0.01417	Non-normal
4	0.00547	Non-normal
5	0.17240	Normal
6	0.02486	Non-normal

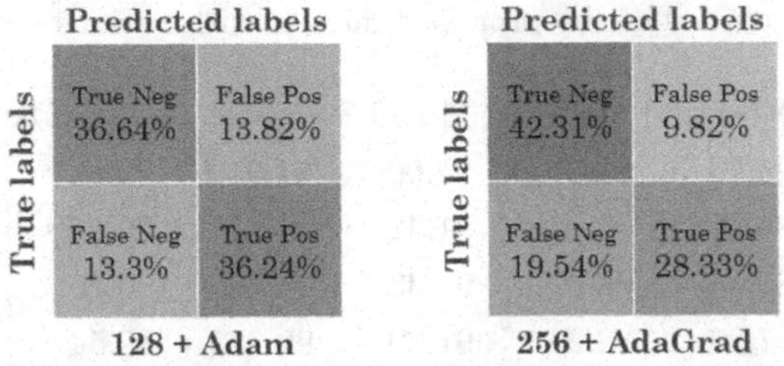

Predicted labels

	True Neg 36.64%	False Pos 13.82%
True labels	False Neg 13.3%	True Pos 36.24%

128 + Adam

Predicted labels

	True Neg 42.31%	False Pos 9.82%
True labels	False Neg 19.54%	True Pos 28.33%

256 + AdaGrad

Fig. 1. Confusion matrix of the *model* 1 and *model* 5

5.3   Emotion Analysis in Texts in Mexican Context

Once the previous experiment was concluded, we continued with the necessary modifications to the design of the computational models, preserving the characteristics of the architectures with the best scores in the Shapiro test in that experiment.

The output of the convolutional neural network is formed by the precision metric, the error and the labels of the emotions detected in each analyzed text. Once the modifications were completed, training and testing of the model was started in 10 epochs from the collected data, obtaining the results in Table 5.

Table 5. Test results of the CNN models with the generated corpus.

Epoch	Model 1		Model 5	
	Accuracy	Loss	Accuracy	Loss
1	0.4750	0.6954	0.4134	0.8500
2	0.5033	0.6928	0.4256	0.7187
3	0.5183	0.6865	0.4025	0.7310
4	0.5783	0.6635	0.4678	0.6954
5	0.7644	0.5364	0.5125	0.5401
7	0.9133	0.3884	0.5965	0.5176
7	0.9572	0.3091	0.6697	0.4943
8	0.9700	0.2721	0.7322	0.3398
9	0.9794	0.2418	0.8200	0.2976
10	0.9828	0.1268	0.8943	0.2387

The computational *model* 1 achieved its best accuracy at epoch 10 after reaching 98.28% accuracy, while the loss was 12.68% at epoch 10 and the computational *model* 5 archived its best accuracy at epoch 10 after reaching 89.43% accuracy and the loss was 23.87% at epoch 10.

With respect to the verification of the performance of the models, the confusion matrix of each of the models for this exercise has been compiled, as shown in Fig. 2.

The confusion matrices have a very similar behavior to the previous experiment, maintaining a balance in the little confusion of classes for *model* 1 and

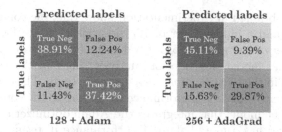

Fig. 2. Confusion matrix of the *model* 1 and *model* 5 in the emotion analysis.

a greater success with the negative class, but there are more errors with the positive class in *model* 2.

The percentages of emotions present in the texts analyzed during the test are shown in Table 6.

Table 6. Percentage of identification of emotions in texts

Emotion	Percentage	
	Model 1	*Model* 5
Joy	0.12	0.21
Sadness	24.20	34.98
Calmness	0.01	0.10
Anger	56.30	52.34
Pleasantness	0.00	1.11
Disgust	13.20	13.25
Expectation	0.72	0.92
Surprise	0.01	0.03
Love	20.12	23.12

6   Conclusions and Future Work

This paper analyzed the state of the art of deep learning applied to affective computing, emotion analysis, and sentiment analysis through the existing literature in Web of Science, IEEE, Scopus, and Google Scholar. In addition, the authors implemented several computational models of convolutional neural networks for emotion analysis and the visualization of the behavior of such models, considering that such work was implemented on texts in the Mexican context, which provides a variant scenario mainly for the preprocessing of the data derived from the many adaptations of the language in the so-called regionalisms.

The findings obtained in experimentation show that the combination of the Adam optimizer and a hidden layer of 128 neurons or the AdaGrad optimizer and a hidden layer of 256 neurons prove to be the best bets in their integration into computational models for emotion analysis, both in the case of polarity classification and in the identification of emotions in texts. These models suggest a very similar behavior in their performance according to the confusion matrices obtained, where the model that integrates the Adam optimizer and 128 neurons in the hidden layer has a more balanced performance in terms of hits and low levels of class confusion. The second model suggests a considerable improvement in the negative class hit, however, the positive class hits are reduced and there is a slight increase in class confusion. The field of affective computation is broad in terms of previous work and research, however, there is a weakness for the study of Spanish languages and that is that very little has been contributed in this area contemplating these languages. One of the main difficulties that have been encountered is the lack of data sets or databases in the Spanish language to support the experimentation of emotion and sentiment analysis in this region. Another of the main challenges faced by the research is the stemming process, in which the diversification due to local adaptations of the Spanish language hinders the process itself, making it difficult to put the words in their canonical form, considering that in addition to the local adaptations of the language, It has been identified that there are words that have been given a diversified meaning according to the geographical areas of the country, which leads to a kind of mixture between polysemous and homonymous words in the same entity. In addition to this, the variation of the verb tenses that are implemented in this language make this task even more difficult. This paper opens the possibility to continue working on the improvement of the computational models of CNNs in affective computing with the manipulation of the activation functions in each layer of the architectures and even in the integration and combination of other techniques with the ideal of perfecting and reaching the highest levels of precision in this task.

References

1. Banafa, A.: ¿Qué es la computación afectiva? (2018). https://www.bbvaopenmind.com/tecnologia/mundo-digital/que-es-la-computacion-afectiva/
2. Bisquerra-Alzina, R.: Psicopedagogía de las emociones. Sintesis, Madrid (2009)
3. Casas García, A., Villena Román, J.: Sistema de Análisis Automático de Sentimientos Basado en Procesamiento del Lenguaje Natural. Ph.D. thesis, Universidad Carlos III de Madrid (2014)
4. Ciresan, D.C., Meier, U., Masci, J., Gambardella, L.M.: Flexible, high performance convolutional neural networks for image classification. In: Proceedings of the Twenty-Second International Joint Conference on Artificial Intelligence Flexible, pp. 1237–1242 (2013). https://www.aaai.org/ocs/index.php/IJCAI/IJCAI11/paper/viewFile/3098/3425
5. Dubiau, L., Ale, J.M.: Análisis de Sentimientos sobre un Corpus en Español: Experimentación con un Caso de Estudio. In: XIV Argentine Symposium on Artificial Intelligence (ASAI)-JAIIO 42 (2013)

6. Fernández, Y.: Qué es IFTTT y cómo lo puedes utilizar para crear automatismos en tus aplicaciones (2019). https://www.xataka.com/basics/que-ifttt-como-puedes-utilizar-para-crear-automatismos-tus-aplicaciones
7. Francisco, V.: Identificación Automática del Contenido Afectivo de un Texto y su Papel en la Presentación de Información. Ph.D. thesis, Universidad Complutense de Madrid (2008)
8. Freire, E., Silva, S.: Redes neuronales. Programa de Visión Artificial (2019). https://bootcampai.medium.com/redes-neuronales-13349dd1a5bb
9. Gavilán, I.: Catálogo de componentes de redes neuronales (y IV): optimizadores (2020). https://ignaciogavilan.com/catalogo-de-componentes-de-redes-neuronales-y-iv-optimizadores/
10. Konate, A., Du, R.: Sentiment analysis of code-mixed Bambara-French social media text using deep learning techniques. Wuhan Univ. J. Nat. Sci. **23**(3), 237–243 (2018). https://doi.org/10.1007/s11859-018-1316-z
11. Lecun, Y., Bottou, L., Bengio, Y., Haffner, P.: Gradient based learning applied to document recognition. IEEE (1998). http://ieeexplore.ieee.org/document/726791/#full-text-section
12. Lee, M.S., Lee, Y.K., Pac, D.S., Lim, M.T, Kim, D.W., Kang, T.K.: Fast emotion recognition based on single pulse PPG signal with convolutional neural network. Appl. Sci. (Switz.) **9**(16), 3355 (2019)
13. PlatoAiStream: BigData: Una guía completa sobre optimizadores de aprendizaje profundo (2021). https://zephyrnet.com/es/a-comprehensive-guide-on-deep-learning-optimizers/
14. Plutchik, R.: The nature of emotions: human emotions have deep evolutionary roots, a fact that may explain their complexity and provide tools for clinical practice. Am. Sci. **89**(4), 344–350 (2001). http://www.jstor.org/stable/27857503
15. Poincaré: Análisis de sentimiento de texto basado en CNN - programador clic (2020). https://programmerclick.com/article/71111623446/
16. Sarin, E., Vashishtha, S., Kaur, S., et al.: SentiSpotMusic: a music recommendation system based on sentiment analysis. In: 2021 4th International Conference on Recent Trends in Computer Science and Technology (ICRTCST), pp. 373–378 (2022)
17. Sarmiento-Ramos, J.L.: Aplicaciones de las redes neuronales y el deep learning a la ingeniería biomédica. Revista UIS Ingenierías **19**(4), 1–18 (2020)
18. Shi, S., Zhao, M., Guan, J.U.N., Huang, H.: Multi-features group emotion analysis based on CNN for Weibo events. DEStech Trans. Comput. Sci. Eng. **L**(cii), 358–368 (2017)
19. Softtek: Las CNN mejoran el análisis de imágenes (2021). https://softtek.eu/tech-magazine/artificial-intelligence/las-redes-neuronales-de-convolucion-cnn-mejoran-el-analisis-de-imagenes/
20. Tripto, N.I., Ali, M.E.: Detecting multilabel sentiment and emotions from Bangla YouTube comments. In: 2018 International Conference on Bangla Speech and Language Processing (ICBSLP), pp. 1–6 (2018)
21. Valle-Cruz, D., Lopez-Chau, A., Sandoval-Almazan, R.: Impression analysis of trending topics in Twitter with classification algorithms. In Proceedings of the 13th International Conference on Theory and Practice of Electronic Governance, pp. 430–441 (2020)
22. Velasco, L.: Optimizadores en redes neuronales profundas: un enfoque práctico (2020). https://velascoluis.medium.com/optimizadores-en-redes-neuronales-profundas-un-enfoque-práctico-819b39a3eb5

23. Yaqub, M., et al.: State-of-the-art CNN optimizer for brain tumor segmentation in magnetic resonance images. Brain Sci. **10**(7), 1–19 (2020)
24. Yudita, S.I., Mantoro, T., Ayu, M.A.: Deep face recognition for imperfect human face images on social media using the CNN method. In: 2021 4th International Conference of Computer and Informatics Engineering (IC2IE), pp. 412–417. IEEE (2021)
25. Zatarain Cabada, R., Barrón Estrada, M.L., Cárdenas López, H.M.: Reconocimiento multimodal de emociones orientadas al aprendizaje. Res. Comput. Sci. **148**(7), 153–165 (2019)

Proficient Annotation Recommendation in a Biomedical Content Authoring Environment

Asim Abbas, Steve Mbouadeu, Avinash Bisram, Nadeem Iqbal, Fazel Keshtkar, and Syed Ahmad Chan Bukhari[✉]

Collins College of Professional Studies, Division of Computer Science, Mathematics and Science, St. John's University, 8000 Utopia Parkway, Queens, NY 11439, USA
{abbasa,iqbaln,keshtkaf,bukharis}@stjohns.edu,
{steve.mbouadeu19,avinash.bisram19}@my.stjohns.edu

Abstract. Given the ubiquity of unstructured biomedical data, significant obstacles still remain in achieving accurate and fast access to online biomedical content. Accompanying semantic annotations with a growing volume biomedical content on the internet is critical to enhancing search engines' context-aware indexing, improving search speed and retrieval accuracy. We propose a novel methodology for annotation recommendation in the biomedical content authoring environment by introducing the socio-technical approach where users can get recommendations from each other for accurate and high quality semantic annotations. We performed experiments to record the system level performance with and without socio-technical features in three scenarios of different context to evaluate the proposed socio-technical approach. At a system level, we achieved 89.98% precision, 89.61% recall, and an 89.45% F1-score for semantic annotation recollection. Similarly, a high accuracy of 90% is achieved with the socio-technical approach compared to without, which obtains 73% accuracy. However almost equable precision, recall, and F1- score of 90% is gained by scenario-1 and scenario-2, whereas scenario-3 achieved relatively less precision, recall and F1-score of 88%. We conclude that our proposed socio-technical approach produces proficient annotation recommendations that could be helpful for various uses ranging from context-aware indexing to retrieval accuracy.

Keywords: Annotation recommendation · Automate semantic annotation · Biomedical semantics · Biomedical content authoring · Peer-to-peer · Annotation ranking

1 Introduction

The timely dissemination of information from the scientific research community to peer investigators and other healthcare professionals requires efficient methods for acquiring biomedical publications. The rapid expansion of the biomedical field has led researchers and practitioners to a number of access-level challenges.

This work is supported by the National Science Foundation grant ID: 2101350.

Due to the lack of machine-interpretable metadata (semantic annotation), essential information present in web content is still opaque to information retrieval and knowledge extraction search engines. Search engines require this metadata to effectively index contents in a context-aware manner for accurate biomedical literature searches and to support ancillary activities like automated integration for meta-analysis [1]. Ideally, biomedical content should include machine-interpretable semantic annotations during the pre-publication stage (when first drafting), as this would greatly advance the semantic web's objective of making information meaningful [2]. However both of these procedures are complicated and require in-depth technical and/or domain knowledge. Therefore, a cutting-edge, publicly available framework for creating biological semantic content would be revolutionary.

The generation or processing of textual information through a semantic amplifying framework is known as semantic content authoring. Primary elements of this process include ontologies, annotators, and a user interface (UI). Semantic annotators are built to facilitate tagging/annotating their encompassing ontology concepts using pre-defined terminologies, whether it being through a manually, automatically, or through a hybrid approach [3]. As a result, users create information that is more semantically rich when compared to typical writing utilities such as word processors [4]. Furthermore, it is categorized that the two semantic content writing methodologies used today can be categorized as either bottom-up or top-down. In a bottom-up approach, a collection of ontologies is utilized to semantically enrich or annotate the textual content of a document [5]. Semantic MediaWiki [6], SweetWiki [7], and Linkator [8] are a few examples of bottom-up-designed semantic content production tools. However, these tools have significant drawbacks. The bottom-up approach is an offline, non-collaborative, and application-centric way of content authoring. Additionally, it cannot be used with the most recent version of Microsoft Word because it was created more than eleven years ago. Top-down approaches were developed to add semantic information to existing ontologies, each of which being extended or populated using a particular template design. Hence, this approach is sometimes referred to as an ontology population approach to content production. Top-down approaches do not improve the non-semantic components of text by annotating them with the appropriate ontology keywords. Instead, they use ontology concepts as fillers while authoring content. Examples of top-down approaches are OntoWiki [9], OWiki [10], and RDFAuthor [11].

Over the years, the development of biomedical semantic annotators has received significant attention from the scientific community due to the importance of the semantic annotation process in biomedical informatics research and retrieval. Biomedical annotators can be further classified into a) general-purpose annota tors for biomedicine, which assert to cover all biomedical subdomains, and b) use case-specific biomedical annotators, which are developed for a specific sub-domain or annotate specific entities like genes and mutations in a given text. Biomedical annotators primarily use term-to-concept matching with or without machine learning-based methods, in contrast to the general purpose non-biomedical semantic annotators, which combine NLP (Natural Language Processing) techniques, ontologies, semantic similarity algorithms, machine learning (ML) models, and graph manipulation techniques [12]. Biomedical annotators such as NOBLE Coder [13], ConceptMapper [14], Neji [15], and Open Biomedical Annotator

[16] use machine learning and annotate text with an acceptable processing speed. However, they lack a strong disambiguation capacity i.e., the ability to distinguish the proper biomedical concept for a particular piece of text among several candidate concepts. Whereas NCBO Annotator [17] and MGrep services are quite slow. Similarly Rysan nMd annotator asserts that it can balance speed and accuracy in the annotation process. On the other hand, its knowledge base is restricted to specific UMLS (Unified Medical Language System) ontologies and does not fully cover all biomedical subdomains [18].

To solve the aforementioned constraints, we designed and developed "Semantically" a publicly available interactive systems that enables users with varying levels of biomedical domain expertise to collaboratively author biomedical semantic content. To develop a robust Biomedical Semantic Content Editor, balancing between speed and accuracy is the key research challenge. Finding the proper semantic annotations in real time during content authoring is particularly challenging since a single semantic annotation is frequently available in multiple biomedical ontologies with various connotations. To balance the efficiency and precision of the current biomedical annotators, we present an unconventional socio-technical method for developing a biomedical semantic content authoring system that involves the original author throughout the annotation process. Our system enables users to convert their content into a variety of online, interoperable formats for hosting and sharing in a decentralized fashion. We conducted a series of experiments using biomedical research articles obtained from Pubmed.org [19] to demonstrate the efficacy of the proposed system. The results show that the proposed system achieves a better accuracy when compared to the existing systems used for the same task in the past, leading to a considerable decrease in annotation costs. Our method also introduces a cutting-edge socio-technical approach to leveraging semantic content authoring to enhance the FAIRness [18] of research literature.

2 Proposed Methodology

This section introduces "Semantically," a web-based, open source, and accessible systems for creating biomedical semantic content that can be used by authors with varying levels of expertise in the biomedical domain. An authoring interface resembling the Microsoft Word editor is provided for end users to write and compose biomedical semantic contents including research papers, clinical notes, and biomedical reports. We leveraged Bioportal [17] endpoint APIs to cater the initial layer of semantic annotation and automate the configuration process for authors. Subsequently, the annotated terms or concepts are highlighted to improve visibility. The system provides a social-collaborative environment in "Semantically Knowledge Cafe" which allows authors to receive assistance from domain experts and peer reviewers to assist in the appropriate annotation of a specific phrase or the entire text. The proposed methodology consists of two primary sections: Base Level Annotations and Annotations generated through Socio-Technical Approach (as shown in Fig. 1).

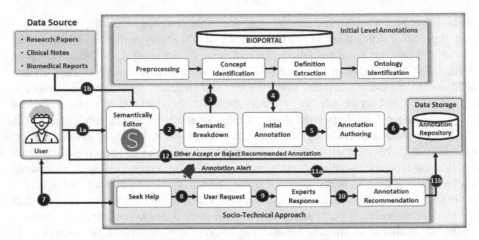

Fig. 1. Proposed methodology workflow

2.1 Base Level Annotation

This semantic annotation approache, a top-down approach, is meant to annotate existing texts using a collection of established ontologies. A biomedical annotator is a key component of a content-level semantic enrichment and annotation process [3]. These annotators leverage publicly accessible biomedical ontology systems like Bioportal [17] and UMLS [5] to assist the biomedical researcher in structuring and annotating their data with ontology concepts that enhance information retrieval and indexing. However, the semantic annotation and enhancement process cannot be easily automated and often requires expert curators. Furthermore, the lack of a user-friendly framework makes the semantic enrichment process more difficult to non-technical individuals. To address this barrier, we used an NCBO Bioportal [17] web-service resource which analyzes the raw textual data, tags it with pertinent biomedical ontology concepts, and provides a basic set of semantic annotations without the requirement of technical expertise.

A. **Semantically Workspace:** The Semantically framework was developed for a wide range of users, including bench scientists, clinicians, and casual users involved in medical journalism. The users initially have the choice to start typing directly in the Semantically text editor or import pre-existing content from research articles, clinical notes, and biomedical reports. Afterward, based on their level of expertise and knowledge with particular ontologies, authors are provided a few annotation options to choose from. Users without a technical background may readily traverse a simplified interface, while more experienced users can employ advanced options to exert more control over the semantic annotation process, as depicted in Fig. 1.

B. **Semantic Breakdown:** Semantic breakdown is the process of retrieving semantic information for a biomedical content utilizing the NCBO Bioportal [17]annotator at a granular level. Term based matching as an ontology approach is supported by the NCBO web services and a set of semantic information is returned. The semantic breakdown process consists mainly of (*i*) Biomedical concept or terminology

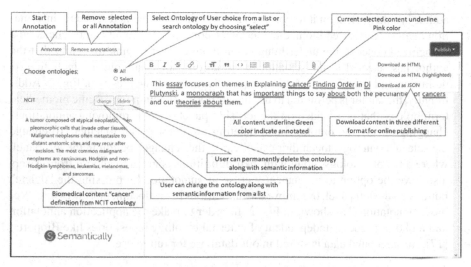

Fig. 2. Proposed biomedical semantic content authoring interface

identification in a series of biomedical data, (*ii*) Ontological semantic information extraction.

i.   *Biomedical Concept Identification:* We make use of the NCBO BioPortal Annotator API [17] to identify biomedical terms. The NCBO annotator accepts the user's free text which is then mapped using a concept recognition tool. The concept recognition tool leverages Mgrep services developed by NCBO and an ontology-based dictionary created from UMLS [5] and Bioportal [17]. The tool identifies the biomedical terms or concepts following a string matching approach and the result is the collection of biomedical terminologies and ontologies to which the terms belong.

ii. *Ontological Semantic Information Extraction:* An ontology is a collection of concepts and the semantic relationships among them. In the proposed methodology, we leverage the Bioportal Ontology web service [17], a repository containing around 1018 ontologies in the biomedical domain. The biomedical text provided to "Semantically" is routed to Bioportal [17] to retrieve the relevant ontologies, acronyms, definitions, and ontology links for individual terminologies. This semantic information is displayed in an annotation panel for user interpretation and comprehension. "Semantically" allows users to author this semantic information based on their knowledge and experience by picking the appropriate ontology from the provided list, suitable acronyms, or eliminating semantic information and annotations for specific terminology.

C. **Semantic Annotation Authoring Process:** During the semantic breakdown process, a set of base level semantic information is acquired from Bioportal. The associated terminology is then underlined with a green color as illustrated in Fig. 2. The author is able to click on it to observe and examine the obtained semantic information against recognized biomedical terms or concepts. When an author clicks on a recognized concept, the underlining color changes over to pink to signify that the author has selected it; the semantic information about that concept, including its definition and associated ontology, appear on the left panel, as is in Fig. 2. Additionally, the author is permitted to use the authoring function for specific biomedical terminology that is provided on the left side panel, such as removing annotations, changing underlying ontology, and deleting the ontology all together. To change the selected ontology, the author can click on the "change" button in the left panel, where a list of ontology along with their definitions is given to choose from. Users also have the option to permanently erase an annotation by pressing the "delete" buttons on the left panel, or remove all annotations in the document by pressing "Remove annotations" as shown in Fig. 2. In order to make the application annotation and authoring process independent of external ontology repositories like Bioportal [17], the annotated data is stored in our database for future use.

2.2   The Socio-Technical Approach

After achieving base level semantic annotation, "Semantically" offers an novel socio-technical environment, where the author is able to discuss and receive recommendations from peers to achieve more precise and high quality annotations. Three different scenarios, as given in the Table 1, are used to assess the proposed approach. In order to assess and recommend the appropriate annotation to the author, we use a statistical technique upon gathering the features as given in Table 2. The setting is known as "Semantically Knowledge Cafe," where the author may post their query, peers or domain experts can react with some self-confidence score, and other community users can credit their reply by up-voting or down-voting in a collaborative and constructive manner. The post author is notified of any recommended annotations with the option of accepting or rejecting the

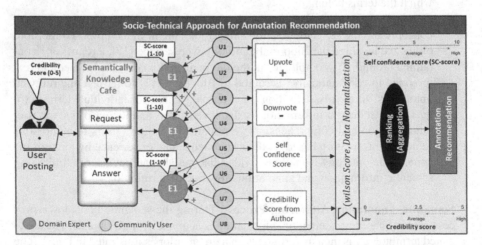

Fig. 3. Proficient annotation recommendation evaluation process

suggestion; only the finally accepted annotation is recorded as an answer in the database. The statistical process of the annotation recommendation is described in Fig. 3.

To evaluate the suggested socio-technical approach, we have devised three scenarios as illustrated in Table 1. These scenarios are related to the query regarding semantic annotation for biomedical content. If we examine the existing question answering platform such as Stackoverflow, such kind of scenarios or query can be found. An example is presented for a Case:1 as below:

The author is asked to identify an accurate ontology annotation from experts for the biomedical term "breast-conserving surgery". The "Scmantically" framework provides an interface where users can input such a query.

Table 1. Semantically annotation recommendation scenarios

Scenario no.	Scenario description
Scenario: 1	Which ontology should i use?
Scenario: 2	What is the suitable ontology vocabulary?
Scenario: 3	Does this ontology best describe this terminology?

For Example:

"Which ontology should I use for "breast-conserving surgery"?
Submitting the above query will create a new post on the "Semantically Knowledge Cafe" forum for experts E_i response where $E_i = e_1, e_2, e_3 \ldots e_n$. Similarly, the platform also provides an interface for the expert E_i to streamline the response process in the forum. For example the expert can describe the suggested annotation, rate their level of self-confidence, and quickly find the precise ontology they mean by utilizing NCBO otology tree widget tool. Whenever the expert E_i submits their response to the post author for Case.1, other community users $U_i = u_1, u_2, u_3 \ldots u_n$ do respond to the expert reply in the form, upvote as $+V$ and downvote as V as shown in Fig. 3. A statistical measure is taken for expert self-confidence score, upvote $(+V)$, downvote $(-V)$ and credibility score from the author by applying Wilson formula and data normalization techniques. Finally an optimal recommendation of annotation is suggested to the author.

2.3 Semantically Recommendation Features (SR-FS)

In order to find more optimal and high quality annotation recommendations, we addressed several features in the proposed socio-technical approach as given in Table 2. These features are produced by and gathered from community users, who actively participated in the socio-technical environment. We presented features as upvote $(+V)$, downvote $(-V)$, expert confidence score (ECS) and user credibility score (UCS) in the following way.

Table 2. Semantically recommendation features (SR-f) descriptions

Features no.	Features name	Feature descriptions
SR.f1	Upvotes (+), Downvotes (−)	Votes from community users
SR.f2	Self confidence score	Score from responder
SR.f3	Credibility score from author	Score from authors

(i) **Upvote and downvote (SR-f1):** In the social network environment upvotes (+V) and downvotes (−V) play a crucial role, whereas +V implies the usefulness or quality of a response or answer while −V point to irrelevance or low quality. This feature measures the quality of domain expert responses to the post. A higher amount of up-votes and small amount of down votes indicates better quality annotation recommendations. These features are further processed by Wilson's score confidence interval for a Bernoulli parameter see Eq. 1 to determine the expert suggested annotation quality score.

$$\text{Wilsonscore} = \left(\hat{p} + \frac{Z_{\alpha/2}^2}{2n} \pm Z_{\alpha/2} \sqrt{\left[\hat{p}(1 - \hat{p}) + Z_{\alpha/2}^2/4n \right]/n} \right) / \left(1 + Z_{\alpha/2}^2/n \right)$$

(1)

where,

$$\hat{p} = \left(\sum_{n=1}^{N} +V \right) / (n)$$

(2)

$$n = \sum_{i=0}^{N} \sum_{j=0}^{M} (+V_i, -V_j)$$

(3)

and, $Z_{\frac{\alpha}{2}}$ is the $\left(1 - \frac{\alpha}{2}\right)$ quantile of the standard normal distribution.

In Eq. 1. $\hat{p}$ is the sum of upvotes (+V) of a community user's U_i to the expert E_i response for a post from an author for correct annotation divided by overall votes (+V, V). Likewise n is the sum of number of upvote and downvote (+V, V) and α is the confidence refers to the statistical confidence level: pick 0.95 to have a 95% chance that our lower bound is correct. However the z-score in this function never changes.

(ii) **Self Confidence Score (SR-f2):** In Psychology a self-confidence score is defined as an individual's trust in their own skills, power, and decision, that they can effectively make. In the proposed approach we allow the expert to give a confidence score for their decision made for annotation recommendation. As shown in Fig. 4, the experts can choose a self confidence level in a range between 1 to 10, indicate how they feel about annotation recommendation. For instance, if an expert believes that their recommendation is slightly above average, they might rate them as a 6, but if an expert feels more confident, rate them as an 8.

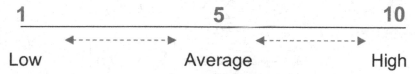

Fig. 4. Self confidence score selection level

$$z_i = (x_i - min(x))/(max(x) - min(x)) * Q \qquad (4)$$

where z_i is the $i_t h$ normalized value in the dataset. Where x_i is the $i_t h$ value in the dataset, e.g. the user confidence score. Similarly $min(x)$ is the minimum value in the dataset, e.g. the minimum value between 1 and 10 is 1, so the min(x) = 1 and $max(x)$ is the maximum value in the dataset, e.g. the maximum value between 1 and 10 is 10, so the max(x) = 10. Finally Q is the maximum number wanted for normalized data value, e.g. we normalized the confidence score between 0 and 1, the maximum value between 0 and 1 for Q is 1.

(iii) **Credibility Score from a User (SR-f3):** The credibility is defined as the ability to be trusted or acknowledged as genuine, truthful or honest. As an attribute, credibility is crucial since it helps to determine the domain expert knowledge, experience and profile. Therefore if a domain expert profile is not credible, others are less likely to believe what is being said or recommend. Subsequently annotation recommendation is received to the author post from "Semantically Knowledge Cafe", the author is allowed to either accept or reject the recommended annotation. Whenever an author accepts a user's recommended annotation, a credibility score between 1–5 is added to the user's profile. Similarly, if the author rejects the annotation recommendation, a negative credibility score of 0 is added to their profile (Fig. 5).

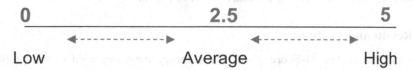

Fig. 5. Credibility score from an author selection level

All the features (SR-f1, SR-f2 and SR-f3) are equally contributed and deeply correlated with one another for final annotation recommendation. Though the final output of SR-f1 is between 0 and 1, Therefore kept the process consistent and features dependencies, we normalize the self-confidence score (SR-f2) between 0 and 1 utilizing Eq. 4. Finally all the SR-FS (Semantically Ranking Feature Score) for each expert E_i annotation recommendation is computed and aggregated using Eq. 5.

$$Sr - Fs = \sum_{j=1}^{m} \sum_{i=1}^{n} \sum_{k=0}^{p} (Fj, Ei, Ak) \tag{5}$$

$$final - score = arg\ max \left[\sum_{i=1}^{N} (Sr - Fs) \right] \tag{6}$$

where F_i is feature score for Annotation A_i and Expert E_i. The final decision or ranking happens based on maximum feature scoring gained by the Expert E_i response to the author post or query see Eq. 6.

3 Experiments

3.1 Datasets

In the proposed methodology a total of 30 people participated. A set of 30 biomedical articles where chosen from Pubmed.org [19] and distributed to the participants randomly. The participants where split into three groups with ten members being assigned to each scenario as we already discussed in proposed methodology Sect. 2.2. Though each user is guided and instructed to process the given documents on "Semantically". Subsequently processing the individual documents by participants a set of base annotations is obtained as outlined in proposed methodology Sect. 2.1. Meanwhile each participant is encourage to post the query or question on the "Semantically Knowledge Cafe" forum for the biomedical content, which they have doubts of for their initial annotation and a number of 645 total posts is generated. Whereupon, all the participants are allowed to respond to the posts with a confidence score ranging from 1 to 10. Cumulatively 2845 number of answers are given by the expert E_i from the participant and other users participate to weigh the reply post in the form of upvotes and downvotes, an average number of 6056 upvotes and 7942 downvotes are recorded. Subsequently, received the recommended annotation, the author is enabled to give credibility score between 0 and 5. Finally, we reviewed the results without socio-technical and with socio-technical approach by three domain specific experts from academia at professor level.

3.2 Results and Analysis

Precision, Recall, and f1-Score where the accuracy measures used in our proposed methodology. Precision counts the number of valid instances(TP) in the set of all retrieved instances (TP + FP). Recall measures the number of valid instances (TP) in the intended class of instances (TP + FN). Finally the F1-score is the harmonic mean between precision and recall, used to obtain the adjusted F measure. The results where computed using the following equations:

$$Precision = \frac{TP}{TP + FP} \tag{7}$$

$$Recall = \frac{TP}{TP + FN} \tag{8}$$

$$f1 - score = 2 * \frac{Precision * Recall}{Precision + Recall} \tag{9}$$

$$Accuracy = \frac{TP + TN}{TP + FN + TN + FP} \tag{10}$$

A. **System Level Performance:** A system level performance is constructed by averaging the scenario level results. Initially we find the results at the document level for each scenario. We than combine the results at scenario level and finally averaged the results. After obtaining the results for each scenario, a domain expert from academia, all being domain professors, were engaged to manually assess the results. Thereafter we manually compared the results of domain experts to the socio-technical approach and calculated the precision, recall and f1-score to find the system efficacy. As a result, the system demonstrated almost identical performance of 90% for an annotation recommendation in a socio-technical environment to the manual expert review as shown in Fig. 6a. Additionally, we used Eq. 10 to compare the performance of the system with and without socio-technical approach. A document's level accuracy is determined with or without the socio-technical approach. In Fig. 6b, the X-axis represents the number of documents processed, while on Y-axis at left represent the level of accuracy without socio-technical approach and on Y-axis at right represent the level of accuracy with socio-technical approach. Consequently, having socio-technical approach is more effective compared not having it at the individual document level. An accuracy of over 90% was achieved by nine documents and the majority of the rest had an accuracy ranging between 87% and 90% using the socio-technical approach. In contrast, only a single document yielded an accuracy above 73% and the rest of the documents had an accuracy between 65% to 73% without the socio-technical approach see Fig. 6b. Overall the proposed socio-technical approach remains the winner by obtaining 89.98% precision, 89.61% recall, and an 89.45% F1-score.

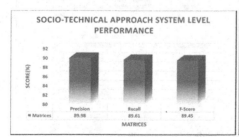

(a) System Level Performance

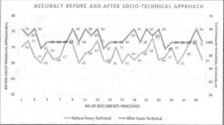

(b) Without and with socio-technical

Fig. 6. (a) presented the System Level Performance of a Socio-technical approach, (b) presented the accuracy of system without and with socio-technical approach

B. **Scenarios Level Performance:** This section presented the scenario level performance of the proposed socio-technical approach, where distinguishing results have been produced by these scenarios taking into account the above three matrices precision, recall and f-score. However scrutinizing the results, immense performance is shown by scenario-1 with a 91.14% precision, 90.85% recall and an F-score of 90.53%, and lower efficiency is achieved by scenario-3 with precision of 88.17%, recall 87.82% and f1-score 87.89%. Similarly, an acceptable performance is earned by scenario-2 upon a precision score of 90.63%, recall 90.15% and f1-score 89.9%, which is near to the performance of scenario-1 see Fig. 7. As we have already discussed in the proposed methodology the different working scenarios. However, after carefully examining the outcomes of each scenario, we came to a conclusion: why are distinctive results produced? Whenever the author posted taking scenario-1, the Expert E_i or responder is open and free to suggest the appropriate ontology, and multiple options are available. On the other hand taking scenario-3 author is bounded to choose the appropriate ontology, only from the list suggested by the automatic annotator of NCBO [7]. Also the author explicitly mentioned a list of ontology to the expert in the post to suggest to ontologies only from the given list. Similarly, considering scenario-2, the expert is able to suggest the appropriate ontology vocabulary class by utilizing the NCBO ontology tree widget tool [4] and the expert E_i is not bounded to recommend the ontology vocabulary class from a user specific vocabulary list. Consequently due to bounding the expert E_i to recommend me ontology from author defined vocabulary list scenario-3 achieve less accuracy compared to scenario-1 and scenario-2 that gain an acceptable level of accuracy as shown in Fig. 7.

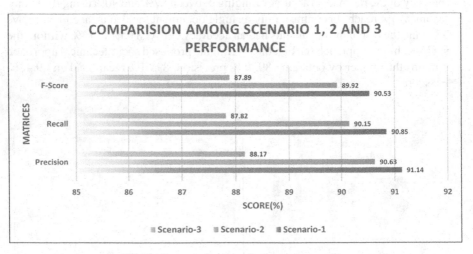

Fig. 7. Scenario level performance comparison

4 Conclusion

One of the main reasons why semantic content authoring is still in its infancy and researchers have not been able to achieve its desired objectives is because researchers did not realize the importance of the original content creator (author). Involvement them rather than heavily focusing on technological sophistication is extremely consequential. The objective of this study is to develop an open source interactive system that empowers individual authors at different levels of expertise in the biomedical domain to generate accurate annotations. Therefore, we proposed a novel socio-technical approach to develop a biomedical semantic con tent authoring system that balances speed and accuracy by keeping the original author in loop through the entire process. Similarly the "Semantically Knowledge Cafe" is a forum style extension authors can post their query for annotation recommendations. Our work provides a stepping stone towards optimizing biomedical information on the web for search engines, making it more meaningful and useful. The application is available at https://gosemantically.com.

References

1. Bukhari, S.A.C.: Semantic enrichment and similarity approximation for biomedical sequence images. University of New Brunswick, Canada (2017)
2. Warren, P., Davies, J.: The semantic web– from vision to reality. ICT Futures, pp. 53–66. https://doi.org/10.1002/9780470758656.ch5
3. Abbas, A., et al.: Clinical concept extraction with lexical semantics to support automatic annotation. Int. J. Environ. Res. Public Health **18**(20), 10564 (2021)
4. Abbas, A., et al.: Biomedical scholarly article editing and sharing using holistic semantic uplifting approach. Int. FLAIRS Conf. Proc. **35** (2022)
5. Abbas, A., et al.: Meaningful information extraction from unstructured clinical documents. Proc. Asia Pac. Adv. Netw. **48**, 42–47 (2019)
6. Laxström, N., Kanner, A.: Multilingual semantic MediaWiki for Finno-Ugric dictionaries. Septentrio Conf. Ser. **2**, 75 (2015). https://doi.org/10.7557/5.3470
7. Buffa, M., et al.: SweetWiki: a semantic Wiki. SSRN Electron. J. https://doi.org/10.2139/ssrn.3199377
8. Araujo, S., Houben, G.-J., Schwabe, D.: Linkator: enriching web pages by automatically adding dereferenceable semantic annotations. In: Benatallah, B., Casati, F., Kappel, G., Rossi, G. (eds.) ICWE 2010. LNCS, vol. 6189, pp. 355–369. Springer, Heidelberg (2010). https://doi.org/10.1007/978-3-642-13911-6_24
9. Auer, S., Dietzold, S., Riechert, T.: OntoWiki – a tool for social, semantic collaboration. In: Cruz, I., et al. (eds.) ISWC 2006. LNCS, vol. 4273, pp. 736–749. Springer, Heidelberg (2006). https://doi.org/10.1007/11926078_53
10. Iorio, A.D., et al.: OWiki: enabling an ontology-led creation of semantic data. In: Hippe, Z.S., Kulikowski, J.L., Mroczek, T. (eds.) Human – Computer Systems Interaction: Backgrounds and Applications 2. AISC, vol. 99, pp. 359–374. Springer, Heidelberg (2012). https://doi.org/10.1007/978-3-642-23172-8_24
11. Tramp, S., Heino, N., Auer, S., Frischmuth, P.: RDFauthor: employing RDFa for collaborative knowledge engineering. In: Cimiano, P., Pinto, H.S. (eds.) EKAW 2010. LNCS (LNAI), vol. 6317, pp. 90–104. Springer, Heidelberg (2010). https://doi.org/10.1007/978-3-642-16438-5_7

12. Jovanović, J., et al.: Semantic annotation in biomedicine: the current landscape. J. Biomed. Semant. **8**(1), 44 (2017)

13. Tseytlin, E., et al.: NOBLE – flexible concept recognition for large-scale biomedical natural language processing. BMC Bioinform. **17**(1) (2016). https://doi.org/10.1186/s12859-015-0871-y

14. Funk, C., et al.: Large-scale biomedical concept recognition: an evaluation of current automatic annotators and their parameters. BMC Bioinform. **15**, 59 (2014)

15. Campos, D., Matos, S., Oliveira, J.L.: A modular framework for biomedical concept recognition. BMC Bioinform. **14**(1) (2013). https://doi.org/10.1186/1471-2105-14-281

16. Shah, N.H., et al.: Comparison of concept recognizers for building the Open Biomedical Annotator. BMC Bioinform. **10**(S9) (2009). https://doi.org/10.1186/1471-2105-10-s9-s14

17. Jonquet, C., et al.: NCBO annotator: semantic annotation of biomedical data. In: International Semantic Web Conference, Poster and Demo session, Washington DC, USA, vol. 110, October 2009

18. Mbouadeu, S.F., et al.: Towards structured biomedical content authoring and publishing. In: 2022 IEEE 16th International Conference on Semantic Computing (ICSC). IEEE (2022)

19. PubMed Help: PubMed Help. National Center for Biotechnology Information, USA (2020)

DKMI: Diversification of Web Image Search Using Knowledge Centric Machine Intelligence

S. Mohnish[1], Gerard Deepak[2(✉)], S. V. Praveen[3], and J. Sheeba Priyadarshini[4]

[1] Madras Institute of Technology, Chennai, India
[2] Manipal Institute of Technology Bengaluru, Manipal Academy of Higher Education, Manipal, India
`gerard.deepak.christuni@gmail.com`
[3] National Institute of Technology, Tiruchirappalli, Tiruchirappalli, India
[4] CHRIST (Deemed to be University), Bangalore, India

Abstract. Web Image Recommendation is quite important in the present-day owing to the large scale of the multimedia content on the World Wide Web (WWW) specifically images. Recommendation of the images that are highly pertinent to the query with diversified yet relevant query results is a challenge. In this paper the DKMI framework for web image recommendation has been proposed which is mainly focused on ontology alignment and knowledge pool derivation using standard crowd-sourced knowledge stores like Wikipedia and DBpedia. Apart from this the DKMI model encompasses differential classification of the same dataset using the GRU and SVM, which are two distinct differential classifiers at two different levels. GRU being a Deep Learning classifier and the SVM being a Machine Learning classifier, enhances the heterogeneity and diversity in the results. Semantic similarity computation using Cosine Similarity, PMI and SOC-PMI at several phases ensures strong relevance computation in the model. The DKMI model yields overall Precision of 97.62% with an accuracy of 98.36% along with the lowest FDR score of 0.03 and is much better than the other models that are considered to be the baseline models.

Keywords: Web image recommendation · Ontology alignment · Knowledge pool · Differential classification · Crowd-sourced knowledge stores

1 Introduction

The world as we know it is changing every single day. Anything and everything are being transformed into a digital medium. Gone are the days when people

© The Author(s), under exclusive license to Springer Nature Switzerland AG 2022
B. Villazón-Terrazas et al. (Eds.): KGSWC 2022, CCIS 1686, pp. 163–177, 2022.
https://doi.org/10.1007/978-3-031-21422-6_12

used to buy music CDs and cameras. Now everything is digitally available on our smartphones and on the internet. Different kinds of data are being mined and generated are stored on computers, which makes it easily accessible. The rapid or exponential increase in the amount of information is mainly attributed to the fact that now more people have more tools to create and share information than ever before. The internet is being filled with large amounts of data, which includes texts, images, videos, documents, etc. daily. Due to the substantial increase in information, the effort required by the user to find information relevant to their search query, is also increased. In particular, when a user searches for an image on the internet, he/she is presented with thousands of images. But only a subset of these images is actually relevant to the user's query. Web Image Retrieval or Recommendation proves to be a useful technique to counter this problem. A special kind of data search used to find images from a large data store or database, such as the internet, is known as Web Image Retrieval. A user may usually provide query terms such as keywords or might click on some image, to look for images, and the system will provide images "similar" to the input query. The search criteria use similarity between anything that ranges from meta tags to the query word itself. To perform retrieval over the annotation words, most of the traditional and popular methods of image retrieval often use some kind of method to adding metadata such as keywords, captioning, title or descriptions to the images. Manual image annotation is laborious, expensive and inefficient as it is time-consuming. There has been a large amount of research done on automatic image annotation, to perform annotation efficiently.

Motivation: Due to the information overload, and in particular multimedia content, and absence of a semantic approach for web image recommendation, it's getting harder and harder for a user to find relevant information that is related to the query.

Contribution: The incorporation of standard knowledge stores like Wikidata and DBpedia for knowledge pool generation and ontology alignment with static domain ontologies for the categories in the dataset. A differential classification using a Deep Learning classifier, namely the GRU, and a feature controlled Machine Learning classifier, namely the SVM, have been hybridized for ensuring diversity in the results and incorporating heterogeneity in classification. Encompassing Cosine Similarity, SOC-PMI and the PMI measure for computing the Semantic Similarity ensures a strong relevance computation. The precision, accuracy, recall and F-Measure are increased when compared to the baseline models. Organization: The remaining paper is organized as follows. Section 2 provides a summary of the related works. Section 3 illustrates the system architecture of the proposed system. Section 4 presents the results achieved and evaluates the performance of the proposed model. Section 6 contains the conclusion of the paper.

2   Related Works

Otani et al. [1] have proposed a methodology that provides a way to represent videos and sentences using Web Image search. They use web image search in sentence embedding process, to disambiguate fine-grained visual concepts. Xie et al. [2] have proposed CARM, which is a context aware re-ranking model for web image search results for a query that is based on previous interaction. CARM outperformed state-of-the-art models, in terms of personalized evaluation metrics. Guo et al. [3] have proposed a new scheme for re-ranking images based on estimation of concept relevance. This facilitates image search for complex queries. Sejal et al. [4] have proposed a system that can retrieve images relevant to a user's input query, called IRAbMC. Computation of keyword relevance probability between annotated keywords and keywords of user input query is used to rank images.

Deepak et al. [5] have proposed a dynamic ontology alignment technique that can recommend relevant webpages. This technique provides an accuracy of 87.73%. Nguyen et al. [6] have presented an approach for image recommendation that is personalized and content-aware, which combines historical image-based features and tagging information. Mishra et al. [7] have developed a novel model that along with content information also takes into account the sequential information which is present in web navigation patterns. Rawat et al. [8] has proposed a deep neural network that can predict multiple tags for an image based on the context in which the image is captured as well as the content. Chen et al. [9] has proposed a model that semantically describes the images by extracting the text information on the web pages. Pang et al. [10] has proposed to model text matching as a problem of image recognition. Zhang et al. [11] has proposed a new model based on Bayesian Personalized Ranking which combines photo importance along with user-item interactive attention. Jere et al. [12] has proposed a model that focuses on image selection by harnessing keyword extraction. Xu et al. [13] has proposed a model called Semi-supervised Multi-concept Retrieval to semantic image retrieval via Deep Learning (SMRDL) which considers multiple concepts as a holistic scene for multi-concept scene learning of uni-modal image retrieval. Bouchakwa et al. [14] has proposed a tag based query reformulation process, by using a set of predefined ontological semantic rules.

Fernando et al. [15] have proposed a framework for improving electronic communication for government agencies targeting several ethic groups using ontologies in multi-faceted dialect encompassing Language Processing Techniques. Ayush et al. [16] have proposed a formal Ontology Model by studying concepts of heat transfer and applying them on several experimentations and formalized an inter-operable Ontology for Heat Transfer as a Primary Domain. Fernando et al. [17] put forth an approach for sharing, retrieval and exchange of Legal Documents in support of e-governmental policies using Ontology Driven Methods. Tiwari et al. [18] have formulated methodologies for representing domain knowledge focusing on Web of Things. Haribabu et al. [19] have encompassed Ontological Semantic Model for facilitating focused Web Search using semantically inclined set expansion. Surya et al. [20] have put forth a model for socially aware Web Page Recommendation using term aggregation and semantic technologies. In [21] Tiwari et al. have formalized the study of knowledge graph construction and put forth their opinion in building knowledge graphs across several interrelated domains in the real-world Web. Deepak et al. [22] have proposed a semantically inclined approach for diversified integration of knowledge encompassing role-based chunking and linguistic processing techniques. Ronak et al. [23] have proposed a framework which is an Ontology Driven Semantic Model in support of higher education in public universities

3   Proposed System Architecture

The Fig. 1 illustrates the proposed system architecture of the semantically inclined knowledge driven web image diversification framework. This model is divided into couple of phases, namely Phase 1 and Phase 2. Phase 1 is dataset driven and Phase 2 is user query driven. In Phase 1, the dataset, which is tagged, annotated and categorical in nature, is subjected to preprocessing to extract the annotations from the images. The dataset is compiled in such a way that each image has got at least one label or annotation and one category. Once the annotations are initially extracted, they are subjected to ontology alignment with static domain ontologies. The static domain ontologies are modelled or crawled using crawlers from structural meta data and is formalized as an ontology using Protégé/Stardog. In order to achieve ontology alignment, Lin Similarity with a threshold of 0.5 is considered. A Lin Similarity Measure, depicted in Eq. (1), is defined as a Node-based Semantic Similarity Measure which is based on the information content of the least common subsumer.

$$\frac{2 * ResnikSimilarity(c_1,\ c_2)}{IC\,(c_1) + IC(c_2)} \tag{1}$$

Reason for choosing 0.5 as a threshold value is mainly due to the fact that at this phase a greater number of relevant instances are needed. The entities aligned from ontologies are sent to Wikidata through Wikidata API and to DBpedia through SPARQL querying. Both Wikidata and DBpedia are knowledge stores which ensures that auxiliary knowledge is included into the framework in order to minimize the cognitive gap between the localized framework and global knowledge available in the real world. As a result, a knowledge pool is synthesized. This knowledge pool is used as a feature pool where features are extracted arbitrarily and is given to the Support Vector Machine (SVM) classifier in order to classify the categorical data.

Support Vector Machine (SVM) is a very popular Supervised Machine Learning algorithm, whose main purpose is to classify different data points into appropriate classes. The main objective of this algorithm is to find the best hyperplane/decision boundary that separates the data points in n-dimensional space into different classes. The dimension of the hyperplane depends on the features present in the data. SVM chooses the first closest point on either side from the hyperplane. These points are called as support vectors. There will a positive support vector on one side and a negative support vector on another side. The equation of the positive and negative support vectors, are represented by Eq. (2).

$$W^T x + b = 1 \text{ and } W^T x + b = \text{-}1 \tag{2}$$

respectively, where W represents the slope. The central hyperplane is represented by the Eq. (3).

$$W^T x + b = 0 \tag{3}$$

The distance between the two support vectors is termed as the marginal distance, and is depicted by Eq. (4).

$$\frac{2}{\|W\|}, s.t. \ Y_i * W^T x_i + b_i \geq 1 \tag{4}$$

This marginal distance is actually the optimization function in SVM. The goal of SVM is to maximize this marginal distance, since more the marginal distance, more the linear separation of data points. But this only holds good for an ideal case, where the classes are clearly separable, without any errors. In reality the data that we use might not be clearly separated and will have some errors. So, to counter this the optimization function can be modified to Eq. (5),

$$(w^*, b^*) = min \left(\frac{\|W\|}{2} + c_i \sum_{i=1}^{n} \varepsilon_i \right) \tag{5}$$

where w* and b* represent the values for the optimal hyperplane, W represents the slope of the hyperplane, c_i represents the number of errors (data points that fall on the wrong side of the hyperplane) and ε_i represents the value of the error (distance of the error from the hyperplane).

Apart from this the dataset is also subjected to classification using a Gated Recurrent Unit (GRU). GRU being a Deep Learning classifier, it doesn't require any external features, and instead, implicit and auto handcrafted feature selection takes place in GRUs. GRUs classify a dataset based on the features extracted by themselves.

Gated Recurrent Unit (GRU) is an improved version of the vanilla Recurrent Neural Networks (RNN), which is very much similar to Long Short-Term Memory (LSTM) neural network. The most important differentiating factor between GRU and LSTM is that the latter comprises of three gates namely - Output gate, Input gate and Forget gate, whereas the former has only two - Reset and Update gate. The reset gate is responsible for the short-term memory of the cell as it determines how much information from previous state needs to be forgot. It is calculated using Eq. (6).

$$r^{(t)} = \sigma \left(W_r \ x^{(t)} + U_r \ h^{(t-1)} \right) \tag{6}$$

The Update gate is responsible for the long-term memory of the cell. It determines how much information from the previous state needs to be retained to be passed on to the future timestamps. It is calculated using the Eq. (7),

$$u^{(t)} = \sigma \left(W_u \ x^{(t)} + U_u \ h^{(t-1)} \right) \tag{7}$$

where σ - sigmoid activation function, x(t) - inputs at timestamp t, h(t-1) - previous state, Wu and Uu - weights for the input and the previous state. The new value of the candidate state, after using the reset gate to forget any irrelevant information, can be represented by Eq. (8),

$$\hat{h}^{(t)} = tanh \ (W_h \ x^{(t)} + (r^{(t)} \circ h^{(t-1)})U_h \tag{8}$$

where $\circ$ - Hadamard Product. The model now uses the update gate to determine whether the candidate state is to be passed on or not. The final state or output state can be represented by Eq. (9).

$$h^{(t)} = u^{(t)} \ \hat{h}^{(t)} + [1 - u^{(t)}] \circ h^{\ (t-1)} \tag{9}$$

So, according to Eq. (9), if the update gate has a value of 0, then the state is not updated and the preceding state [h(t)] is retained. If the update gate has a value of 1, then the new state [h(t)] is the updated candidate state [ĥ(t)], which was in turn calculated by discarding some unwanted information of the previous state.

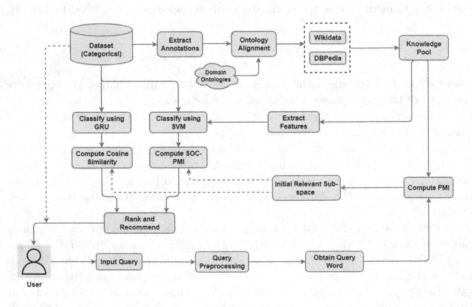

Fig. 1. Proposed system architecture.

Figure 1 illustrates the system architecture of the proposed model. At this stage, we have two distinct classified sets of the same dataset. One has been classified using the SVM classifier, which is a very basic classifier, with features from the knowledge pool and the second one has been classified using GRU, which is a robust classifier.

The reason why this type of architecture is chosen over other typologies is because the proposed architecture is a hybrid ensemble framework which encompasses the strategy of both learning as well as inferencing. Learning using several learning models and algorithms like the GRUs which is a deep learning strategy and the SVM, and inferencing using several semantic similarities schemes like Cosine similarity, SOC-PMI, PMI with differential thresholds and different configuration types, which helps in inferencing and filtering out the irrelevant as well as the least relevant and yielding the most relevant entities. So, the classifiers do the first level of simplification and then the inferencing schemes i.e., the cosine similarity, SOC-PMI and PMI ensures further filtering of the least relevant among the most relevant entities and yield a much focused data point. Apart from this encompassment of auxiliary knowledge through the Wikidata, DBPedia and other domain ontologies is chosen a lateral knowledge accumulation is achieved as well as auxiliary knowledge can increase the density of relevant entities. Due to all these reasons, this kind of a architecture is chosen because it has a hybridized learning scheme, hybridized semantic inferencing scheme along with sources like Wikidata and DBPedia to encompass auxiliary knowledge so as to enhance the density of knowledge.

The reason why the combination of GRU and SVM is used is because one is a very strong deep learning classifier and the other is light weight bi-linear machine learning classifier. In the GRUs the feature selection is automatic which means the classifier itself will take control of the features, whereas SVM being a machine learning classifier model where the features are selected and fed into the classifier. The reason why a light weight machine learning classifier with no auto handcrafted feature selection is used and also the reason why the GRUs with an auto handcrafted and automatic feature selection is used is mainly due to the fact that heterogeneity has to be improved. If two strong deep learning classifiers are used, there won't be much heterogeneity because, outlier based learning would be encountered as feature selection is auto handcrafted. If two machine learning algorithms are used, the feature selection load will become higher and the classification accuracy cannot be achieved. So due to these reasons the differential classifiers are used, one being a very strong deep learning classifier and the other being a light weight machine learning classifier. Moreover, this approach ensures that the computational complexity is not very high and it doesn't become computationally expensive.

The second phase or Phase 2 involves querying. Basic query is put through pre-processing, which involves Named Entity Recognition (NER), Lemmatization, Tokenization and Stop Word Removal to yield the individual query words. These individual query words are mapped with the instances in the knowledge pool, by computing the Pointwise Mutual Information (PMI) with a threshold of 0.5. PMI is depicted in Eq. (10),

$$PMI(x_1, x_2) = log_2 \frac{P(x_1, x_2)}{P(x_1) P(x_2)} \tag{10}$$

where P(x1) and P(x2) is the probability that x1 and x2 appear independently respectively, P (x1, x2) is the probability that x1 and x2 appear together, and x1 and x2 are two words in the text corpus. Here a very weak mutual information scheme is ensured in order to increase the entities at this phase to yield the Initial Relevant Sub-space. The entities in this sub-space are further used to compute the semantic similarity, using the Second Order Cooccurrence - PMI (SOC-PMI), with the classified entities from the SVM classifier. SOC-PMI uses PMI to sort the important neighbouring words of the two words. The semantic PMI similarity function between two words, w1 and w2, is depicted by Eq. (11),

$$Sim(w_1, w_2) = \frac{f\,(w1,\ w2,\ \beta_1)}{\beta_1} + \frac{f\,(w1,\ w2,\ \beta_2)}{\beta_2} \tag{11}$$

where the function f (w1, w2, β) sums the positive PMI values of all the words of w2 that are semantically close and which are also common to list of words that are semantically close to w1. Similarly, the entities in the sub-space are again used to compute the semantic similarity with the classified entities from GRU classifier, using Cosine Similarity. Cosine similarity is used as a measure to compare how similar two data points are, irrespective of their size. In mathematical terms, it computes the similarity of two vectors in an inner product space and is calculated by using the cosine of the angle between the two vectors that are projected in multi-dimensional space. Both SOC-PMI and Cosine Similarity in this phase have a threshold of 0.5. Since the Initial Relevant Sub-space is already a relevant set, relevant amongst the relevant is considered and hence the threshold is relaxed.

Finally, we are ranking and recommending in the increasing order of SOC-PMI and Cosine Similarity for the instances to the user. Initially only terms are given to the user and based on the user's click of the term along with the categories of the dataset, the images are loaded and yielded to the user. If the user is satisfied, the recommendation stops here, else the current user clicks are recorded and submitted as input query. This process carries on until no further clicks are recorded.

4   Results and Performance Evaluation

The performance of the proposed DKMI (Diversification of Web Images using Knowledge Centric Machine Intelligence) framework for diversified web image recommendation which is a knowledge centric approach is evaluated using Precision, Recall, Accuracy, F-Measure percentages. The False Discovery Rate (FDR) and Normalized Discounted Cumulative Gain (nDCG) are also used as standard and potential metrics. Precision, Recall, Accuracy and F-Measure quantify the relevance of the results provided by the framework. The False Discovery Rate indicates the number of false positives returned by the model and the nDCG value quantifies the variety in the results. From Table 1 it is clear that the proposed DKMI model provides the highest percentage of Precision, Recall, Accuracy and F-Measure, the lowest value of FDR and the highest value of the nDCG. The DKMI framework yields 97.62% of average Precision, 99.10% of average Recall, 98.36% of average Accuracy, 98.35% of F-Measure with an FDR of 0.03 and nDCG of 0.98.

Table 1. Comparison of Performance of the proposed DKMI with other approaches

Model	Precision %	Recall %	Accuracy %	F-Measure %	FDR	nDCG
PIR [11]	84.29	85.07	84.68	85.31	0.16	0.85
MLIRNA [12]	87.24	89.03	88.14	88.13	0.13	0.84
SIRDL [13]	92.18	94.06	93.12	93.11	0.08	0.92
MLDA [14]	93.04	95.12	94.08	94.59	0.07	0.94
Proposed DKMI model	97.62	99.1	98.36	98.35	0.03	0.98

To compare the performance of the DKMI model, it has been baselined with PIR [11], MLIRNA [12], SIRDL [13] and MLDA [14] respectively. Testing was done for 4418 queries for which the ground truth has been collected from the web usage data, the web browsing data as well as consultation of users, who were between the ages of 18 and 54, for over a period of 7 months. The number of users who participated in the ground truth collection were 724. To assess the performance of the DKMI model, the baseline models and DKMI model were evaluated in the exact same environment for the exact same number of queries for the exact same dataset and ground truth. Hybridization of two distinct datasets namely the Kaggle's corel-images (https://www.kaggle.com/datasets/elkamel/corel-images) dataset and the 59K OCR dataset was incorporated. For experimentations however, these two datasets could not overlap each other. The 59K OCR snippets from 95K annotated Twitter images (https://doi.org/10.34740/kaggle/ds/732777) dataset was already annotated and further annotations were done automatically using customized Python annotators. The corel-images dataset was further annotated and these two datasets were integrated into a single large dataset and were used as the main dataset.

Table 1 indicates that PIR [11] yields 84.29% of average Precision, 85.07% of average Recall, 84.68% of average Accuracy, 85.31% of F-Measure along with FDR of 0.16 and nDCG value of 0.85. Similarly, MLIRNA yields 87.24% of average Precision, 89.03% of average Recall, 88.14% of average Accuracy, 88.13% of F-Measure along with FDR of 0.13 and nDCG value of 0.84. On the other hand, SIRDL yields 92.18% of average Precision, 94.06% of average Recall, 93.12% of average Accuracy, 93.11% of F-Measure along with FDR of 0.08 and nDCG value of 0.92. It is also seen that MLDA yields 93.04% of average Precision, 95.12% of average Recall, 94.08% of average Accuracy, 94.59% of F-Measure along with FDR of 0.07 and nDCG value of 0.94.

Table 2 depicts the results for two example queries - Real Estate and Wedding Jewellery. For the proposed model, the various facets yielded are displayed and for the other models, there were no facets but the consolidated categories of

Table 2. Results for two example queries

Query	TIR	MLIRNA	SIRDL	MLTA	Proposed DKMI
Real Estate	Real Estate Agents, Several Agents and their Organizations Alone, House for Sale Hoardings	Real Estate Agents, Real Estate Around, Real Estate and Loan, Only Several Real Estate Brokers Contact Details were yielded	Real Brand Logo, Estates like Farmlands and Grasslands were depicted, pictures of Brokers and Agents and several pictures of Landscape were yielded	Real Life Hiking Advertisements, Brokering and Agents for Real Estate were alone in the Top-10Search Results	Real Estate Advertisements, Classifieds Related to Property Real Estates, Real Estate of Houses, Brokering, Real Estate on Farm Lands and Agricultural Lands, Logos of Real Estate, Real Estate Courses, Books on Real Estate, Real Estate and Economics, Real Estate Agents, Real Estate Companies Logos
Wedding Jewellery	Wedding Jewels and their design, Wedding Rituals, Wedding Necklace	Wedding Bridal Collections, Gold Ring for Weddings, Several Cluttered Jewellery Images were displayed	Random Wedding Jewellery Pictures, Queen of Rings Wedding Collections, Ring Size for men, Ring Designs for Men and Women	Images of Several Wedding Jewels were displayed without any specific labels. The results were cluttered. The designs included Chains, Watches, Bracelets, Lockets and other Jewellery	Wedding Jewellery Designs, Collections from Several Wedding Brands, Indian Wedding Jewels, Bridal Jewels for Christian Wedding, Gothic Style Wedding Jewellery, Wedding Jewellery Around the Globe, Ruby and Diamond Bridal Collection, Christian Wedding, Collections, Temple Jewellery Wedding Brandline, Muslim Weddings Bridal Jewellery, Rings for Wedding, Wedding Jewel Making, Emerald Collections for Bridal Wedding, Jewels for the Bridegroom

the top ten results are displayed. The proposed DKMI model has a wide diversity of results which is a full cover search because of the auxiliary knowledge augmented and a very strong ecosystem of semantic similarity as well as auxiliary knowledge. That is reason why there are a lot of facets generated and the respected images were yielded. Whereas for the other models, the results were quite shallow, cluttered, random, weren't diverse enough and the top ten results itself only contained many instances of very few categories. The diversity of the results of the proposed model is mainly due to the fact that auxiliary knowledge, strong semantic similarity ecosystem, strong relevance computation mechanism as well as strong learning is incorporated in the model.

The reason why the proposed DKMI attains the highest Precision, Recall, Accuracy, F-Measure percentages, with the highest nDCG and lowest FDR is due to the reason that it is a Knowledge Centric approach which uses knowledge from heterogeneous sources. Although it is dataset driven the annotations are extracted and further aligned with domain ontologies using strategic schemes, selective strategic alignment and apart from this formulation of knowledge pool by using two distinct strong knowledge stores, namely Wikidata and DBpedia, ensures large diversity of entities anchored into the model as the subsequent auxiliary knowledge. Apart from this the usage of GRUs for implicit classification of the dataset, classification of the dataset using a subsequent SVM, which is a feature controlled machine learning model, computation of Cosine Similarity and SOC-PMI for semantic similarity computation at different levels with differential thresholds, and the usage of PMI for further computation of semantic similarity between the initial knowledge pool and the relevant subspace ensures large amount of auxiliary knowledge and heterogeneous entities. Also, there is a high degree of regulatory mechanism in terms of relevance computation with differential semantic similarity measures. Amalgamation of Deep Learning based GRU classifier and a feature controlled Machine Learning classifier improves heterogeneity in classification. Due to all the above results the proposed framework gives the highest Precision, Recall, Accuracy, F-Measure percentages, with the highest nDCG and lowest FDR. Also, due to large amount of auxiliary knowledge with extremely controlled similarity computation mechanisms, the proposed DKMI framework outperforms the baseline models and enhances the diversity in the results.

The reason why the PIR [11] model doesn't perform as expected, even though it is a personalized image recommendation model, is because it uses only user intents along with user-item interactive attention and photo importance. Importance of user interaction in the current user clicks alone is captured. Using a Bayesian personalized ranking model with some weighting factor, the user-item interactive attention is computed here. So, in this model there are neither strong learning mechanisms nor strong learning paradigms. Also, there is no auxiliary knowledge to diversify the results and no presence of strong relevance computation mechanism. Due to all the above factors, the PIR model lags to a large extent.

The MLIRNA [12] model, a Machine Learning model for image recommendation for news articles, is a highly domain centric model. It is a Machine Learning model with statistical computational schemes with Open Calais entity detection. Due to usage of Open Calais, there is some amount of knowledge incorporated but there is no splicing, ranking and relevance computation mechanisms. Although the knowledge is attenuated there is no organization of knowledge. As a result, this model also does not perform as expected.

The reason why the SIRDL [13] framework doesn't outperform the proposed model is because, even though it is a semantic image recommendation model which uses Deep Learning, it is a uni-modal image retrieval modal with limited labelled examples. The assumption is that, limited examples with a strong learning environment can accelerate the quality of the results. However, in this case there is a lot of underfitting of relevant auxiliary knowledge and overfitting of unilateral knowledge. This clearly indicates that there is neither diversity nor diversification, and also there is high number of recommendations with similar content. As a result, although it performs well it doesn't outperform the proposed model.

The MLDA [14] model proposes tag based query reformulation process. Due to the integration of tags for the query where multi-label image diversification takes place, it proves to be a very good model. But the only drawback of this model is that even though the query is grown, which in turn enhances diversity, it relies only on tags. High grade auxiliary knowledge from heterogeneous sources is required, rather than depending only on the tags. Apart from this, this model also lags in relevance computation mechanisms. As a result, this model performs very well but again doesn't outperform the proposed model.

Due to all the above reasons the proposed DKMI model proves to be a better model. Large quantities of heterogeneous entities from several varied entity sources, with strong relevance computation mechanisms and with proper thresholds at every instance, along with usage of two distinct classifiers ensures large number of variations and variety in the classification itself, thereby increasing the variety of the results. Hence, the proposed DKMI model outperforms all the baseline models.

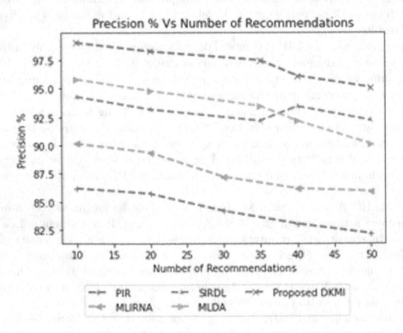

Fig. 2. Precision % vs number of recommendations

Figure 2 depicts the Precision percentage vs the number of recommendations distribution line graph for the proposed DKMI model along with the baseline models. From Fig. 2 it is clear that the DKMI model occupies the highest position in the hierarchy which is immediately followed by MLDA [14], SIRDL [13], MLIRNA [12] and PIR [11] respectively.

Since the proposed DKMI model uses two distinct classifiers namely GRU and SVM, and also due to the fact that it is a Knowledge Centric model, it has the best performance amongst all other baseline models. As the MLDA [14] model is highly dependent on the tags, it doesn't perform as good when compared to the proposed DKMI model. Limited labelled examples, which essentially leads to less diversity, hampers the performance of the SIRDL [13] model. Since the MLIRNA [12] model is a highly domain centric model, which lacks splicing, ranking and relevance computation mechanisms, it doesn't outperform the proposed model. Due to the absence of auxiliary knowledge and strong relevance computation mechanisms, the PIR [11] model turns out to be the least performing model among all the baseline models.

5 Conclusions

In this paper a Knowledge Centric framework that uses Machine Intelligence for Web Image Recommendation has been proposed. Due to the enormous increase in the information, specifically images, available on the World Wide Web, it is becoming increasingly inconvenient for users to find information that is relevant to their query. As a result of this Web Image Recommendation has become very important. Recommendation of diversified yet relevant, results that are relevant to the query still pose a challenge The proposed DKMI model is primarily focused on ontology alignment and knowledge. The knowledge pool is derived using standard crowd-sourced knowledge stores like Wikipedia and DBpedia. In order to obtain heterogeneity and diversity in the results the proposed model incorporates two distinct differential classifiers, namely GRU which is a Deep Learning based classifier and SVM which is a Machine Learning based classifier, at two different levels. A strong relevance computation in the model is ensured by the computation the Semantic Similarity using Cosine Similarity, PMI and SOC-PMI. The proposed model yields overall Precision of 97.62% with an accuracy of 98.36% along with the lowest FDR score of 0.03. As a result, it outperforms all the baseline models.

References

1. Otani, M., Nakashima, Y., Rahtu, E., Heikkilä, J., Yokoya, N.: Learning joint representations of videos and sentences with web image search. In: Hua, G., Jégou, H. (eds.) ECCV 2016. LNCS, vol. 9913, pp. 651–667. Springer, Cham (2016). https://doi.org/10.1007/978-3-319-46604-0_46
2. Xie, X., et al.: Improving web image search with contextual information. In: Proceedings of the 28th ACM International Conference on Information and Knowledge Management, pp. 1683–1692 (2019)
3. Guo, D., Gao, P.: Complex-query web image search with concept-based relevance estimation. World Wide Web 19(2), 247–264 (2016)

4. Sejal, D., Rashmi, V., Venugopal, K.R., Iyengar, S.S., Patnaik, L.M.: Image recommendation based on keyword relevance using absorbing markov chain and image features. Int. J. Multimedia Inf. Retr. **5**(3), 185–199 (2016)
5. Deepak, G., Ahmed, A., Skanda, B.: An intelligent inventive system for personalised webpage recommendation based on ontology semantics. Int. J. Intell. Syst. Technol. Appl. **18**(1–2), 115–132 (2019)
6. Nguyen, H.T.H., Wistuba, M., Schmidt-Thieme, L.: Personalized tag recommendation for images using deep transfer learning. In: Ceci, M., Hollmén, J., Todorovski, L., Vens, C., Džeroski, S. (eds.) ECML PKDD 2017. LNCS (LNAI), vol. 10535, pp. 705–720. Springer, Cham (2017). https://doi.org/10.1007/978-3-319-71246-8_43
7. Mishra, R., Kumar, P., Bhasker, B.: A web recommendation system considering sequential information. Decis. Support Syst. **75**, 1–10 (2015)
8. Rawat, Y.S., Kankanhalli, M.S.: ConTagNet: exploiting user context for image tag recommendation. In Proceedings of the 24th ACM international conference on Multimedia, pp. 1102–1106 (2016)
9. Chen, Z., Wenyin, L., Zhang, F., Li, M., Zhang, H.: Web mining for web image retrieval. J. Am. Soc. Inf. Sci. Technol. **52**(10), 831–839 (2001)
10. Pang, L., Lan, Y., Guo, J., Xu, J., Wan, S., Cheng, X.: Text matching as image recognition. In Proceedings of the AAAI Conference on Artificial Intelligence, vol. 30 (2016)
11. Zhang, W., Wang, Z., Chen, T.: Personalized image recommendation with photo importance and user-item interactive attention. In 2019 IEEE International Conference on Multimedia and Expo Workshops (ICMEW), pp. 501–506. IEEE (2019)
12. Jere, R., Pandey, A., Shaikh, H., Nadgeri, S., Chandankhede, P.: Using machine learning for image recommendation in news articles. In: Shetty D., P., Shetty, S. (eds.) Recent Advances in Artificial Intelligence and Data Engineering. AISC, vol. 1386, pp. 215–225. Springer, Singapore (2022). https://doi.org/10.1007/978-981-16-3342-3_18
13. Haijiao, X., Huang, C., Wang, D.: Enhancing semantic image retrieval with limited labeled examples via deep learning. Knowl. Based Syst. **163**, 252–266 (2019)
14. Bouchakwa, M., Ayadi, Y., Amous, I.: Multi-level diversification approach of semantic-based image retrieval results. Prog. Artif. Intell. **9**(1), 1–30 (2020)
15. Ortiz-Rodriguez, F., Tiwari, S., Panchal, R., Medina-Quintero, J.M., Barrera, R.: MEXIN: multidialectal ontology supporting NLP approach to improve government electronic communication with the Mexican ethnic groups. In DG. O 2022: The 23rd Annual International Conference on Digital Government Research, pp. 461–463 (2022)
16. Kumar, A., Deepak, G., Santhanavijayan, A.: HeTOnto: a novel approach for conceptualization, modeling, visualization, and formalization of domain centric ontologies for heat transfer. In: 2020 IEEE International Conference on Electronics, Computing and Communication Technologies (CONECCT), pp. 1–6. IEEE (2020)
17. Ortiz-Rodriguez, F., Medina-Quintero, J.M., Tiwari, S., Villanueva, V.: EGODO ontology: sharing, retrieving, and exchanging legal documentation across e-government. In: Futuristic Trends for Sustainable Development and Sustainable Ecosystems, pp. 261–276. IGI Global (2022)
18. Tiwari, S., Siarry, P., Mehta, S., Jabbar, M.A.: Tools, Languages, Methodologies for Representing Semantics on the Web of Things. Wiley, New York (2022)
19. Haribabu, S., Siva Sai Kumar, P., Padhy, S., Deepak, G., Santhanavijayan, A., Kumar, N.: A novel approach for ontology focused inter-domain personalized search based on semantic set expansion. In: 2019 Fifteenth International Conference on Information Processing (ICINPRO), pp. 1–5. IEEE (2019)

20. Surya, D., Deepak, G., Santhanavijayan, A.: KSTAR: a knowledge based approach for socially relevant term aggregation for web page recommendation. In: Motahhir, S., Bossoufi, B. (eds.) ICDTA 2021. LNNS, vol. 211, pp. 555–564. Springer, Cham (2021). https://doi.org/10.1007/978-3-030-73882-2_50

21. Tiwari, S., Gaurav, D., Srivastava, A., Rai, C., Abhishek, K.: A preliminary study of knowledge graphs and their construction. In: Tavares, J.M.R.S., Chakrabarti, S., Bhattacharya, A., Ghatak, S. (eds.) Emerging Technologies in Data Mining and Information Security. LNNS, vol. 164, pp. 11–20. Springer, Singapore (2021). https://doi.org/10.1007/978-981-15-9774-9_2

22. Deepak, G., Kumar, N., Santhanavijayan, A.: A semantic approach for entity linking by diverse knowledge integration incorporating role-based chunking. Procedia Comput. Sci. **167**, 737–746 (2020)

23. Panchal, R., Swaminarayan, P., Tiwari, S., Ortiz-Rodriguez, F.: AISHE-Onto: a semantic model for public higher education universities. In: DG. O2021: The 22nd Annual International Conference on Digital Government Research, pp. 545–547 (2021)

Does Wikidata Support Analogical Reasoning?

Filip Ilievski[✉], Jay Pujara, and Kartik Shenoy

Information Sciences Institute, University of Southern California, Los Angeles, USA
{ilievski,jpujara,kshenoy}@isi.edu

Abstract. Analogical reasoning methods have been built over various resources, including commonsense knowledge bases, lexical resources, language models, or their combination. While the wide coverage of knowledge about entities and events make Wikidata a promising resource for analogical reasoning across situations and domains, Wikidata has not been employed for this task yet. In this paper, we investigate whether the knowledge in Wikidata supports analogical reasoning. Specifically, we study whether relational knowledge is modeled consistently in Wikidata, observing that relevant relational information is typically missing or modeled in an inconsistent way. Our further experiments show that Wikidata can be used to create data for analogy classification, but this requires much manual effort. To facilitate future work that can support analogies, we discuss key desiderata, and devise a set of metrics to guide an automatic method for extracting analogies from Wikidata.

Keywords: Wikidata · Analogical reasoning · Ontologies · User experience

1 Introduction

Cognitive science research has provided rich evidence that humans use analogical reasoning to understand, explain, or imagine novel situations within or across domains [13]. Analogical thinking can connect the Great Depression and the financial crisis based on *causal* knowledge [21], or compare the Sun and the Earth to the Earth and the Moon based on the *revolves* relation [11]. Corresponding cognitive systems have been build to realize and test this skill algorithmically, such as the Structured Mapping Engine [9] and the Companion architecture [10]. Natural Language Processing research on analogy has been popularized through the proportional analogy task, illustrated through the famous example of *man:woman-king:queen* by the word2vec system [19]. Recognizing the gap between the large-scale word analogy systems and the expressive cognitive systems, recent AI research has focused on integrating neural (language) models with cognitive systems to solve tasks like sketch object recognition [3], product innovation [14], narrative understanding [21], and moral decision making [5].

As it can be expected, these analogical reasoning efforts are often centered around a knowledge base that enables models to understand implicit relations, such as *causes* or *revolves*. Curiously, despite the large quality and richness of Wikidata, and its increasing adoption for many knowledge-intensive tasks [16, 20, 23], prior work on analogical reasoning has not considered leveraging its ontology or its factual knowledge to reason by analogy. Instead, existing systems have typically leveraged publicly available parts of Cyc, semantic lexical resources, language models [27], or their combination [10].

Considering the coverage of millions of ontological concepts and instances, intuitively, Wikidata could serve as a valuable resource for analogical reasoning. In this paper, we perform an initial study on whether the Wikidata knowledge supports analogical reasoning. Specifically, we focus on three key questions:

1. *Does Wikidata express relational information consistently?* Analogical reasoning revolves around relational similarity, therefore, consistency in the knowledge modeling of relational knowledge is crucial to enable reasoning systems to connect between two situations or domains. We investigate whether relational information is consistently modeled in Wikidata.
2. *Does Wikidata support extraction of analogy classification data?* Given its wide coverage, Wikidata may have the potential to generate large-scale analogy detection tasks automatically. We investigate how much manual effort is required to create such benchmarks for subclass-of relations in Wikidata, and we evaluate the performance of state-of-the-art NLP systems.
3. *Which desiderata and metrics can guide automatic generation of analogies using the Wikidata structure?* Considering that the Wikidata ontology is not uniform in terms of its granularity and expressivity, it is important to define desired properties for analogical reasoning and design automated metrics that can quantify this variation and enable the selection of potential analogical correspondences automatically.

2 Does Wikidata Express Relational Information Consistently?

2.1 Data and Setup

We sample 20 *subclass-of (P279)* relations from Wikidata, whose subject label is a superstring of its object label. For example, we keep the subclass pair *red wine - wine*, while we discard *dog - pet*. We prioritize Qnodes with low identifiers as a simple proxy for well-known entities and concepts. Example pairs include *stellar atmosphere - atmosphere* and *computer keyboard - keyboard*. We analyze whether the subclass-of relation is complemented by additional information that can help us categorize the nature of the inheritance relation, expressed either as other relations of the subject or qualifiers on the subclass-of relation. Based on prior work on categorization of semantic relations for noun compounds [12], we define an initial set of seven inheritance categories: PURPOSE, PROPERTY, LOCATION, OWNERSHIP, MATERIAL, INSTANCE, and TEMPORAL, and

Table 1. Ten exemplar compound noun pairs.

Subject	Object	Category	Qualifiers	Statements	Siblings
Computer keyboard (Q250)	Keyboard (Q1921606)	PURPOSE	follows: mobile phone	Computer keyboard - part of - computer	Typewriter keyboard, Braille keyboard, musical keyboard
Natural science (Q7991)	Science (Q336)	PROPERTY	–	–	Human science, information science, modern science, Ancient Egyptian science, ...
Beach volleyball (Q4543)	Volleyball (Q1734)	LOCATION	–	–	Snow volleyball, women's volleyball, men's volleyball
Fairy tale (Q699)	Tale (Q17991521)	PROPERTY	–	–	Old-fashioned tale, cumulative tale, urbain tale, German folk tale
Shia Islam (Q9585)	Islam (Q432)	INSTANCE	–	–	Sunni Islam, Islam in Denmark, Islamic eschaetology, Gospel in Islam
Stellar atmosphere (Q6311)	Atmosphere (Q8104)	LOCATION	of: star	–	Extrasolar atmosphere, extraterrestrial atmosphere, Reducing atmosphere
Electric charge (Q1111)	Charge (Q73792)	PROPERTY	of: electromagnetic field	–	Magnetic charge, color charge, weak hypercharge
Red wine (Q1827)	Wine (Q282)	PROPERTY	–	Red wine - color - red	White wine, Mexican wine, Polish wine, straw wine, de-alcoholised wine, ...
Day sky (Q4812)	Sky (Q527)	PROPERTY	–	–	Blue sky, morning sky, Velazquez sky, ...
Animal rights (Q426)	Right (Q2386606)	PROPERTY	of: nonhuman animal	–	Hunting rights, women's rights, right to property,

annotate each pair with one category. Besides obtaining relations for the original pair, we obtain siblings of the subject (other Qnodes that are direct children of the same object) and investigate their structures seeking for regularities.

2.2 Findings

We show ten out of the twenty pairs in Table 1. Overall, we find that Wikidata describes relations sparsely, which does not help us identify the compound relation category. Specifically, out of 20 pairs, we found 4 cases where Wikidata provided a qualifier to further specify the relation. Among these four qualifiers, three were expressed with *of* (e.g., *stellar atmosphere - atmosphere* is further specified by the qualifier *of - star*) and a single case used the *follows* qualifier (*computer keyboard - keyboard* is specified by *follows - mobile phone*). In addition to being sparse, we find the qualifier information to correlate weakly with our semantic categories, as *of* corresponds to both LOCATION (in *stellar atmosphere - atmosphere - of: star*) and to PROPERTY (in *electric charge - charge - of: electromagnetic field*).

Comparing the subject to its siblings also reveals a wide diversity in semantics that is not explicitly modeled. The siblings of *natural science* (Q7991) with a parent *science* (Q336) sometimes imply a property relation (e.g., human science, information science), and in others imply a spatial (science in Ivano-Frankivsk), temporal (modern science), or spatio-temporal (Ancient Egyptian science) specification. Some of the siblings with a PROPERTY relation use the *studies* property in Wikidata to indicate the subject of the science (e.g., *human science - studies - humans*). However, this property is again not consistently applied across the different siblings.

3 Does Wikidata Support Extraction of Analogy Classification Data?

3.1 Dataset Construction

Given that Wikidata does not support analogical reasoning directly through its relational modeling, we next investigate the possibility of creating a dataset of analogical pairs with Wikidata. Using the word pairs from the previous section as a seed set, we create analogies manually by searching for more P279 relations in Wikidata. We form 200 such quadruples, each consisting of two word pairs. The 200 quadruples are split evenly into analogical and non-analogical pairs.

Within the 100 positive cases (analogical pairs), we systematically sample three sets of cases: 1) 25 direct analogies (Pos-direct), which are pairs with a common object node, e.g., `computer science-science: food science-science`. 2) 50 parent-based analogies (Pos-parent), consisting of pairs of words where the objects have a common parent. For example, we sample the pair indoor `indoor golf-golf: road tennis-tennis`. 3) 25 distant analogies (Pos-distant), where the two objects have a more distant, yet semantically meaningful, common ancestor. An example is `Computer keyboard-keyboard: text display-display device`, where both keyboard and display device are products. For all three categories, we make sure that the analogical pairs are connected with the same relation, e.g., MATERIAL. To sample negative cases (Neg), we sampled random pairs and validated that their relation is not analogical. An example negative analogy is `Shia Islam-Islam: ancient music-music`. The sampling procedure of the positive and the negative analogical pairs included a manual postprocessing step, as Wikidata does not provide a direct way to filter the data.

The resulting 100 positive analogies are distributed as shown in Table 2. This Table shows that some categories, most notably PROPERTY, dominate the dataset, suggesting that this category may need to be further refined in future work. In this Table, we also provide an example analogical pair for each of the seven categories an each of the three levels of analogy. We show negative

Table 2. Distribution of the 100 positive analogies, with representative examples.

Category	Count	Example
PROPERTY	53	`fairy tale-tale: protestant cathedral-cathedral`
LOCATION	19	`sitting volleyball-volleyball: backyard cricket-cricket`
MATERIAL	11	`carbon fibers-fiber: iron wire-wire`
INSTANCE	5	`Shia Islam-Islam: Antiochian Greeks-Greek diaspora`
OWNERSHIP	5	`fetal liver-liver: elephant skin-skin`
PURPOSE	4	`computer keyboard-keyboard: scoreboard-display device`
TEMPORAL	3	`day sky-sky: winter garden-garden`

Table 3. Examples of non-analogical pairs.

Example	Category 1	Category 2
`beach basketball-basketball: Middle English-English`	LOCATION	TEMPORAL
`electronic music-music: Government of France-government`	PROPERTY	LOCATION
`carbon fibers-fiber: private bank-bank`	MATERIAL	PROPERTY
`mammal tooth-tooth - slang dictionary-dictionary`	OWNERSHIP	PROPERTY
`stellar atmosphere atmosphere: marker pen-pen`	LOCATION	PURPOSE
`animal rights-right: cash-money`	PROPERTY	INSTANCE

examples in Table 3. Here, again the data is dominated by the PROPERTY relation.

Table 4. Baseline results on the analogical benchmark.

Model	Supervised	Accuracy
GloVe	No	0.72
Word2Vec	No	0.74
BERT	No	0.775
BERT	Yes	**0.815**

3.2 Baseline Models

Given that the dataset consists of proportional analogies on which language models have already been evaluated, we experiment with three models and to evaluate their performance: GloVe [24], Word2Vec [18], and BERT [6]. For GloVe and Word2Vec, we compute analogical similarity as $sim = cosine(emb(child1) - emb(parent1), emb(child2) - emb(parent2))$. In the case of BERT, we compute similarity as $sim = cosine(emb(sentence1), emb(sentence2))$, where sentence1 and sentence2 are template-based sentences constructed following {subject-label} *is {subject-description}*. This means that for BERT, we do not consider the parent nodes explicitly, they are only indirectly covered by the Wikidata descriptions of the subjects (children).

For all models, we define a threshold t, such that the pairs are analogical if $sim > t$, and non-analogical otherwise. To rule out the impact of the threshold, we perform grid search and report the results with the optimal t.

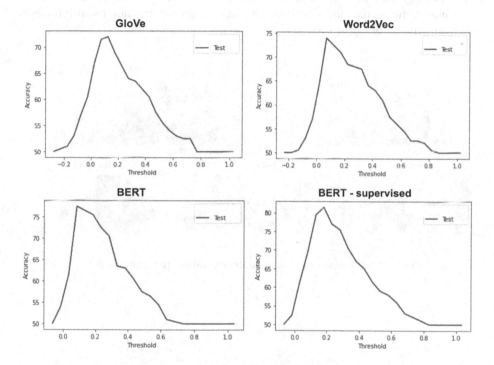

Fig. 1. Optimal thresholds for each model.

3.3 Results

The best obtained results per model in are shown in Table 4. Among the unsupervised models, we observe that the performance of BERT is better than GloVe and Word2Vec. We also evaluate a supervised version of the BERT model, which is an SVM model trained with cross-validation. We observe that with this training, the BERT model is able to improve its performance to 81.5%.

The impact of the different thresholds for each of the four models is shown in Fig. 1. We observe that the optimal threshold for most models is around 0.1. Interestingly, when we distinguish between the similarities for the negative and the different kinds of positive cases (shown for BERT-supervised in Fig. 2), we observe that the similarities for the non-analogies are often between 0 and 0.1. The analogies for the positive class are more uniformly distributed, and peak between 0.2 and 0.3. Interestingly, the similarity values for both analogies and non-analogies are typically low, which signals that simple embedding calculations cannot be used to extract analogical links from language models, but also that the wide range of domains covered make this task even more challenging.

Among the different positive classes, we observe that the direct analogies have highest similarities, and the similarity decreases with the ontological distance. These findings can be expected, and they indicate that language models contain some useful signals for identifying analogical pairs, yet they require further improvements to perform robustly.

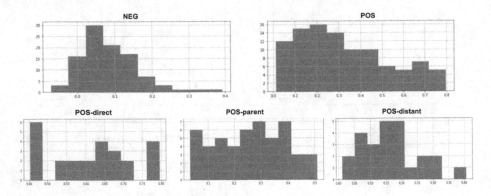

Fig. 2. Distribution of cosine similarity values per analogy set.

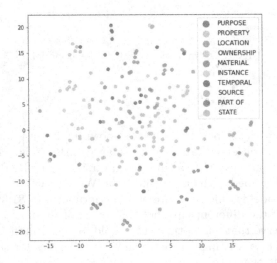

Fig. 3. BERT-supervised embeddings for each pair, colored with its category.

Considering these results, we investigate whether the models are able to distinguish between the different semantic categories of properties. Figure 3 shows a t-SNE plot of the BERT-supervised embeddings for each of the data points, colored by their category. The lack of same-colored visual clusters for BERT-supervised shows that overall BERT is unable to clearly distinguish between the semantic categories. Some exceptions exist, for instance, the MATERIAL nodes are often found together in the bottom-right part of the plot. We obtained similar or worse results for the other models as well. This shows that even with supervision, language models are not able to directly understand analogical categories derived from Wikidata. These results are consistent with findings on existing datasets for analogical reasoning in the NLP domain [27].

4   Towards Automated Analogy Generation Using Wikidata Structure

The experiments in the prior section were evaluated based on a small, manually curated set of analogical entities. To move beyond such manually curated resources, we explored some key desiderata for meaningful analogical relationships. Using the Wikidata ontology and a set of analytical metrics of ontological structure, we evaluated several techniques for filtering ontological classes to identify candidate analogical relationships. In this section, we describe the important properties of analogies, how such properties can be captured using metrics from the Wikidata ontology, and provide illustrative examples from five such approaches for analogy generation.

4.1   What Makes a Good (Wikidata) Analogy?

Our exploration focused on classes in Wikidata, with the goal of finding meaningful child-parent relationships. We hypothesize that parent-child relationships provide a foundation for analogy generation since (a) child classes specialize from parent classes by some criteria and (b) sibling classes must differ from each other by some criteria. For example the class of houses is a subclass of buildings, and the differentiation between buildings and houses occurs based on purpose – houses are buildings that are used for residential purposes, rather than commercial, governmental, or industrial purposes. Similarly, houses and office buildings are both types of buildings, but again differ based on purpose. To better understand the differences between parent and child classes and sibling classes, we consider entities that are instances of those classes.

Generality: Ontologies can contain very specific classes that are limited to a specialized domain, and used for technically precise concepts. For example, a butter mill is a type of building, but very few people have direct experience or understanding of butter mills. While such classes may still provide meaningful analogies, their specialization makes such analogies difficult to judge and less accessible. Our experiments exclude classes that may be too niche to provide a large set of meaningful analogies.

Selectivity: Useful parent-child analogical class pairs must show a clear selective criteria for distinction between parent and child classes. If such class pairs are too similar, the conceptual differences between their constituent entities may be hard to characterize. For example, if most tools are devices, the distinction between tool and device diminishes. Non-selective entity pairs may result in confusing analogies, so useful analogical classes will show selective differences between classes.

Salience: Similar to selectivity, salience is the ability to differentiate between two related classes. If two class pairs have largely overlapping constituent entities, the criteria to distinguish between classes may not be apparent. If many houses are also office buildings, the distinction between these two classes becomes difficult to justify. We choose salient pairs to ensure clear analogies.

Diversity: Interesting analogies may stem from parent classes that have several different subclasses that have significant instances. A structure with many meaningful subclasses can suggest a diverse set of concepts that can easily be used to create analogies. For example, by having many distinct building types with many different specific building instances, generating diverse analogies with buildings is more feasible. In contrast, if one building type dominates over others, or there are myriad, sparse building types analogy generation may become difficult.

4.2   Metrics on the Wikidata Ontology

We pair the qualities of desirable ontological properties with specific metrics to realize these properties. By defining such metrics, we are able to analytically derive candidate classes and their associated instances for use in analogy generation. When paired with large scale validation data, these metrics can allow tailored analogy generation based on a specific type of analogy. We focus on two specific types of Wikidata relationships parent-child `subclass of` relationships (P279) and class-instance `is a` relationships (P31).

Generality → Instance Count: To operationalize the generality of a Wikidata class, we introduce a metric of instance counts. Formally,

$$IC = |C_i|$$

where IC is the instance count, C_i is the set of instances of the class C and $|C_i|$ is the cardinality of set C_i. A class with very few instances signals a specific or potentially undefined concept class. Filtering out classes with few instances can help avoid analogies that are difficult to understand.

Selectivity → Reduction Ratio: We determine how selective a parent-child relationship is based on the ratio of child instances to parent instances, which we refer to as the reduction ratio. Formally,

$$RR = 1 - \frac{|C_i|}{|P_i|}$$

where RR is the reduction ratio, C_i is the set of instances of child class C and P_i is the set of instances of parent class P.

If all parent instances are also child instances, the reduction ratio will be low. Low reduction ratios mean that a subclass is largely co-referent with the parent class and not very selective. In contrast, high reduction ratios mean that a subclass has very few instances of the parent class, and only represents a small portion of the concept space. Ideally, reduction ratios would be in between these extremes, identifying classes that still have a meaningful portion of the parent class without being too general or too specific.

Salience → Class Overlap: To identify salience, we employ a ratio corresponding to class overlap as measured by the Jaccard set similarity. For each class, we measure the overlap with the largest sibling class. Formally,

$$CO = \frac{|C_i \cap S_i|}{|C_i \cup S_i|}$$

CO is the class overlap metric, C_i are instances of class C and S_i are instances of sibling class S. The numerator designates the number of entities that are instances of both C_i and S_i, while the denominator is the number of distinct entities that are instances of either C or S.

If the class overlap is high, the two classes share many instances and will be difficult to distinguish. If the class overlap is low, the two classes are largely distinct. Desirable analogical relationships between classes will require low class overlap to ensure that the classes are easily distinguished.

Diversity → Entropy: We measure diversity of a parent class with respect to its subclasses by measuring the entropy. Formally, we first define the probability of a class as $P(C) = \frac{|C_i|}{|P_i|}$, where P_i are instances of parent class P and C_i are instances of child class C. Next, we can formulate the entropy as

$$H(P) = \prod_{C \subset P} -P(C) \cdot log(P(C))$$

$H(P)$ is the entropy of the parent class defined in terms of its subclasses C.

If the entropy of a parent class is very low, the class hierarchy is likely to be dominated by a single, large subclass. Conversely, if the entropy is very high the hierarchy may contain a large number of small classes. A moderate value for entropies can achieve a balance between a dominating class and many small classes.

4.3 Illustrative Examples of Wikidata Relationships

We provide some illustrative examples found by filtering Wikidata with metrics above, using filters such as $IC > 1000, RR < .85, CO < 0.5, H(P) > 1$. One such example is the class of *visual artwork* (Q4502142), which includes two sibling classes of *painting* (Q3305213) and *film* (Q11424). An example analogical pair at an instance level that can be generated from these two sibling classes is The

Bohemian (Q1000128) :: Amélie (Q484048), the former depicting and the latter having a narrative location of Paris (Q90).

Another such example of sibling analogical relationships is for the parent class *building* (Q3305213), which has subclasses *residential building* (Q11755880) and *religious building* (Q16970). An analogical pair at the instance level of these classes might be Fallingwater (Q463179) and Unity Temple (Q1680814), both of which were designed by architect Frank Lloyd Wright (Q5604).

An example of a parent-child analogical relationship is for the parent class *computer* (Q68) and the child class *supercomputer* (Q121117). Instances that form an analogical pair are the IBM 3790 (Q11223800) and Watson (Q12253), both of which were manufactured by IBM (Q37156).

These examples suggest that filtering the Wikidata ontology to identify classes containing high analogical potential is possible and can allow large-scale, automated analogies that capture interesting real-world relationships. In our ongoing work, we hope to develop fully automated analogies that can help explain the deeper structural relationships in the world using the immensity of knowledge in public knowledge graphs.

5   Related Work

Analogical Reasoning. To perform analogical mapping and reasoning over structures, prior research has used publicly available portions from Cyc [17]. For instance, the Companion architecture [10] operates over a subset of Cyc integrated with several other resources, primarily stemming from the Natural Language Processing (NLP) domain. Most works on analogical reasoning in NLP have considered the task of establishing proportional analogies (a:b - c:d), where language models and word embeddings have been the dominant resource [18,27]. Nagarajah et al. [21] studied the possibility of using language models or frames in FrameNet to perform various kinds of analogical reasoning over narratives. In [28], the authors investigate distinguishing features that enable models to distinguish between weak and strong analogies in STEM. While Wikidata has not been leveraged for analogical reasoning, recent work [4] has proposed a case-based reasoning method for Knowledge Base Question Answering that learns relational patterns in Wikidata between questions and their answers. Our study is complementary to prior work, revealing that Wikidata has much useful knowledge that can support analogies, yet it may not be ready yet for analogical reasoning at scale. Much of the knowledge that facilitates analogies is *commonsense knowledge* (e.g., polio vaccine cures polio), whose coverage in Wikidata can be further improved [15]. Interestingly, unlike other general-domain KGs like FreeBase,[1] Wikidata's notability clause is flexible and supports the inclusion of commonsense knowledge.[2]

---

[1] http://videolectures.net/iswc2017_taylor_applied_semantics/, accessed October 2, 2022.
[2] https://www.wikidata.org/wiki/Wikidata:Notability.

Quality and User Experience. Chen et al. [2] devise a framework for evaluating the fitness of knowledge graphs for downstream applications. Piscopo and Simperl [25] survey dozens of papers and provide three categories of metrics: intrinsic (accuracy, trustworthiness, and consistency of entities), contextual (completeness and timeliness of resources), and representation (i.e., understanding, interoperability of entities). Enhanced metrics for timeliness of knowledge have been proposed by [8] and illustrated on a case study with Wikidata. Noy et al. [22] describe that large graph systems have three key determinants of quality and usefulness: coverage, correctness, and freshness, and investigate the approach towards achieving those at five major technical companies. The authors recommend consolidating descriptions of people, places, and other entities in Wikidata as a common core. While Wikidata has been shown to be more reliable and expressive than other public knowledge graphs (e.g., DBpedia) [7], prior work has recognized challenges with its quality. Shenoy et al. [26] study over 300 Wikidata dumps and reveal that establishing identity, semantic typing, and satisfying semantic constraints are thorny issues that need further consideration. Recognizing that correctness, completeness, and freshness are difficult challenges, Wikidata has several tools to monitor, analyze, and surface issues with quality. These include the Objective Revision Evaluation Service (ORES) for vandalism detection,[3] ReCoIn [1] for completeness estimation, and the Primary Sources Tool (PST) for curation of the contribution process.[4] Our paper is orthogonal to prior efforts to organize, measure, or improve intrinsic aspects of Wikidata quality, as our goal is to investigate its fitness for analogical reasoning.

6 Conclusions

Recognizing that prior work on analogical reasoning has not considered the vast knowledge in Wikidata, this paper presented an initial study of the ability of Wikidata to support analogical reasoning. Our experiments with compositional subclass-of relations showed that relational knowledge in Wikidata is not consistently modeled, which can be expected given that Wikidata is created in a collaborative manner following the wisdom-of-the-crowd idea. Follow-up efforts to generate analogical classification data from Wikidata resulted in a dataset with 200 analogical quads (pair-of-pairs); yet, selecting these pairs required a substantial manual effort. To facilitate automatic analogy generation using the KG structure, we discussed what makes a good (Wikidata) analogy, suggesting four key desiderata: generality, selectivity, salience, and diversity. We paired these desirable ontological properties with specific metrics to realize them, and provided illustrative examples of Wikidata relationships sampled through these metrics. Future work will investigate how to apply the findings from this paper, and operationalize the desiderata and the metrics into a method that can sample analogical pairs. Future work should also investigate reframing and normalizing

---

[3] https://www.wikidata.org/wiki/Wikidata:ORES.
[4] https://www.wikidata.org/wiki/Wikidata:Primary_sources_tool#References.

of the knowledge in Wikidata to allow for more direct support for analogical reasoning. We release the data and the code supporting our analysis.[5],[6]

References

1. Balaraman, V., Razniewski, S., Nutt, W.: Recoin: relative completeness in Wikidata. In: Companion Proceedings of the Web Conference 2018, WWW 2018, Republic and Canton of Geneva, CHE, pp. 1787–1792. International World Wide Web Conferences Steering Committee (2018). https://doi.org/10.1145/3184558. 3191641
2. Chen, H., Cao, G., Chen, J., Ding, J.: A practical framework for evaluating the quality of knowledge graph. In: Zhu, X., Qin, B., Zhu, X., Liu, M., Qian, L. (eds.) CCKS 2019. CCIS, vol. 1134, pp. 111–122. Springer, Singapore (2019). https://doi.org/10.1007/978-981-15-1956-7_10
3. Chen, K., Rabkina, I., McLure, M.D., Forbus, K.D.: Human-like sketch object recognition via analogical learning. In: Proceedings of the AAAI Conference on Artificial Intelligence, vol. 33, pp. 1336–1343 (2019)
4. Das, R., et al.: Knowledge base question answering by case-based reasoning over subgraphs. arXiv preprint arXiv:2202.10610 (2022)
5. Dehghani, M., Tomai, E., Forbus, K.D., Klenk, M.: An integrated reasoning approach to moral decision-making. In: AAAI, pp. 1280–1286 (2008)
6. Devlin, J., Chang, M.W., Lee, K., Toutanova, K.: BERT: pre-training of deep bidirectional transformers for language understanding. arXiv preprint arXiv:1810.04805 (2018)
7. Färber, M., Bartscherer, F., Menne, C., Rettinger, A.: Linked data quality of DBpedia, Freebase, OpenCyc, Wikidata, and YAGO. Semant. Web 9(1), 77–129 (2018)
8. Ferradji, M.A., Benchikha, F.: Enhanced metrics for temporal dimensions toward assessing Linked Data: a case study of Wikidata. J. King Saud Univ. Comput. Inf. Sci. 34, 4983-4992 (2021)
9. Forbus, K.D., Ferguson, R.W., Lovett, A.M., Gentner, D.: Extending SME to handle large-scale cognitive modeling. Cogn. Sci. 41(5), 1152–1201 (2017)
10. Forbus, K.D., Hinrichs, T.R.: Analogy and qualitative representations in the companion cognitive architecture (2017)
11. Gentner, D., Brem, S., Ferguson, R., Wolff, P.: Analogy and creativity in the works of Johannes Kepler (1997)
12. Girju, R., Moldovan, D., Tatu, M., Antohe, D.: On the semantics of noun compounds. Comput. Speech Lang. 19(4), 479–496 (2005)
13. Holyoak, K.J., Thagard, P., Sutherland, S.: Mental leaps: analogy in creative thought. Nature 373(6515), 572 (1995)
14. Hope, T., Chan, J., Kittur, A., Shahaf, D.: Accelerating innovation through analogy mining. In: Proceedings of the 23rd ACM SIGKDD International Conference on Knowledge Discovery and Data Mining, pp. 235–243 (2017)
15. Ilievski, F., Szekely, P., Schwabe, D.: Commonsense knowledge in Wikidata. arXiv preprint arXiv:2008.08114 (2020)

---

[5] https://drive.google.com/file/d/1jJz4yAyBKjq4Mm47eMKD12w-DH5uugnN.
[6] https://github.com/usc-isi-i2/analogical-transfer-learning/blob/main/Analogical %20Proj%20Experiments.ipynb.

16. Klein, N., Ilievski, F., Szekely, P.: Generating explainable abstractions for wikidata entities. In: Proceedings of the 11th on Knowledge Capture Conference, pp. 89–96 (2021)
17. Lenat, D.B.: CYC: a large-scale investment in knowledge infrastructure. Commun. ACM **38**(11), 33–38 (1995)
18. Mikolov, T., Chen, K., Corrado, G., Dean, J.: Efficient estimation of word representations in vector space. arXiv preprint arXiv:1301.3781 (2013)
19. Mikolov, T., Yih, W.t., Zweig, G.: Linguistic regularities in continuous space word representations. In: Proceedings of the 2013 Conference of the North American Chapter of the Association for Computational Linguistics: Human Language Technologies, pp. 746–751 (2013)
20. Möller, C., Lehmann, J., Usbeck, R.: Survey on English entity linking on Wikidata. arXiv preprint arXiv:2112.01989 (2021)
21. Nagarajah, T., Ilievski, F., Pujara, J.: Understanding narratives through dimensions of analogy. In: Workshop on Qualitative Reasoning (QR) (2022)
22. Noy, N., Gao, Y., Jain, A., Narayanan, A., Patterson, A., Taylor, J.: Industry-scale knowledge graphs: Lessons and challenges: Five diverse technology companies show how it's done. Queue **17**(2), 48–75 (2019)
23. Oguz, B., et al.: UniK-QA: unified representations of structured and unstructured knowledge for open-domain question answering. arXiv preprint arXiv:2012.14610 (2020)
24. Pennington, J., Socher, R., Manning, C.D.: Glove: global vectors for word representation. In: Proceedings of the 2014 Conference on Empirical Methods In Natural Language Processing (EMNLP), pp. 1532–1543 (2014)
25. Piscopo, A., Simperl, E.: What we talk about when we talk about Wikidata quality: a literature survey. In: Proceedings of the 15th International Symposium on Open Collaboration, pp. 1–11 (2019)
26. Shenoy, K., Ilievski, F., Garijo, D., Schwabe, D., Szekely, P.: A study of the quality of Wikidata. J. Web Semant. (Community Based Knowl. Bases) **72**, 100679 (2022)
27. Ushio, A., Espinosa Anke, L., Schockaert, S., Camacho-Collados, J.: BERT is to NLP what AlexNet is to CV: can pre-trained language models identify analogies? In: Proceedings of the 59th Annual Meeting of the Association for Computational Linguistics and the 11th International Joint Conference on Natural Language Processing (Volume 1: Long Papers), pp. 3609–3624. Association for Computational Linguistics, August 2021. https://doi.org/10.18653/v1/2021.acl-long.280. https://aclanthology.org/2021.acl-long.280
28. Wijesiriwardene, T., Wickramarachchi, R., Shalin, V.L., Sheth, A.P.: Towards efficient scoring of student-generated long-form analogies in stem (2022)

Flexible Queries over Knowledge Graphs

José Félix Yagüe[1], Ignacio Huitzil[2], Carlos Bobed[1,3],
and Fernando Bobillo[1,3(✉)]

[1] University of Zaragoza, Zaragoza, Spain
{cbobed,fbobillo}@unizar.es
[2] Artificial Intelligence Research Institute (IIIA), CSIC, Bellaterra, Spain
[3] Aragon Institute of Engineering Research (I3A), Zaragoza, Spain

Abstract. The increasing interest in Knowledge Graphs to represent real-world knowledge and the common need to manage imprecise knowledge in many real-world applications demand the study of approaches to solve flexible queries over Knowledge Graphs. In this paper, we propose a novel approach to solve that problem which reuses Semantic Web standards (RDF and SPARQL) and builds a fuzzy layer on top of them.

Keywords: Knowledge graphs · Fuzzy logic · Flexible querying

1 Introduction

Semantic Web technologies are receiving a lot of attention to represent the knowledge in many domains and applications. Such Semantic Web technologies include ontologies, or formal and shared specifications of the vocabulary of a domain of interest [14], Knowledge Graphs [6], a graph-based model to capture data at large scale, and Linked Data, a set of best practices for publishing and connecting data on the Web [2]. Knowledge Graphs, Linked Data, and even ontologies are usually expressed in RDF language[1].

In many real-world applications, there is a need to manage imprecise knowledge. In such cases, Fuzzy Set Theory and Fuzzy Logic have proved to be useful for more than half a century [9,17]. Fuzzy sets allow to represent the partial membership of an element to a set, and Fuzzy Logic allows to manage propositions which are partially true and make deductions through approximate reasoning.

Therefore, fuzzy extensions of Semantic Web technologies have been proposed. Most of the literature has focused on fuzzy ontologies [10,18] or fuzzy Description Logics [3], as the main formalism behind fuzzy ontologies. Unfortunately, the combination of Fuzzy Logic and Knowledge Graphs has not received a similar attention.

---

[1] https://www.w3.org/TR/rdf11-primer/, last accessed on 26th July 2022.

B. Villazón-Terrazas et al. (Eds.): KGSWC 2022, CCIS 1686, pp. 192–200, 2022.
https://doi.org/10.1007/978-3-031-21422-6_14

In this paper, we will propose a novel approach to answer flexible queries over classical Knowledge Graphs. While previous work focuses on the support of fuzzy axioms (using non-standard RDF), we instead restrict to fuzzy datatypes. This way, we are able to stick to the Semantic Web standards, using standard RDF and SPARQL query endpoints, and building the fuzzy layer on top of them as a series of additional steps.

The remainder of this paper is structured as follows. Section 2 provides some background on Knowledge Graphs and Fuzzy Logic. Then, Sect. 3 discusses our novel approach to answer flexible queries over Knowledge Graphs. Finally, Sect. 4 overviews some related work, and Sect. 5 sets up some conclusions and ideas for future work.

2 Background

Knowledge Graphs. Although they have been there for quite a long time under different names (e.g., Semantic Networks), Knowledge Graphs have lately received a lot of attention since the adoption of the term by Google in 2012. There is not a consensual definition of what a Knowledge Graph is, but we could consider the view on them as labeled directed graphs the most spread one. In this view, a Knowledge Graph is a labeled directed graph (E, R, L), with E being entities, R being relations between such entities, and L a labeling function that maps each element in the graph to its name/type. Influenced indeed by the RDF data model, this definition derives into regarding them as sets of triples subject-predicate-object (SPO) triples. However, as noted in [6], the notion of Knowledge Graph is broader than that. Without binding the definition to any particular data model, Hogan et al. [6] adopt the following definition: *a Knowledge Graph is a graph of data intended to accumulate and convey knowledge of the real world, whose nodes represent entities of interest and whose edges represent relations between these entities* (the interested reader can find pointers to alternative definitions in [6]).

Included in the previous definition, we would find RDF graphs, which are labeled directed graphs. RDF is the W3C standard language for representing information in the Web. It is based on triples of the form $\langle s, p, o \rangle$, where s is the subject, p is the property, and o is the object. That is, $\langle s, p, o \rangle$ states the s is related with o via the property p. In general, such relationships are not symmetrical.

To define the schema that the RDF graph follows, we can use different languages with different expressivities, ranging from RDF-S (RDF-Schema[2]), which allows us to define hierarchies of concepts and properties, the domain and ranges of properties, and typing the resources asserting the concepts they belong to, to the different profiles of OWL[3], which extends RDF-S allowing to model expressive ontologies in different fragments of Description Logics [1].

---

[2] https://www.w3.org/TR/rdf-schema/, last accessed on 26th July 2022.
[3] https://www.w3.org/TR/owl2-overview/, last accessed on 26th July 2022.

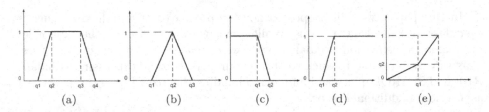

Fig. 1. (a) Trapezoidal; (b) Triangular; (c) Left-shoulder; (d) Right shoulder; (e) Linear fuzzy membership functions.

SPARQL[4] is its standard query language. It supports four different types of queries (namely, SELECT, ASK, CONSTRUCT, and DESCRIBE) whose body is described in terms of basic graph patterns (BGPs) composed using free variables. Such BGPs are to be matched crisply to the underling graph in order to provide possible mappings that fulfill the query conditions.

Fuzzy Logic. Fuzzy Logic is widely used to manage imprecise and vague knowledge. It is a generalization of classical logic proposed by Zadeh where statements are not necessarily either true or false, but hold to some degree of truth [17].

The cornerstone of Fuzzy Logic is the concept of fuzzy set, which is a generalization of a classical set where elements can have a partial membership. A fuzzy set A is characterized by a membership function $\mu_A(x)$ which associates with each object x a real number in $[0,1]$ representing the membership degree of x in A. As in classical sets, 0 means no-membership and 1 full membership, but now an intermediate value between 0 and 1 denotes partial membership to F.

To build fuzzy membership functions, common options are the trapezoidal (Fig. 1(a)), triangular (Fig. 1(b)), left-shoulder (Fig. 1(c)), right-shoulder (Fig. 1(d)), and linear (Fig. 1(e)) membership functions.

Example 1. The fuzzy set of beers with LowAlcohol can be defined using a a triangular function **triangular**$(0.5, 3.1, 6.55)$. If the alcohol of Ambar_Especial beer is $5.2°$, its membership degree to LowAlcohol can be evaluated as: $\mu_{\mathsf{LowAlcohol}}(\mathsf{Ambar\_Especial}) = (\mathbf{triangular}(0.5, 3.1, 6.55))(5.2) = 0.39$.

Fuzzy Logic enables approximate reasoning. Logical operations over classical sets are also generalized to the fuzzy case. To compute the conjunction, disjunction, complement and implication over fuzzy sets one can use different families of functions, namely a *t-norm* function $\otimes$, a *t-conorm* function $\oplus$, a *negation* function $\ominus$ and an *implication* function $\Rightarrow$ (see [9] for details). For instance, the minimum and the product are t-norms and the maximum is a t-conorm.

There are other ways to combine fuzzy sets. For example, an aggregation operator is a function that takes n values in $[0,1]$(possibly representing the membership degrees to n fuzzy sets) and returns a single value in $[0,1]$. Some

---

[4] https://www.w3.org/TR/sparql11-overview/, last accessed on 26th July 2022.

examples are the weighted mean (WMEAN) or the *Ordered Weighted Averaging* (OWA) operator.

To conclude this section, a *fuzzy modifier* (also called fuzzy hedge) modifies the shape of a fuzzy set by altering its membership function. Two common examples are the weakening modifier **very**, characterized by the function $\mathsf{very}(x) = x^2$, and the increasing modifier **few**, defined as $\mathsf{few}(x) = \sqrt{x}$.

Example 2. If we apply **very** to the fuzzy set LowAlcohol, for each beer x the degree of being a very low alcoholic beer can be computed as: $\mu_{\mathsf{VeryLowAlcohol}}(x) = \mathsf{very}(\mu_{\mathsf{LowAlcohol}}(x)) = (\mu_{\mathsf{LowAlcohol}}(x))^2$. Note that as the degree is in $[0,1]$, by squaring it, we reinforce the need to belong "very" to the set.

3   Flexible Querying

Representing the Queries. Given a Knowledge Graph $\mathcal{K}$, a flexible query is characterized by the following parameters:

- A set of classes $C_1, C_2, \ldots, C_N$
- A set of functional numerical data properties $P_1, P_2, \ldots, P_n$
- A list of fuzzy datatypes $D_1, D_2, \ldots, D_n$
- An optional fusion operator @: $[0,1]^n \to [0,1]$
- An optional fuzzy hedge $h\colon [0,1] \to [0,1]$

Possible fusion operators are t-norms, t-conorms, weighted mean, or OWA.

Example 3. Given a Knowledge Graph about beers, a possible flexible query to retrieve "very [relevant] Spanish Lager beers with low alcohol and bitterness levels" might be described as:

- Classes: Lager, SpanishBeer
- Data properties: alcohol, bitterness
- Fuzzy datatypes: LowAlcohol, defined as **triangular**$(0.5, 3.1, 6.55)$, and Low-Bitterness, defined as **triangular**$(15, 27, 41)$
- Fusion operator: product t-norm
- Fuzzy hedge: $\mathsf{very}(x) = x^2$

We propose to rely on Fuzzy OWL 2 datatypes [4], which includes trapezoidal, triangular, left-shoulder, and right-shoulder datatypes. In Fuzzy OWL 2, a datatype declaration can be associated with an OWL 2 annotation encoding a fuzzy membership function, using an XML-like syntax proposed in [4]. Interestingly, such annotations can be encoded as triples in the Knowledge Graph.

Example 4. To express that fuzzy datatype *LowAlcohol* corresponds to a triangular function **triangular**$(0.5, 3.1, 6.55)$, we use the following set of RDF triples:

```
@prefix ex: <http://www.example.org/beer/> .
@prefix owl: <http://www.w3.org/2002/07/owl#> .
@prefix rdf: <http://www.w3.org/1999/02/22-rdf-syntax-ns#> .
@prefix rdfs: <http://www.w3.org/2000/01/rdf-schema#> .
ex:fuzzyLabel rdf:type owl:AnnotationProperty .
ex:LowAlcohol rdf:type rdfs:Datatype .
ex:LowAlcohol ex:fuzzyLabel """<fuzzyOwl2 fuzzyType=\"datatype\">
 <Datatype type=\"triangular\" a=\"0.5\" b=\"3.1\" c=\"6.55\" />
 </fuzzyOwl2>""" .
```

Solving the Queries. To solve the query, we perform the following steps:

1. The first step is to retrieve, for each member of all the classes, the values of the data properties. This can be done using the following SPARQL query:

```
SELECT ?x ?Y₁ ... ?Yₙ
WHERE {
    ?x rdf:type/rdfs:subClassOf* C₁ .
    ...
    ?x rdf:type/rdfs:subClassOf* Cₙ .
    ?x P₁ ?Y₁ .
    ...
    ?x Pₙ ?Yₙ .
}
```

The result of this step is a list of tuples $\langle x, y_1^x, \ldots, y_n^x \rangle$. Note that this query not only retrieves the direct instances of the concepts $C_1, \ldots, C_n$, but also their indirect instances (i.e., including the instances of some of their subclasses). It could also be extended to retrieve values Y_i linked to ?x via a subproperty of P_i, or to infer the classes of ?x via some domain or role restrictions.

2. Next, for each $i \in \{1, \ldots, n\}$, we retrieve the definition F_i of the input fuzzy datatype D_i, using the following SPARQL query:

```
SELECT ?Fᵢ
WHERE {
    Dᵢ rdf:type rdfs:Datatype .
    Dᵢ fuzzyLabel Fᵢ .
}
```

3. The next step is to compute, for each individual ?x and each data property P_i, the membership degree of the value Y_i to the fuzzy datatype F_i:

$$z_i^x = F_i(y_i^x), \ \forall i \in \{1, \ldots, n\} \tag{1}$$

4. Then, we aggregate all the values z_i^x corresponding to an individual ?x into a single result r_x using the input fusion operator

$$r_x = @_{i \in \{1, \ldots, n\}}(z_i^x) \tag{2}$$

5. Now it is time to use the fuzzy hedge to modify the membership degree for each ?x:

$$\alpha_x = h(r_x) \tag{3}$$

6. Finally, the answer to the query is a list of values $\langle ?x, \alpha_x \rangle$, sorted in decreasing order according to the value α_x.

Example 5. Let us solve the query in Example 3 given the following set of triples on a beer domain:

```
@prefix ex: <http://www.example.org/beer/> .
@prefix rdf: <http://www.w3.org/1999/02/22-rdf-syntax-ns#> .
@prefix rdfs: <http://www.w3.org/2000/01/rdf-schema#> .
@prefix xsd: <http://www.w3.org/2001/XMLSchema#> .
ex:ImperialPilsStrongPaleLager rdfs:subclassOf ex:Lager .
ex:PaleLager rdfs:subclassOf ex:Lager .
ex:Alhambra_Reserva_1925 rdf:type ex:SpanishBeer .
ex:Alhambra_Reserva_1925 rdf:type ex:ImperialPilsStrongPaleLager .
ex:Alhambra_Reserva_1925 ex:alcohol "6.4"^^xsd:decimal.
ex:Alhambra_Reserva_1925 ex:bitterness "25"^^xsd:decimal .
ex:Ambar_Especial rdf:type ex:SpanishBeer .
ex:Ambar_Especial rdf:type ex:PaleLager .
ex:Ambar_Especial ex:alcohol "5.2"^^xsd:decimal .
ex:Ambar_Especial ex:bitterness "25"^^xsd:decimal .
ex:BrewDog_Punk_IPA ex:rdf:type ex:IndiaPaleAleIPA .
ex:BrewDog_Punk_IPA ex:alcohol "6"^^xsd:decimal .
ex:BrewDog_Punk_IPA ex:bitterness "60"^^xsd:decimal .
```

The first SPARQL query retrieves both Alhambra_Reserva_1925 and Ambar_Especial, as they are both Spanish beers of a (subclass of) Lager style, together with their values of alcohol and bitterness. After that, the algorithm computes z_1 and z_2, and then $\alpha_x = (z_1 \cdot z_2)^2$. More precisely, the obtained values are:

?x	?Y1	?Y2	z1	z2	α_x
Alhambra_Reserva_1925	6.4	25	0.04	0.83	0.001
Ambar_Especial	5.2	25	0.39	0.83	0.11

Finally, the result is the following ordered list of pairs:

$\langle$Ambar_Especial, 0.11$\rangle$

$\langle$Alhambra_Reserva_1925, 0.001$\rangle$

The main advantage of this algorithm is that it is possible to reuse standard RDF language and SPARQL query endpoints, similarly as the authors in [7,8] do for fuzzy ontologies. Note that it is trivial to extend the characterization of the query to include an additional parameter k so that only the top-k results (i.e., the individuals with the k highest values of α_x) are retrieved. Note as well that some fusion operators and fuzzy hedges can be expressed using SPARQL built-in functions (or inner expressions), allowing to perform more computations without relying on external processing. However, the resulting SPARQL query could be quite complex so as to be included here, and there are operators that cannot be expressed in standard SPARQL, such as the geometric mean (involving n-th roots).

Implementation. We are currently implementing a prototype tool. It is developed in Java. The SPARQL queries are solved using Apache Jena and Jena Java API[5] through a (local or remote) SPARQL endpoint. The definitions of the fuzzy datatypes are parsed (to retrieve the function type and its parameters) using Fuzzy OWL 2 API.

The idea is develop a graphical user interface so that the user can easily provide all the relevant information:

- Providing the URL of the endpoint storing the Knowledge Graph.
- Providing the names of the concepts, data properties, and fuzzy datatypes. Rather than allowing the user to select a concept/property/datatype from the Knowledge Graphs (which requires showing all possible options, overwhelming the user in large Knowledge Graphs), some auto-complete mechanism seems more appropriate.
- Selecting the fusion and hedge operators from a list of options built into the implementation.

4   Related Work

Some previous works consider Fuzzy Logic and RDF language. The oldest proposal was made by Vaneková [16] et al., who propose to represent a fuzzy fact of the form "a resource s belongs to a fuzzy set f with degree α", with $\alpha \in (0, 1]$, using an RDF triple $\langle s, f, \alpha \rangle$. While this approach makes it possible to use a single triple reusing standard RDF, there is an implicit relationship type which is not being represented. For example, the authors do not represent the fact that item002 is of type (rdf:type) cheap, or the fact item002's price is cheap. M. Mazzieri and A. F. Dragoni consider statements of the form $\langle s, p, o, \alpha \rangle$ [11]. The idea is to represent, for example, that the triple $\langle$Zaragoza, isCloseTo, Huesca$\rangle$ holds with degree 0.7. Depending on the property p, we can have concept assertions (rdf:type), subclass axioms (rdfs:subclassOf), subproperty axioms (rdfs:subPropertyOf) ore, more generally, property assertions. A. E. A. Djebri discusses different approaches to annotate such statements: reification, n-ary properties, single named graph, singleton properties, and RDF-star [5]. However, these approaches do not consider fuzzy datatypes, as we do.

---

[5] http://jena.apache.org.

U. Straccia proposed a reasoning algorithm for a fuzzy extension of ρDF (a fragment of RDF Schema) with statements of the form $\langle s, p, o, \alpha \rangle$ but no fuzzy datatypes [15]. He also proposed fuzzy conjunctive queries over a fuzzy graph, which can include fuzzy triples, and assignments involving fuzzy membership functions and fusion operators, e.g., $q(x, s) \leftarrow \langle x, rdf : type, SportsCar \rangle \land \langle x, hasPrice, y \rangle \land \{ s := s1 \cdot cheap(y) \}$ [15]. This approach does not detail how to represent the syntax of the fuzzy datatypes (we instead use Fuzzy OWL 2 datatypes already represented in the RDF graph) and does not consider fuzzy hedges. Our approach could be friendlier to the user as there is no need to write such complex queries.

J. Z. Pan et al. proposed Fuzzy SPARQL, an extension of SPARQL designed to query fuzzy ontologies (but not fuzzy Knowledge Graphs) [12] with statements of the form $\langle s, p, o, \alpha \rangle$, but it does not consider fuzzy datatypes or fuzzy hedges.

O. Pivert et al. proposed another fuzzy extension of SPARQL called FURQL (Fuzzy RDF Query Language) [13], allowing fuzzy properties (that partially hold) and fuzzy datatypes in queries over a fuzzy RDF graph with statements of the form $\langle s, p, o, \alpha \rangle$. However, although fuzzy datatypes are used in the examples, it is not discussed how to represent them in the graph, it is not possible to specify the fusion operator, and fuzzy hedges are not supported.

5 Conclusions and Future Work

This paper has proposed a Fuzzy Logic based approach to answer flexible queries over Knowledge Graphs. Our approach makes it possible to reuse existing RDF graphs and SPARQL endpoints, building a fuzzy layer on top of them. It also supports Fuzzy OWL 2 datatypes to describe fuzzy membership functions.

The main direction for future work is the completion of the prototype implementation, with a modular design so that new fusion operators can be added, and an intuitive user graphical interface. After finishing the implementation, we would like to evaluate the performance on some real scenarios.

Acknowledgments. C. Bobed and F. Bobillo were supported by the I+D+i project PID2020-113903RB-I00, funded by MCIN/AEI/10.13039/501100011033, and by DGA/FEDER.

References

1. Baader, F., Calvanese, D., McGuinness, D., Nardi, D., Patel-Schneider, P.F.: The Description Logic Handbook: Theory, Implementation, and Applications. Cambridge University Press, Cambridge (2003)
2. Bizer, C., Heath, T., Berners-Lee, T.: Linked Data - the story so far. Int. J. Semant. Web Inf. Syst. 5(3), 1–22 (2009)
3. Bobillo, F., Cerami, M., Esteva, F., García-Cerdaña, À., Peñaloza, R., Straccia, U.: Fuzzy description logics. In: Cintula, P., Fermüller, C., Noguera, C. (eds.) Handbook of Mathematical Fuzzy Logic Volume III, Studies in Logic, Mathematical Logic and Foundations, Chap. XVI, vol. 58, pp. 1105–1181. College Publications (2015)

4. Bobillo, F., Straccia, U.: Fuzzy ontology representation using OWL 2. Int. J. Approx. Reason. **52**(7), 1073–1094 (2011)
5. Djebri, A.E.A.: Uncertainty management for linked data reliability on the Semantic Web. Ph.D. thesis, Université Côte D'Azur (2022). https://hal.archives-ouvertes.fr/tel-03679118
6. Hogan, A., et al.: Knowledge Graphs, vol. 22. Morgan & Claypool, San Rafael (2021)
7. Huitzil, I., Alegre, F., Bobillo, F.: GimmeHop: a recommender system for mobile devices using ontology reasoners and Fuzzy Logic. Fuzzy Sets Syst. **401**, 55–77 (2020)
8. Huitzil, I., Molina-Solana, M., Gómez-Romero, J., Bobillo, F.: Minimalistic fuzzy ontology reasoning: an application to building Information Modeling. Appl. Soft Comput. **103**, 107158 (2021)
9. Klir, G.J., Yuan, B.: Fuzzy Sets and Fuzzy Logic: Theory and Applications. Prentice-Hall, Englewood Cliffs (1995)
10. Lukasiewicz, T., Straccia, U.: Managing uncertainty and vagueness in Description Logics for the Semantic Web. J. Web Semant. **6**(4), 291–308 (2008)
11. Mazzieri, M., Dragoni, A.F.: A fuzzy semantics for the resource description framework. In: da Costa, P.C.G., d'Amato, C., Fanizzi, N., Laskey, K.B., Laskey, K.J., Lukasiewicz, T., Nickles, M., Pool, M. (eds.) URSW 2005-2007. LNCS (LNAI), vol. 5327, pp. 244–261. Springer, Heidelberg (2008). https://doi.org/10.1007/978-3-540-89765-1_15
12. Pan, J.Z., Stamou, G., Stoilos, G., Thomas, E., Taylor, S.: Scalable querying service over fuzzy ontologies. In: Proceedings of the 17th International World Wide Web Conference (WWW 2008), pp. 575–584 (2008)
13. Pivert, O., Slama, O., Thion, V.: An extension of SPARQL with fuzzy navigational capabilities for querying fuzzy RDF data. In: Proceedings of the 2016 IEEE International Conference on Fuzzy Systems (FUZZ-IEEE 2016), pp. 2409–2416. IEEE (2016)
14. Staab, S., Studer, R.: Handbook on Ontologies. International Handbooks on Information Systems, Springer, Heidelberg (2004)
15. Straccia, U.: A minimal deductive system for general fuzzy RDF. In: Polleres, A., Swift, T. (eds.) RR 2009. LNCS, vol. 5837, pp. 166–181. Springer, Heidelberg (2009). https://doi.org/10.1007/978-3-642-05082-4_12
16. Vaneková, V., Bella, J., Gurský, P., Horváth, T.: Fuzzy RDF in the Semantic Web: deduction and induction. In: Proceedings of the 6th Workshop on Data Analysis (WDA 2005), pp. 16–29
17. Zadeh, L.A.: Fuzzy sets. Inf. Control **8**, 338–353 (1965)
18. Zhang, F., Cheng, J., Ma, Z.: A survey on fuzzy ontologies for the Semantic Web. Knowl. Eng. Rev. **31**(3), 278–321 (2016)

Knowledge Graphs for Community Detection in Textual Data

Federica Rollo[✉] and Laura Po

"Enzo Ferrari" Engineering Department, University of Modena and Reggio Emilia,
Modena, Italy
{federica.rollo,laura.po}@unimore.it

Abstract. Online sources produce a huge amount of textual data, i.e.,
freeform text. To derive insightful information from them and facilitate
the application of Machine Learning algorithms textual data need to
be processed and structured. Knowledge Graphs (KGs) are intelligent
systems for the analysis of documents. In recent years, they have been
adopted in multiple contexts, including text mining for the development
of data-driven solutions to different problems. The scope of this paper
is to provide a methodology to build KGs from textual data and apply
algorithms to group similar documents in communities. The methodology
exploits semantic and statistical approaches to extract relevant insights
from each document; these data are then organized in a KG that allows
for their interconnection. The methodology has been successfully tested
on news articles related to crime events occurred in the city of Modena,
in Italy. The promising results demonstrate how KG-based analysis can
improve the management of information coming from online sources.

Keywords: Knowledge graph · Entity linking · Keyphrase extraction ·
Graph analysis · Newspaper · Community detection · Neo4j

1 Introduction

The amount of heterogeneous data produced every day by online sources grows
ever more. The problem of information overload stimulates the need of more efficient and effective methods to integrate and analyze the content published across
multiple sources. In most cases, traditional information retrieval approaches rely
on Natural Language Processing techniques and statistical methods to generate
information from textual data. In this context, the use of semantics could allow a
more accurate data extraction and improve downstream tasks such as document
classification or topic detection. The objective is to transform heterogeneous and
unstructured data (i.e., freeform text) into useful and meaningful information.
In recent years, Knowledge Graphs (KGs) have emerged as a compelling tool
for organizing the structured knowledge. Their application in several contexts
allows to solve a plenty of different problems, i.e., question answering over natural language text, development of recommendation systems, data enrichment,

B. Villazón-Terrazas et al. (Eds.): KGSWC 2022, CCIS 1686, pp. 201–215, 2022.
https://doi.org/10.1007/978-3-031-21422-6_15

document classification, fake news detection, topic prediction [7,10–12,14,17,22]. In particular, we explored the use of KGs to identify similar documents through community detection. The problem of community detection is to group nodes of a graph into communities that are closely related within and weakly across communities. Community detection is useful for topic detection in a large amount of data. Different approaches can be followed for building KGs from textual data. First of all, it is necessary to define which are the entities (nodes) to represent, their attributes and how they are related each other. Statistical methods can be applied to calculate the probability of relationships for KG completion. Methodologies for the construction of a KG can be domain-agnostics or based on domain-specific ontologies. Also, recent methods exploit the numerical representations (embeddings) for improving semantic relationships between nodes. In [3], the authors highlight the importance to include contextual information in the KG for improving the community detection task. In particular, they add attributes to the KG nodes, integrate multiple domain-specific concept graphs to widen the semantic meaning of the attributes, and, finally, apply community detection algorithms to discover cluster of similar nodes. The authors of [2] experimented the use of KGs for the analysis of co-authorship in scientific researches, while in [25] KGs are employed to understand the evolution of a scientific topic over time. In [16,18], a graph analytical approach is proposed to identify the main topics published on social media. The graph is based on the co-occurrence of entities and keywords across the news articles, however, semantic relationships are not included. In [8], the authors propose a framework to support the Legal Enforcement Agencies (LEAs) to detect criminal networks and prevent cybercrime on social networks. The nodes of the graph are the users of the social network and the relationships between users are weighted considering the frequency of private messages they exchanged. Community detection is applied to identify the key members of different crimes. In [24], millions of sexual ads are crawled from the web to provide LEAs a KG tool to fight human trafficking and support victims. Data from the web are organized in a pre-defined ontology, then, the authors address the problem of duplicates through text similarity and entity resolution.

In this paper, we propose a domain-agnostic methodology to identify similar documents through the application of community detection algorithms to KGs. We combine statistical and semantic methods, and exploit existing Knowledge Base for entity resolution and duplicate detection among entities. Some experiments are conducted on an openly available dataset of Italian news articles related to real-world crime events. The use of such a methodology in the specific context of the experiments could allow for crime analysis through newspapers, i.e., discovering the number of crimes of the same type [15,20]. The remainder of the paper is organized as follows: Sect. 2 describes the methodology to construct the KG and apply the Louvain algorithm for community detection. In Sect. 3, we present our experiments on a dataset of Italian news articles related to real-world crime events and discuss the results. In the end, Sect. 4 outlines some conclusions and future works.

2   Methodology

The methodology used to construct the KG from textual documents is illustrated in Fig. 1. Firstly, named entities and keyphrases are extracted from each document. These data are used to construct the KG where each document is a node connected to other nodes which represent the entities occurred in its text and its keyphrases. The same entity or the same keyphrase can be connected to more than one document node. Then, other operations are performed on the graph to create additional connections between the nodes representing the documents. At this point, community detection algorithms can be applied to the graph to group similar documents in the same cluster. The Python code used for these steps in available at https://github.com/federicarollo/CrimeKG.

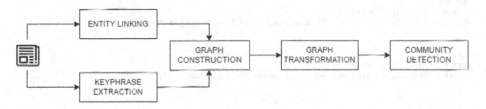

Fig. 1. The methodology implemented to construct and analyze a Knowledge Graph from textual documents.

2.1   Entity Linking

Entity Linking (EL) consists on the identification of named entities in a text and their mapping w.r.t. a Knowledge Base. In our case, we are interested in matching the entities mentioned in the documents with the correspondent DBpedia resource[1]. In particular, we want to detect entities which refer to person, organization, place, and building, i.e., the types *dbo:Location, dbo:PopulatedPlace, dbo:Organisation, dbo:Person, dbo:Building* in the DBpedia ontology (*dbo*).

DBpedia spotlight[1] [13] was developed for this scope. Its python implementation in spacy library[2] is able to disambiguate and annotate the mentions of DBpedia resources in natural language text. It takes in input the text to annotate and returns the URIs of annotated entities and other data in json format, e.g., the value of support (how prominent is the entity in Lucene Model, i.e., the number of inlinks in Wikipedia) and the similarity score. However, the category of the DBpedia ontology each entity belongs to is often unavailable in the request response. For this purpose, some additional SPARQL queries are performed. Firstly, a SPARQL query is in charge to retrieve the value of the property *rdf:type* of the extracted URIs. If the property is not available or none of the results refer to *dbo*, the properties *rdf:subClassOf, owl:equivalentClass* and

---

[1] https://www.dbpedia-spotlight.org/.
[2] https://github.com/MartinoMensio/spacy-dbpedia-spotlight.

owl:sameAs are exploited in combination with *rdf:type*, as shown in Queries 1, 2, 3 of Listing 1.1. Let consider in these exemplar queries < http://dbpedia.org/resource/Modena> the resource of DBpedia representing the city of Modena. If Queries 1, 2 and 3 do not return any result, the property *wikiPageRedirects* of DBpedia ontology is used to find the resource which the annotated entity is redirected to (Query 4). The result of EL is a list of DBpedia URIs associated to each document with a score that is the similarity score returned by DBpedia spotlight.

Listing 1.1. SPARQL queries used to retrieve the type of annotated entities.

```
Prefix rdfs: <http://www.w3.org/2000/01/rdf-schema#>
Prefix rdfs: <http://www.w3.org/1999/02/22-rdf-syntax-ns#>
Prefix dbo: <http://dbpedia.org/ontology/>
Prefix owl: <http://www.w3.org/2002/07/owl#>

# QUERY 1
select distinct ?type where {
   <http://dbpedia.org/resource/Modena> rdfs:subClassOf{0,5}/a ?type .
}

# QUERY 2
select distinct ?type where {
   <http://dbpedia.org/resource/Modena> owl:equivalentClass{0,1}/a ?type .
}

# QUERY 3
select distinct ?type where {
   <http://dbpedia.org/resource/Modena> owl:sameAs{0,1} ?x .
   ?x a ?type
}

# QUERY 4
select distinct ?type where {
   <http://dbpedia.org/resource/Modena> dbo:wikiPageRedirects ?new_resource .
   ?new_resource a ?type .
}
```

2.2 Automatic Keyphrase Extraction

Automatic Keyphrase Extraction (AKE) is the task for the automated extraction of relevant single or multiple-token phrases from a textual document. The scope of AKE is to capture the main concepts of a document. Documents sharing the same or similar keyphrases are very likely to be related to the same topic.

We compared the python implementation of three unsupervised approaches for automatic AKE: Rake (Rapid Automatic Keyword Extraction) [23], Yake (Yet Another Keyword Extractor) [6] and KeyBERT [9]. The first two algorithms do not require training, while KeyBERT needs a pre-trained language-specific BERT model. Rake and Yake rely on statistical text features, taking into account the frequency of terms in the document and their co-occurrence. On the other hand, KeyBERT exploits a semantic approach since it identifies keyphrases comparing the contextualized word embeddings to the document embedding through the calculation of cosine similarity. All the algorithms allow for the extraction of keyphrases containing a variable number of words. We made preliminary tests on 50 Italian documents to identify keyphrases of maximum 4 words. Through these tests, we noticed a critical point of KeyBERT. Probably, the algorithm is not able to tokenize sentences correctly. Indeed, in most cases the algorithm returns keyphrases that consist of words belonging to two different sentences. Moreover, the resulting keyphrases are often very similar each other even if the "diversity" parameter of the extractor is set to a high numerical value.[3] Comparing the keyphrases obtained by Rake and Yake, we identified more representative words by using Yake. For this reason, we decided to use Yake in our methodology. In addition to the text of the document and the maximum number of words allowed in the keyphrases, Yake takes in input the specification of the language, the deduplication function and its threshold, the number of keyphrases to return and the number of words in the sliding window used to form the list of candidate keyphrases. The deduplication function is a distance similarity measure that allows to filter similar keyphrases. Three different measures are implemented: the Levenshtein similarity, the Jaro-Winkler similarity, and the sequence matcher. The Yake extractor returns a list of keyphrases with the corresponding score. The lower the score, the more significant the keyphrase will be. At the end of AKE, a list of keyphrases with their score is associated to each document.

2.3 Knowledge Graph Generation

Based on the previously extracted data, the KG will consist of three types of nodes: Document nodes, Entity nodes, and Keyphrase nodes. The Document nodes are identified by the url of the web page containing the document. The text of the document and other relevant data, such as the publication date of the document, are stored as properties of the node. The Entity nodes represent the resources of DBpedia extracted through EL. These nodes are identified by the URIs of each resource. In the end, the Keyphrase nodes contain the keyphrases detected by AKE and are identified by the text of the keyphrases.

A Document node is connected to an Entity node with the relationship "appear" if that entity has been extracted from that document, and the weight of the relationship is the ratio of the frequency of the entity in the document to the frequency of the entity in the whole dataset. By doing so, the weight is always between 0 and 1. The scope of this approach is to give more importance to entity occurring more frequently in a document and less in the dataset. In the same way, a Keyphrase node is linked to a document with the relationship "appear" if it is extracted from that document by AKE, and the weight of the relationship is obtained subtracting from 1 the score returned by the keyphrase extractor, that was previously normalized to be between 0 and 1. This operation allows to reverse the meaning of the Yake score. In this way, the keyphrase with the highest score is the one more representative of the document. The same entity can be connected to multiple Document nodes, as well as the same Keyphrase node can be shared by more than one Document node. Figure 2 shows an exemplar graph of two documents (D_1 and D_2) sharing one keyphrase (K_1).

---

[3] See examples of the "diversity" parameter in KeyBERT at https://github.com/ MaartenGr/KeyBERT.

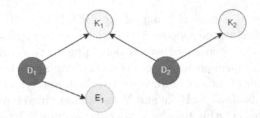

Fig. 2. Exemplar graph of two documents D_1 and D_2 sharing the keyphrase K_1.

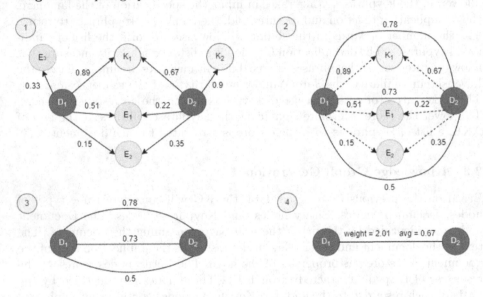

Fig. 3. Four steps for converting Document-Entity and Document-Keyphrase relationships to Document-Document relationships.

The KG is generated in Neo4j.[4] Before the application of community detection algorithms, some operations are made on the KG to integrate undirected Document-Document relationships based on the presence of Document-Entity and Document-Keyphrase relationships. Figure 3 illustrates the steps performed for the transformation of the KG. D_1 and D_2 nodes share 3 nodes (K_1, E_1 and E_2), and each relationship has a weight (Step 1). In Step 2, the weights of the relationships Document-Entity are summed up, while the weights of the relationships Document-Keyphrase are averaged. Again, this operation ensures the weights to be between 0 and 1. The resulting KG is shown in Step 3, where D_1 and D_2 nodes are connected each other through 3 undirected relationships with weights 0.78, 0.73 and 0.5. This operation generates a high number of new

---

[4] https://neo4j.com/.

relationships; indeed, given n Document nodes connected to the same entity or keyphrase, the number of generated relationships for each entity or keyphrase is:

$$\frac{n!}{2*(n-2)!}$$

The modifications to the graph are made using the Cypher query language in Neo4j, as shown in Listing 1.2. The first query creates the relationship "shared_entity", while in the second one the relationship "shared_key" is generated.

Listing 1.2. Cypher queries to add Document-Document relationships based on the presence of Document-Entity and Document-Keyphrase relationships.

```
MATCH (d1:Document)-[r1:appear]->(e:Entity)<-[r2:appear]-(d2:Document)
WHERE d1.id>d2.id
CREATE (d1)-[r3:shared_entity]->(d2)
SET r3.weight=apoc.convert.toFloat(r1.weight)+apoc.convert.toFloat(r2.weight)

MATCH (d1:Document)-[r1:appear]->(k:Keyphrase)<-[r2:appear]-(d2:Document)
WHERE d1.id>d2.id
CREATE (d1)-[r3:shared_key]->(d2)
SET r3.weight=(apoc.convert.toFloat(r1.weight)+apoc.convert.toFloat(r2.weight))/2
```

Then, a single relationship named "connected" is created with two properties: "weight" as the sum of the weights of relationships among the two Document nodes and "median" as their median. We choose the median instead of the arithmetic average since we expect a skewed distribution of the weights and median is not affected by outliers. Since these operations on the graph densely connect nodes each other a pruning method is suggested to remove less important relationships. The selection of relationships can be based on the values of weight and median. We want to give more importance to documents sharing multiple entities/keyphrases w.r.t. documents sharing just one relationship even if the score of that relationship is high. Notice that if two documents share just one relationship, the values of weight and median of the "connected" relationship are the same. In addition, we prefer to maintain relationships with high median score because this indicates that the most shared relationships have high scores and that entity/keyphrase is very representative of that documents. The Cypher queries used for Step 4 are provided in Listing 1.3.

Listing 1.3. Cypher queries to add the connected relationships and remove relationships with median under the pruning threshold.

```
MATCH (d1:Document)-[r]->(d2:Document)
WHERE d1.id>d2.id
WITH d1, d2,
     sum(r.weight) AS totalWeights,
     apoc.agg.median(r.weight) AS medianWeights
CREATE (d1)-[:CONNECTED {weight:totalWeights, median:medianWeights}]->(d2)

MATCH (d1:Document)-[r:CONNECTED]->(d2:Document)
WHERE r.median ≤ pruning_threshold
DELETE r
```

2.4   Community Detection

The Neo4j Graph Data Science library[5] offers some already implemented Community Detection algorithms that can be easily applied to a proper KG. One of these algorithms is Louvain [4]. Due to its increasing popularity and application, we decided to include it in our methodology. Louvain is an iterative heuristic algorithm introduced in 2008. It tries to identify communities in a graph by optimizing the modularity score. Modularity is a numerical value between -0.5 (non-modular clustering) and 1 (fully modular clustering) that quantifies the quality of an assignment of nodes to communities and evaluate how densely connected the nodes in the same community are w.r.t. relationships outside communities. The modularity of a community c is given by the formula:

$$Q_c = \frac{\sum_{in}}{2m} - (\frac{\sum_{tot}}{2m})^2$$

where m is the sum of all of the relationship weights in the graph, $\sum_{in}$ is the sum of relationship weights between nodes within the community c considering each relationship twice; and $\sum_{tot}$ is the sum of all relationship weights of nodes within the community including relationships which link to nodes of other communities. The iterative procedure of Louvain groups nodes into communities based on how closely connected nodes are and calculates the modularity. The nodes are assigned to a different community if this change leads to increased modularity.

The implementation of Louvain in Neo4j allows detecting communities in directed, undirected and weighted graphs. Before the application of the algorithm, it is necessary to project the KG in a graph with the function shown in Listing 1.4. In this context, the direction of the Document-Document relationships is meaningless. For this reason, we convert the "connected" relationships into undirected relationships with the corresponding weights and project just the Document nodes. Then, the Louvain method can be applied with the function *gds.louvain.stream* that takes in input the name of the graph. In the configuration of the algorithm, it is possible to specify the tolerance value, i.e., the minimum change of modularity value to assign the node to a different community.

Listing 1.4. Projection of the graph in Neo4j.

```
CALL gds.graph.create(
      'GRAPH_NAME',
      ['Document'],
      { CONNECTED: { orientation: 'UNDIRECTED' } },
      { relationshipProperties: ['weight', 'median'] }
)
```

3   Experiment

The methodology described in Sect. 2 has been tested on a dataset of Italian news articles. The dataset contains 10,395 news articles from the "Gazzetta di

---

[5] https://neo4j.com/docs/graph-data-science/.

Modena" newspaper.[6] The news articles are related to some crime events occurred in the province of Modena from 2011 to 2021 and cover 13 types/categories of crimes (theft, robbery, murder, sexual violence, mistreatment, aggression, illegal sale, drug dealing, scam, fraud, money laundering, evasion, and kidnapping). The dataset has been obtained by the application of web crawler method along with several semantic approaches for information retrieval [20,21] and is openly available.[7] The dataset is unbalanced on the crime category: the most news articles are related to thefts (70%), while sexual violence, money laundering, evasion and fraud are less than 1% of the dataset. For these experiments we take into account the news articles related to mistreatment, aggression and sexual violence which amount to 803 news articles. The news articles contain on average 314 words (the lowest number of words is 39 and the highest one is 1939). No pre-processing phase was needed on the text of the news articles before the application of our methodology.

3.1 Configuration

Our methodology requires the configuration of a few parameters. After some preliminary experiments, we set the similarity score of DBpedia spotlight to 0.7 for EL. In AKE, we exploited the Jaro-Winkler distance with threshold equal to 0.7 to detect duplicates among keyphrases. Since better results were obtained with keyphrases of just one word (keyword) we chose to add them in the graph and we do not integrate keyphrases of multiple words. We chose the first 5 keyphrases for each document ordered by the Yake score. Considering only the first 5 keyphrases and following the approach described in Sect. 2.2 to normalize the scores, the score of the fifth keyphrase becomes very small. To avoid very small values, firstly, we select the 30 most relevant keyphrases and normalize their scores, then we add to the graph only the 5 (if available) most relevant keyphrases per document. The tolerance value in the Louvain algorithm was left to the default value (0.0001). This is a very small value and we decided to use it since our dataset is related to a restricted domain and we expect to generate a very densely connected graph; therefore, a low tolerance allows for more separation of the graph in smaller communities.

3.2 Results and Discussions

The KG obtained by the 803 news articles consists of 803 Document nodes, 1832 Entity nodes and 1544 Keyphrase nodes. The Document-Entity relationships are 5435, while the Document-Keyphrase relationships are 4015. Analyzing the graphs, we discovered that 1148 entities (63%) are connected to just one document, while the number of keyphrases related to just one document

---

[6] https://gazzettadimodena.gelocal.it.
[7] Italian Crime News dataset: https://paperswithcode.com/dataset/italian-crime-news.

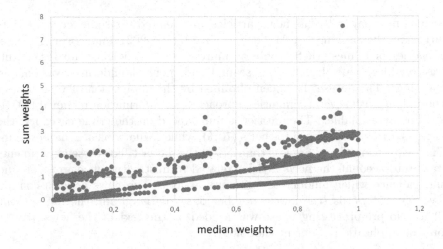

Fig. 4. Median and sum of the weights of "connected" relationships.

is 1063 (69%). Our methodology generates 129041 Document-Document relationships (80588 of them derived from entities - shared_entity - and 48453 from
keyphrases - shared_key) and 56769 "connected" relationships. The top 3 entities that appear more frequently in the dataset are Modena (http://dbpedia.
org/resource/Modena), Carpi (http://dbpedia.org/resource/Carpi) and Bologna
(http://dbpedia.org/resource/Bologna). They are cities in the Emilia-Romagna
region near Modena. Among the most frequent keyphrases there are "cara-
binieri", "polizia", that are the Italian low enforcement agencies. Figure 4 shows
the median and the sum of the weights of the "connected" relationships. As
expected, the median is always a number between 0 and 1. The median is higher
than 0.8 in the 0.67% of the cases and lower than 0.1 in the 0.2% of the cases. The
analysis of the median value could give a suggestion for the value of the threshold
to cut the relationships in the pruning method. Low sum values indicate that
the two connected documents do not have a high similarity/relation.

The Louvain algorithm has been applied to the obtained graph and several
tests have been performed using different values of pruning threshold. Table 1
shows the number of communities and the modularity value for different values of
that threshold. As can be seen, the number of communities is quite low regardless
the value of threshold as well as the value of modularity. This happens because
the graph is very densely connected and the Louvain algorithm is not able to
separate the graph and distinguish small groups of nodes more densely connected
each other. The highest value of modularity is obtained with pruning threshold
0.8, however, the difference with the other modularity values is not significant. As
further step, we decided to remove the "connected" relationships with identical
values of weights sum and weights median. Indeed, as highlighted in Sect. 2.3,
in that cases, the documents share just one entity or keyphrase, thus, their
similarity is low. Focusing on our use case, this means that probably the news
articles are not related to similar crime event. The resulting graph contains 803

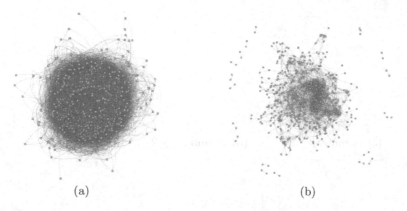

(a)   (b)

Fig. 5. Overview of the initial graph (on the left) and the graph obtained after pruning relationships with median of weights lower than 0.8 or sum of weights equal to their median (on the right).

Table 1 Number of "connected" relationships removed from the graph based on the pruning threshold (T) applied, communities detected by Louvain and modularity value.

T	Deleted relationships	Number of communities	Modularity
0.1	11427 (20%)	9	0.33021
0.2	11861 (21%)	9	0.33239
0.3	12026 (21%)	9	0.33353
0.4	12264 (22%)	9	0.33381
0.5	14020 (25%)	8	0.33776
0.6	15027 (26%)	10	0.33577
0.7	15502 (27%)	10	0.33742
0.8	18966 (33%)	10	0.34788

Document nodes and 2968 "connected" relationships and is shown in Fig. 5(b), while Fig. 5(a) illustrates the initial graph, before all the pruning operations. It is clear that the initial graph is too densely connected for the application of community detection. In the graph of Fig. 5(b), Louvain finds 270 communities with modularity value of 0.56. 29 communities contain just one document, while the number of documents in the other communities is variable, the largest one contains 71 documents.

We calculated the cosine similarity of the text of news articles in the same community. The cosine similarity was applied to the vector representations obtained by Word2Vec, following the approach described in [5,19]. Figure 6 shows the heatmaps of cosine similarity values in the 9 largest communities. The colormap is the same for all the heatmaps, lighter colors indicate higher similarity. As can be seen, similarity within the same community is always high, this confirms the results of community detection. Focusing on a community of 10 news

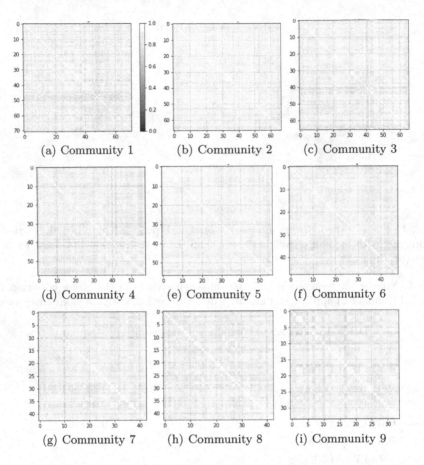

(a) Community 1   (b) Community 2   (c) Community 3

(d) Community 4   (e) Community 5   (f) Community 6

(g) Community 7   (h) Community 8   (i) Community 9

Fig. 6. Cosine similarity of documents in the same community.

articles, we discovered that 3 of them, published in 2 different days, are related to a sexual violence occurred in 2021 in a popular park in Modena. Other 4 news articles of the same community describe aggressions among students happened in the same park of the previous 3 news articles. The 4 aggressions occurred in different years (one in 2016, one in 2017 and two in 2019) but have some common characteristics, i.e., the location, the participants. Analyzing in deep also the other communities, it was possible to verify that the news articles corresponding to the Document nodes in the same community are related to very similar crime events, i.e., aggressions with the same approach, sexual violence against people with similar characteristics, aggressions in the same place. This result demonstrates the usefulness of community detection algorithms, and in particular Louvain, to find similar documents in a collection.

4   Conclusions and Future Work

The paper presented a methodology for the construction of a KG from textual documents and the application of the community detection algorithm Louvain to identify similar documents. We provide the python source code for both the scopes in a GitHub repository. This allows other researchers reproducing our experiments and extending the use of our methodology to other documents. The tool used for the creation and the analysis of the KG is Neo4j. The approach does not depend from the type of documents to analyze and which topics they belong to. Moreover, it can be easily adapted to multi-language analyses. The results of the experiments conducted on an open-source Italian news articles dataset are promising even if the value of modularity over detected communities is low. We demonstrate the effectiveness and efficiency of using KGs to study the similarity of a consistent amount of textual data.

As future work, it could be interesting to explore the possibility of connecting entities and keyphrases to external ontologies, taxonomies or vocabularies such as WordNet, BabelNet, and study the semantic similarity of the nodes. This should allow to link together entities/keyphrases with the same/similar meaning and decrease the number of Entity and Keyphrase nodes connected to only one Document. The integration of semantic technologies should improve the results of community detection and increase the value of modularity. Other experiments for the extraction of keyphrases with multiple words should allow the extraction of more representative concepts. In addition, different pruning techniques could be tested at different level of the KG generation, i.e., after the creation of the "appear" relationships, or the "shared_entity" and "shared_key" relationships. In the end, other community detection algorithms could be applied for comparison.

Acknowledgments. This work is partially supported by the project "Deep Learning for Urban Event Extraction from News and Social media streams" founded by the Engineering Department "Enzo Ferrari" of the University of Modena and Reggio Emilia.

References

1. Auer, S., Bizer, C., Kobilarov, G., Lehmann, J., Cyganiak, R., Ives, Z.: DBpedia: a nucleus for a web of open data. In: Aberer, K., et al. (eds.) ASWC/ISWC -2007. LNCS, vol. 4825, pp. 722–735. Springer, Heidelberg (2007). https://doi.org/10.1007/978-3-540-76298-0_52
2. Aung, T.T., Nyunt, T.T.S.: Community detection in scientific co-authorship networks using Neo4j. In: 2020 IEEE Conference on Computer Applications (ICCA), pp. 1–6 (2020)
3. Bhatt, S.P., et al.: Knowledge graph enhanced community detection and characterization. In: Culpepper, J.S., Moffat, A., Bennett, P.N., Lerman, K. (eds.) Proceedings of the Twelfth ACM International Conference on Web Search and Data Mining, WSDM 2019, Melbourne, VIC, Australia, 11–15 February 2019, pp. 51–59. ACM (2019). https://doi.org/10.1145/3289600.3291031

4. Blondel, V.D., Guillaume, J.L., Lambiotte, R., Lefebvre, E.: Fast unfolding of communities in large networks. J. Stat. Mech. Theory Experiment **2008**(10), P10008 (2008). https://doi.org/10.1088/1742-5468/2008/10/p10008
5. Bonisoli, G., Rollo, F., Po, L.: Using word embeddings for Italian crime news categorization. In: Ganzha, M., Maciaszek, L.A., Paprzycki, M., Slezak, D. (eds.) Proceedings of the 16th Conference on Computer Science and Intelligence Systems, Online, 2–5 September 2021. Annals of Computer Science and Information Systems, 25, pp. 461–470, 2021. https://doi.org/10.15439/2021F118
6. Campos, R., Mangaravite, V., Pasquali, A., Jorge, A., Nunes, C., Jatowt, A.: Yake! keyword extraction from single documents using multiple local features. Inf. Sci. **509**, 257–289 (2020). https://doi.org/10.1016/j.ins.2019.09.013
7. Chen, Q., Wang, W., Huang, K., Coenen, F.: Zero-shot text classification via knowledge graph embedding for social media data. IEEE Internet Things J. **9**(12), 9205–9213 (2022)
8. Elezaj, O., Yayilgan, S.Y., Kalemi, E.: Criminal network community detection in social media forensics. In: Yildirim Yayilgan, S., Bajwa, I.S., Sanfilippo, F. (eds.) INTAP 2020. CCIS, vol. 1382, pp. 371–383. Springer, Cham (2021). https://doi.org/10.1007/978-3-030-71711-7_31
9. Grootendorst, M.: Keybert: minimal keyword extraction with bert (2020). https://doi.org/10.5281/zenodo.4461265
10. Hsu, P.Y., Chen, C.T., Chou, C., Huang, S.H.: Explainable mutual fund recommendation system developed based on knowledge graph embeddings. Appl. Intell. **52**(9), 10779–10804 (2022). https://doi.org/10.1007/s10489-021-03136-1
11. Ji, S., Pan, S., Cambria, E., Marttinen, P., Yu, P.S.: A survey on knowledge graphs: representation, acquisition, and applications. IEEE Trans. Neural Netw. Learn. Syst. **33**(2), 494–514 (2022)
12. Koloski, B., Stepišnik Perdih, T., Robnik-Šikonja, M., Pollak, S., Škrlj, B.: Knowledge graph informed fake news classification via heterogeneous representation ensembles. Neurocomputing **496**, 208–226 (2022). https://www.sciencedirect.com/science/article/pii/S0925231222001199
13. Mendes, P.N., Jakob, M., García-Silva, A., Bizer, C.: Dbpedia spotlight: shedding light on the web of documents. In: Ghidini, C., Ngomo, A.N., Lindstaedt, S.N., Pellegrini, T. (eds.) Proceedings the 7th International Conference on Semantic Systems, I-SEMANTICS 2011, Graz, Austria, 7–9 September 2011, pp. 1–8. ACM International Conference Proceeding Series, ACM (2011). https://doi.org/10.1145/2063518.2063519
14. Nigam, V.V., Paul, S., Agrawal, A.P., Bansal, R.: A review paper on the application of knowledge graph on various service providing platforms. In: 2020 10th International Conference on Cloud Computing, Data Science & Engineering (Confluence), pp. 716–720 (2020)
15. Po, L., Rollo, F.: Building an urban theft map by analyzing newspaper crime reports. In: 13th International Workshop on Semantic and Social Media Adaptation and Personalization, SMAP 2018, Zaragoza, Spain, 6–7 September 2018, pp. 13–18. IEEE (2018). https://doi.org/10.1109/SMAP.2018.8501866
16. Po, L., Rollo, F., Trillo Lado, R.: Topic detection in multichannel Italian newspapers. In: Calì, A., Gorgan, D., Ugarte, M. (eds.) IKC 2016. LNCS, vol. 10151, pp. 62–75. Springer, Cham (2017). https://doi.org/10.1007/978-3-319-53640-8_6

17. Rinaldi, A.M., Russo, C., Tommasino, C.: Web document categorization using knowledge graph and semantic textual topic detection. In: Gervasi, O., et al. (eds.) ICCSA 2021. LNCS, vol. 12951, pp. 40–51. Springer, Web document categorization using knowledge graph and semantic textual topic detection (2021). https://doi.org/10.1007/978-3-030-86970-0_4

18. Rollo, F.: A key-entity graph for clustering multichannel news: student research abstract. In: Seffah, A., Penzenstadler, B., Alves, C., Peng, X. (eds.) Proceedings of the Symposium on Applied Computing, SAC 2017, Marrakech, Morocco, 3–7 April 2017, pp. 699–700. ACM (2017). https://doi.org/10.1145/3019612.3019930

19. Rollo, F., Bonisoli, G., Po, L.: Supervised and unsupervised categorization of an imbalanced Italian crime news dataset. In: Ziemba, E., Chmielarz, W. (eds.) FedCSIS-AIST/ISM -2021. LNBIP, vol. 442, pp. 117–139. Springer, Cham (2022). https://doi.org/10.1007/978-3-030-98997-2_6

20. Rollo, F., Po, L.: Crime event localization and deduplication. In: Pan, J.Z., et al. (eds.) ISWC 2020. LNCS, vol. 12507, pp. 361–377. Springer, Cham (2020). https://doi.org/10.1007/978-3-030-62466-8_23

21. Rollo, F., Po, L., Bonisoli, G.: Online news event extraction for crime analysis. In: Amato, G., Bartalesi, V., Bianchini, D., Gennaro, C., Torlone, R. (eds.) Proceedings of the 30th Italian Symposium on Advanced Database Systems, SEBD 2022, Tirrenia (PI), Italy, June 19–22, 2022. CEUR Workshop Proceedings, vol. 3194, pp. 223–230. CEUR-WS.org (2022). http://ceur-ws.org/Vol-3194/paper28.pdf

22. Rony, M.R.A.H., Chaudhuri, D., Usbeck, R., Lehmann, J.: Tree-KGQA: an unsupervised approach for question answering over knowledge graphs. IEEE Access 10, 50467–50478 (2022)

23. Rose, S., Engel, D., Cramer, N., Cowley, W.: Automatic Keyword Extraction from Individual Documents, chap. 1, pp. 1–20. Wiley (2010). https://onlinelibrary.wiley.com/doi/abs/10.1002/9780470689646.ch1

24. Szekely, P., et al.: Building and using a knowledge graph to combat human trafficking. In: Arenas, M., et al. (eds.) ISWC 2015. LNCS, vol. 9367, pp. 205–221. Springer, Cham (2015). https://doi.org/10.1007/978-3-319-25010-6_12

25. Tosi, M.D.L., dos Reis, J.C.: Understanding the evolution of a scientific field by clustering and visualizing knowledge graphs. J. Inf. Sci. 48(1), 71–89 (2022). https://doi.org/10.1177/0165551520937915

Framework for Author Name Disambiguation in Scientific Papers Using an Ontological Approach and Deep Learning

Lisandra Díaz-de-la-Paz[1,2](✉) iD, Leonardo Concepción-Pérez[1,2] iD,
Jorge Armando Portal-Díaz[1] iD, Alberto Taboada-Crispi[1,2] iD,
and Amed Abel Leiva-Mederos[1,2] iD

[1] Universidad Central "Marta Abreu" de Las Villas (UCLV), Carretera a Camajuaní Km 5 1/2, Santa Clara, Cuba
`{ldp,ataboada,amed}@uclv.edu.cu`, `{lcperez,jportal}@uclv.cu`
[2] UCLV, Centro de Investigaciones de la Informática, Santa Clara, Cuba

Abstract. The aim of this paper is to solve the problem of disambiguation of authors' names in scientific papers. In particular, it focuses on the problem of synonyms and homonyms. Thus, we often find two or more names written in different forms denoting the same person. Moreover, there may be several authors using the same name. To address both the synonym and homonym problems in scientific papers, we propose a framework that uses a hybrid approach of an ontological model and a deep learning model. First, we describe the design of the ontology model, the automatic ontology creation process, and the construction of a weighted co-author network through a set of semantic rules and queries. Second, the selected features are preprocessed during the attribute engineering process to measure the similarity indicator for each feature. Third, the similarity indicators are reduced to a vector space model and used as input to the Deep Learning-based author name disambiguation method to model different types of features. Fourth, the proposed framework is tested on smaller groups of the gold standard large dataset of scientific papers from several international databases named LAGOS-AND and achieves promising results compared to other similar solutions proposed in the literature.

Keywords: Author name disambiguation · Deep learning · Framework · Ontology · Scientific papers

This work is partially supported by Project 3 "ICT supporting the educational processes and the knowledge management in higher education (ELINF)" of the NETWORK University Cooperation "Strengthening of the role of ICT in Cuban Universities for the development of the society". We thank Carlos Alberto Morell for his useful suggestions and ideas and the team of Li Zhang, Wei Lu and Jinqing Yang for providing the corpus used to train the Doc2Vec model of the gold standard dataset LAGOS-AND.

1   Introduction

Author name disambiguation (AND) can occur in two different forms: (1) when two or more names are written in different forms but represent the same person (name variety problem, also called synonyms), and (2) when multiple authors have the same designation name but represent different people (polysemy, also called homonyms). According to [1], the coincidence of both problems is called the name mixture problem and is most common in real-world datasets. In digital libraries, both problems occur together and manifest themselves in the description of scientific papers and in bibliographic metadata. Currently, AND is very common in scientific publication data. With the rapid growth of scientific publications and authors, AND is becoming increasingly important for data cleaning in scientific network analysis and mining [2]. Author names are ambiguous because they may be written in different forms, abbreviations may be used, typos may occur, a person's name may change after marriage in some countries, norms for author names in journals vary, and some people use the same name designation. All these aspects affect the search for information about these author names. To solve the AND problem, there are important contributions that use an author number or code. There are several databases (e.g., Scopus, PubMed, Web of Science), publishers (e.g., Elsevier, PLoS, Thomson Reuters, Nature, Wiley), manuscript submission systems (e.g., ScholarOne), research and professional associations (e.g., ACS, IEEE, AAAS), and others that use a unique researcher number ID to solve the problem of author ambiguity through proprietary identification systems. Some examples of these proprietary identification systems are Scopus Author ID, ResearcherID or Open Researcher and Contributor ID (ORCID). However, there is a large subset of author names that do not show up in any of these systems because a large number of publications and conferences do not yet ask authors for their ORCID ID (or other proprietary identification system), or they have not asked for it in the past (which is obvious for older publications). Authors may also be submitting incorrect metadata information to the system [1]. The AND problem is still being researched to improve the quality measurements of the new solutions. Numerous approaches have been used to solve the AND problem. In order to locate the main approaches in the literature, several authors AND conducted surveys and reviews that classified the different approaches as follows:

- In [3], the authors proposed a AND taxonomy to classify the AND techniques. The taxonomy is divided into two main categories: machine learning techniques and non-machine learning techniques. Machine learning techniques include supervised, unsupervised, and semi-supervised techniques. Non-machine learning techniques include graph-based and heuristic techniques.
- In [4], the authors classified the existing methods of AND into two different categories depending on their main approach: Author Grouping and Author Assignment. Author grouping methods attempt to group author records from the same author based on some sort of similarity in their attributes, including heuristic, graph-based, and methods that use string matching strategies. Framework for AND in Scient. Papers Using an Ontological App. And DL3 Author attribution methods aim to directly attribute authorship to respective authors using either a classification or a clustering technique. Alternatively, the methods can be grouped according to the evidence studied in the

disambiguation task, namely citation attributes (only), web information, or implicit data that can be extracted from the available information. The categories in this taxonomy are not completely disjoint; some methods use two or more types of evidence or mix approaches.

- In [1], the authors categorized the methods of AND into five types: (1) supervised learning, (2) unsupervised learning, (3) semi-supervised learning, (4) graph-based, and (5) ontology-based. They also explained the advantages and disadvantages of using these methods. The authors expressed that the two less explored methods are graph-based and ontology-based, especially the latter one that allows semantics to be added to the disambiguation process. The papers analyzed in this survey are from the period between 2004 and 2016, which means that some important contributions from the last five years are missing.

Despite the different classification methods of the AND techniques, all of them agree in their goal of grouping each author with their corresponding publications, dealing in some way with problems of both synonyms and homonyms. To our knowledge, we prefer a hybrid approach to solve the problem AND through a hybrid solution. Therefore, the main contribution of this work is to develop a framework that combines an ontological model to represent authors, publications, and a weighted co-author network created by semantic rules with deep learning techniques in smaller groups of the gold standard large dataset of scientific papers from several international databases called LAGOS-AND [5].

The rest of this paper is organized as follows. In Sect. 2, we provide an overview of related work. In Sect. 3, we describe our framework in detail and formalize the AND problem from the perspective of the ontological model and the deep learning techniques used. In Sect. 4, we then evaluate the overall framework and validate it in terms of its implications for research and practice. In Sect. 5, we conclude the paper and provide directions for future research.

2   Related Works

In information science, ontology is a set of concepts and categories in a subject area or domain, showing their properties and the relationships between them. In other words, it is the knowledge representation of a domain. Ontologies are a fundamental artificial intelligence tool for knowledge-based systems (KBS) development. With its formal and well-defined structure, an ontology provides a machine-understandable language that enables automatic reasoning for problem solving. Typical KBSs are expert systems and decision support systems [6].

2.1   Ontology-Based AND

Ontology-based AND has been used by many researchers in various fields. Examples: Authority control of individuals and organizations [7], person identities in linked open data [8], scale-free collaboration networks [9], ontology-based crosslanguage information retrieval system Tamil-English [10], ontological framework for information extraction with fuzzy rule base and word sense disambiguation [11], etc. Especially in digital libraries or databases, researchers have used this kind of method less. For example, in [1], the authors present a summary table that analyzes only two works that propose an ontology-based solution for author name disambiguation. The papers are [12] and [13]. According to [1, 12] focuses on entity disambiguation using an ontology-based method, background knowledge, and attributes such as authors, conferences, and journals. In [12], data from DBLP and a corpus from DBWorld were used to prove the results using a largely populated ontology. The main limitation of this work is that it needs to be tested on more robust platforms.

On the other hand, [13] addresses the problem of sharing names through OnCu ontology-based categories using the author ontology and the domain ontology of computer science. In [13], collected contributions from AAAI, ISWC, ESWC, and WWW conference websites were used to perform their evaluation based on category usage over the created ambiguity dataset. The main limitation of this work is that it does not consider property relations.

Table 1. Summary of ontology-based methods for AND in the last 10 years.

Ref	Tool/Method	Features	Findings	Limitations
[14]	Ontology-based personal name disambiguation (OnPerDis)for Chinese personal names on the web	Name, basic information introduction, contact, and personal relationship	The approach achieves good performance in the three categories of disambiguation of personal names. The F-scores of the approach improve by more than 4%, 5.51%, and almost 9.8%, respectively	More instances of person ontology need to be added to the knowledge base of OnPerDis. Also, the mapping relationships between English names and corresponding Chinese names need to be investigated. 3 experiments were conducted, but they focused more on information extraction than disambiguation of person names
[15]	Researcher Name Resolver (RNR) with a web resource	Researcher name and affiliation external direct links and external search links	RNR constructs researcher URIs to display researcher pages with profiles and links to related external resources	Administrative staff appropriately engage with researcher profiles and maintain researcher profiles in their daily work as researchers. The method has not been compared to any other in the literature
[2]	A semi supervised framework for AND in academic social networks that addresses both synonym and homonym issues	Co-authorinformation, title, year, publisher, keywords, affiliation, and topic information	A self-learned method is proposed to solve the ambiguity of co-author information to improve the performance of other models	LDA topic inference is the most time-consuming method in the proposal, about 34 h. The authors tested different combinations of the comparison methods, but were not compared with other similar AND works

(continued)

Table 1. (*continued*)

Ref	Tool/Method	Features	Findings	Limitations
[16]	PDF2TXT, Semantic Finger-print Generator, Comparator, ClaimDecisionMaker, Publica-tion Assignment, Arbiter	Metadata is used to extract information about co-authors and institutions, while text data is used to fingerprint	The method introduced semantic fingerprint integrated with co-author features to AND problem	The size of the dataset was too small and the recall index was low. The method may not work for two authors with the same name and research areasThe method was not compared with other methods from the literature. It was only tested with 7 Chinese author names An ontology-based solution approach for collecting, displaying and managing researcher profiles
[17]	An ontology-based solution approach for capturing, displaying and managing researcher profiles	Publication title, author name, email, department, keywords, publication year, volume, etc	Semantic rules implemented to find collaborations between professors	Similarity indicators between analyzed attributes are not considered, only exact matches. The ontology can be enriched with further semantic relations containing summaries and keywords of publications, researchers and topics of interest
[18]	Rule-based binary Classifier and hierarchical agglomerative clustering approach. Reclassification of existing publications from MAKG into a set of 19 disciplines	Author name, Affiliation, Co-authors, Title,Years, Journals and Conferences, References	The evaluation, showed that ComplEx is the best large scale entity embedding method we could apply to the MAKG	for trained entity embedding, future research could generate em-beddings with higher dimensionality. The main challenge of the task lies in the hardware requirements for training embedding at such a large scale
[19]	Framework Literally Author Name Disambiguation (LAND)	Author names used to get LNFI blocks sorted. Title and Publication date used for comparisons	Benchmark dataset that defines an SCC compliant SCC compliant with the Open Citations Data Model and another SCC (named AMiner-534K) LAND The draft addresses data with in knowledge graphs	Includes author collaboration and network information along with the AND network information along with the topic of interest/expertise, which is obtained by processing the authors' publications extracted using deep learning approaches. With this additional data, they can test whether they can use the results for the task of AND

In summary, the models of ontologies that disambiguate author names have solved many of the problems in isolation and in more specific contexts (see Table 1.). Following this analysis, we believe that the development of a new ontology is necessary that combines the context of research and authority control in libraries.

2.2 Deep Learning-Based AND

On the other hand, [20] investigated how deep neural network (DNN) results can be used to form new clusters and how contributions can be assigned to existing clusters. In the first case, the authors obtained an F1 value of 48.0% for the optimal cut-off in each cluster, while in continuous clustering they obtained an F1 value of 84.3%. The main drawback of their approach is the need for pairwise preprocessing, which scales in quadratic order with the number of contributions in each clustering. This caused most of the computation time in their study. Moreover, [20] focused only on disambiguation of homonymous authors, and when they also consider synonymy of names, the number of pairs increases further.

In contrast to [20, 21] proposed an author identification method in bibliographic data that uses DNN to solve both synonyms and homonyms. The method solves the synonym problem better than the homonym problem; moreover, its performance on the combined synonym-homonym problem is not yet satisfactory. The complexity of detecting and assigning publications to their respective authors is not an easy task. The results show that neural networks with one layer significantly outperform other classical machine learning methods such as Naïve Bayes (NB), Random Forest (RF), and Support Vector Machines (SVM) in average accuracy. Moreover, for homonyms and homonym synonyms, a suitable method should be implemented in other datasets to improve the performance. The use of feature engineering based on a semantic approach for title attributes could improve performance in all cases. In addition, [21] authors confirm that the use of deep neural networks is usually very helpful for working with larger datasets.

Although some work has been developed using the deep learning approach, we believe that this approach has not yet been sufficiently exploited, as well as the possibilities that this approach offers in combination with knowledge extracted from an ontology to solve the AND problem. Therefore, in this paper we describe the development process of the author's name disambiguation ontology (AND ontology) that enables the representation of intelligent system elements using Deep Learning.

3   Materials and Methods

In this paper, we propose a framework for solving the problem of author disambiguation in scientific papers. Fig. 1 shows our framework. In the first phase (1-3), the information from the dataset is imported in CSV format and using R2RML mapping language, the data is transformed into its semantic type corresponding to the ontology model AND previously imported in the W3C Web Ontology Language (OWL). Then, the URI are created and finally the generated triplets are presented in Resource Description Framework (RDF) format as output to be loaded into the AND ontology with all the injected data. The second stage (4) deals with the ontology model AND, the construction of co-author relationships through semantic rules. We also note that attributes such as co-author frequency, total co-author frequency, and normalized weight can be very helpful in supervised disambiguation. The author's full name, title, location, organization, abstract, and co-author information play an important role in solving the problem AND. The third stage (5-6) is used to preprocess the data to calculate the similarity indicators for each attribute. Finally, the fourth stage (7) uses the similarity indicators and other vectorized data to train the deep learning model, which in turn feeds back the AND ontology through a data transformation tool. The process concludes with the feedback of the AND ontology, and the cycle in the framework is run once.

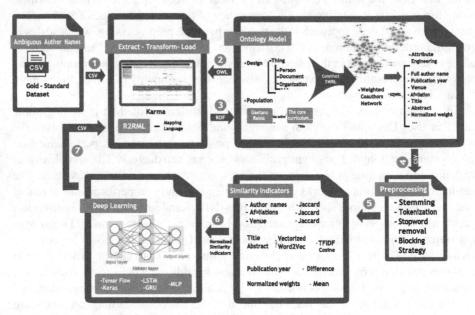

Fig. 1. Framework for author name disambiguation.

3.1 Dataset

We use the LAGOS-AND [5] to support our results, as it accounts for problems with synonyms as well as homonyms, while other datasets are unlikely to provide the former. LAGOS-AND is a recompilation of the following 13 known datasets: Aminer-Rich, Aminer-Simple, Aminer-WhoisWho, Aminer-Zhang, BDBCompCota, DBLP-CiteSeerX, DBLP-GESIS, DBLP-Kim, DBLP-Qian, PubMedGS, PubMed-Kim, REXA-Cullota, and SCAD -zbMATH-Muller. In [5], the authors present a method to automatically generate a large labeled dataset for author name disambiguation (AND) in academia by using authoritative sources, ORCID and DOI. This dataset contains 7.5 m citations from 797,000 unique authors and shows great similarities to the entire Microsoft Academic Graph (MAG) for six gold standard validations. All datasets included in this large gold standard dataset are freely available for academic use without additional restrictions. In this work, the authors investigated the long-standing problem of name synonyms and showed for the first time the degree of variation in surnames. To accurately capture author similarity, the authors in [5] converted "block-based datasets" to "author-pair datasets", and the signal monitored (0/1) is whether the paired instance represents the same author (1) or not (0). Some homonymous authors (0) and synonymous authors (1). Consequently, the "paired dataset" consists of 500K instances, half of which are positive.

3.2 Ontology Model

In this subsection, the ontology model development process is explained with the phases of requirements definition, vocabulary selection for reuse, ontology implementation and integration, ontology evaluation, documentation, and maintenance. The ontology building process followed the regulations for ontology development and the Neon specifications [23, 24] through an application corresponding to an ontology for disambiguation of author names in scientific papers called AND Ontology. One of the goals of the AND ontology is to provide a common representation of data for author name disambiguation. Each element of the ontology (class, property) must be named with only one term to avoid semantic heterogeneity. The language of the ontology will be OWL-2. Scope of ontology. One of the ways to determine the scope of ontology is to make a list of questions that such a system should answer [25]. In the thematic domain of author name disambiguation, some of the possible questions that should be answered by the ontology are the following:

1. In which places has author X published?
2. What is the alias of author X?
3. Which authors collaborate with author X?
4. How many contributions has author X made?
5. What is the affiliation of author X?
6. What is the preferred name of author X?
7. How many authors write a publication P?
8. Did author X write the publication with the title T?
9. What is the exclusivity of publication P?
10. Which authors have a weighted co-author ratio greater than a threshold X?

Judging by this list of questions, the ontology must contain information about the different types of documents and publications, as well as about the people or actors who authored these documents, their organization, and their collaboration through co-author relationships, among others. The development of the AND ontology is based on different vocabularies that allow the use and reuse of classes and properties from other ontologies. In the field of author name disambiguation in scientific papers, there are many applications based on ontologies with few design values. For this reason, languages that provide clues for formalizing ontologies have been selected and are listed below:

- FOAF[1]. It describes the characteristics of people and social groups that are independent of time and technology. FOAF defines the classes for person, organization, and the subclass OrgUnit refers to Department, Faculty, Research center, and College.
- SKOS[2]. It describes simple knowledge organization for the web. SKOS defines the class Concept.
- GEO[3]. It describes the vocabulary for building geographic ontologies and geospatial data. GEO defines the Country class and the Place subclass.

We also reuse several ontology systems developed for the domain, whose conceptual quality is sufficient for the development of other ontological schemes. The ontologies that best describe the domain and are most complete were used for this design. The ontologies selected for reuse are:

- BIBO[4]. It describes the characteristics of bibliographic records such as the class Document and the subclasses Conference Proceeding, Event, Journal, and Publication.
- VIVO[5]. It presents researchers in the context of their experiences, outcomes, interests, accomplishments, and related institutions.
- GND[6]. Used to describe the name of the person and variant names of the person. This ontology takes into account the Authority Resource and the Anglo-American Cataloguing Rules (AACR2).

The concepts we have declared in this section were selected using AgreementMaker, a software tool that allows us to map ontologies and determine whether there is similarity of terms and equality in the order of the hierarchical structure of classes. Agreement-Maker helped us to find not only similar classes, but also properties related to specific concepts that appear in an ontology. This tool allowed us to identify unique terms that are not polysemous or homonymous. The following criteria were used to select the terminological concept base of the ontology:

1. Selection of the class whose hierarchy best describes each concept associated with disambiguation of author names in scientific papers.
2. Selection of classes whose annotations and definitions were accepted by IEEE.
3. The terms of other ontologies associated with the domain are used to establish synonymy relations within the ontology.
4. A base ontology is taken to integrate ontologies into it to build the domain.

This approach to ontology organization uses a mixed solution: symmetric and asymmetric.

Semantic Rules Defined. Following [22], we construct a co-authorship relation between the individuals whose names are the same or very similar to determine whether

---

[1] http://xmlns.com/foaf/0.1/.

[2] http://www.w3c.org/2004/02/skos/.

[3] http://www.w3c.org/2003/01/geo/.

[4] https://purl.org/ontology/bibo/.

[5] https://bioportal.bioontology.org/ontologies/VIVO.

[6] https://d-nb.info/standards/elementset/gnd.

or not they are the same person. We examined the six semantic rules proposed by [22] and found several drawbacks, such as that Rule 2, Rule 4, and Rule 5 have conceptual flaws because they use SQWRL to satisfy the rule and then assign the result to an object property or a data property of their ontology, which is not allowed in this language. The result could be a query but not a rule and they are expressed like semantic rules. If we type it in SWRL tab in Protégé, the system immediately recognizes it as a query. Taking this into account, we adapt their idea and change some rules and queries to construct the co-authorship relations.

Rule 1. Calculate the Co-authorship Relationship.

bibo : Document(?d) $\wedge$ and:nrAuthors (?d, ?nr) $\wedge$ swrlb:greaterThanOrEqual (?nr, 2)$\wedge$ and:hasAuthor(?d, ?a1) $\wedge$ and:hasAuthor(?d, ?a2) $\wedge$ and:hasURI(?a1, ?a1URI) $\wedge$ and:hasURI(?a2, ?a2URI) $\wedge$ swrlb:notEqual(?a1URI, ?a2URI) $\wedge$ sameAs(?a1, ?a1) $\wedge$ sameAs (?a2, ?a2)$\wedge$ swrlx:makeOWLThing(?rel, ?a1, ?a2) $\rightarrow$ and : Co $-$ authorRelation(?rel) $\wedge$ and:hasCoauthor (? a1, rel) $\wedge$ and:hasCoauthorValue(?rel, ?a2)

Rule 3. Calculate the Exclusive Co-authorship for a Given Document.

bibo : Document(?d) $\wedge$ and:nrAuthors(?d, ?nr) $\wedge$ swrlb:greaterThanOrEqual(?nr, 2) $\wedge$ swrlm:eval(?e, "1/(nr-1)", ?nr) $\rightarrow$ and : hasExclusivity(?d, ?e)

Rule 6. Calculate the Co-authorship Weight.

and : hasCoauthorV alue(?rel, ?a2) $\wedge$ swrlm:eval(?w,"a12f/totalFa1", ?a12f, ?totalFa1) $\wedge$ and:hasCoauthor(?a1,?rel) $\wedge$ and:hasTotalFrequency(?a1,?totalFa1) $\wedge$ and : hasCoauthorFrequency(?rel, ?a12f)
$\rightarrow$ and : hasCoauthorWeight(?rel, ?w)

Query 1. Show the Co-authorship Relationship Between Pairs of Authors.

bibo : Document(?p) $\wedge$ and:nrAuthors(?p, ?nr) $\wedge$ swrlb:greaterThanOrEqual (?nr, 2) $\wedge$ and:hasAuthor(?p, ?a1) $\wedge$ and:hasAuthor(?p, ?a2) $\wedge$ and:hasURI(?a1, ?a1URI) $\wedge$ and:hasURI(?a2, ?a2URI) $\wedge$ swrlb:notEqual(?a1URI, ?a2URI) $\wedge$ sameAs(?a1, ?a1) $\wedge$ sameAs(?a2, ?a2) $\wedge$ swrlx:makeOWLThing(?rel, ?a1, ?a2) $\rightarrow$ sqwrl : select(?a1, ?a2, ?rel)

Query 2. Show the Number of Authors for Each Document.

bibo : Document(?d)$\wedge$ and:hasAuthor (?d, ?a)·sqwrl : makeSet(?s, ?a) $\wedge$ sqwrl:groupBy(?s,?d) ·sqwrl : size(?size, ?s) $\rightarrow$ sqwrl : select(?d, ?size)

Query 3. Show the Exclusive Co-authorship for a Given Document.

bibo : Document(?d) $\wedge$ and:nrAuthors (?d, ?nr) $\wedge$ swrlb:greaterThanOrEqual (?nr, 2) $\wedge$ swrlm:eval(?e," 1/(nr-1)",?nr) $\rightarrow$ sqwrl : select(?d ?e)

Query 4. Show the Frequency of Co-authorship.

bibo:Document(?p) $\wedge$ and:nrAuthors (?p, ?nr) $\wedge$ swrlb:greaterThanOrEqual(?nr, 2) $\wedge$ and:hasAuthor(?p,?a1) $\wedge$ and:hasAuthor (?p,?a2) $\wedge$ and:hasURI(?a1, ?a1URI) $\wedge$ and:hasURI(?a2, ?a2URI) $\wedge$ swrlb:notEqual(?a1URI, ?a2URI) $\wedge$ sameAs(?a1, ?a1) $\wedge$ sameAs(?a2, ?a2) $\wedge$ and:hasExclusivity(?p, ?e) ·sqwrl : makeSet(?s, ?e) $\wedge$ sqwrl:groupBy (?s, ?a1, ?a2) · sqwrl:sum (?f, ?s) $\rightarrow$ sqwrl :select(?a1, ?a2, ?f)

Query 5. Show the Overall Frequency of Co-authorship.

foaf : Person(?a1) $\wedge$ and:hasCoauthor(?a1, ?rel) $\wedge$ and:hasCoauthorValue (?rel, ?a2) $\wedge$ and:hasCoauthorFrequency(?rel,?f)·sqwrl : makeBag(?s, ?f) $\wedge$ sqwrl:groupBy(?s,?a1)·sqwrl : sum(?totalFa1, ?s) $\rightarrow$ sqwrl : select(?a1, ?totalFa1)

Query 6. Show the Co-authorship Weight.

and : hasCoauthor(?a1, ?rel) ∧ and:hasCoauthorValue (?rel, ?a2)

∧ and:hasTotalFrequency (?a1, ?totalFa1) ∧ and:hasCoauthorFrequency(?rel,

?a12f) ∧ swrlm:eval(?w,"a12f/totalFa1", ?a12f, ?totalFa1)

→ sqwrl : select(?rel?, ?w)

For the construction of the graph of co-authorship relations, we refer to the definitions of the directed weighted co-authorship graph model presented in [22], which we have adopted for our framework to determine the weights and other key metrics of co-authorship relations. Let $A = a_1, ..., a_n$ denote the set of n authors. Let the set of m publication be denoted as $P = p_1, ..., p_k, ...p_m$. Let f (p_k) define the number of authors of publication p_k. Then we used the definitions introduced by [22].

Definition 1 (Exclusivity per publication). If authors a_i and a_j are co-authors in publication p_k, then $g_{(i, j, k)} = 1/(f$ (p_k) $- 1)$. $g_{(i, j, k)}$ represents the degree to which authors a_i and a_j have exclusive co-authorship for a given publication. In this definition, the relationships between co-authors are weighted more heavily for publications with a smaller total number of co-authors than for publications with a large number of co-authors.

Definition 2 (Co-authorship frequency). Another important metric for a pair of authors a_i and a_j, is the frequency of co-authorship: $c_{(i,j)} = \sum_{(k=1)}^{m} g_{(i,j,k)}$ it sums the exclusivity values $g_{(i,j,k)}$ for the same pair i, j over all publications k, ($k = 1..m$) in which they appear as co-authors. This gives more weight to authors who publish more publications jointly and exclusively.

Definition 3 (Total co-authorship frequency). Consists of the sum of all co-authorship frequency values c_{ik} over a given author a_i and all his co-authors a_k ($k = 1..n$) in in all publications in which a_i appears as an author $c_i = \sum_{(k=1)}^{n} c_{i,k}$.

Definition 4 (Normalized weight). To obtain a normalized value for the weight of co-authorship between two authors, the following normalization step should be performed, in which the total co-authorship frequency of a given author is taken into account when calculating the co-authorship frequency between that author and every other co-author by him: $w_{ij} = c_{ij}/c_i$. This ensures that the weights of an author's relationships sum to one.

Weighted Co-authorship Network. We have defined the co-author network as a set of nodes vi and edges ei, where the node vi represents an author or coauthor name and ei represents the co-author relationship in each document. The co-author network is constructed using the semantic rule R1. Then, we calculate the weight of each ei using the semantic rule R6 which depends on the other rules (R2, R3, R4 and R5). Since R2, R4 and R5 are queries, we need to compute these values using an external programming language or shop the results in a CSV file and insert the results into the ontology using the Karma integration tool. Either variant is a viable option. Part of the co-author relationship in the AND Ontology graph is shown in Fig. 2.

3.3 Preprocessing

The preprocessing step aims to prepare the data for the next steps. This usually includes standardization of author names as well as removal of stop words, tokenization, and stemming of work titles and abstracts. Then, the attributes are vectorized following the procedure of [5], using the Doc2Vec model of [27]. A similar procedure is performed for the venue and affiliation attributes: First, they are converted to lowercase, stop words and special characters are removed, and the Bag of Words is extracted. In the LAGOS-AND dataset, all ambiguous authors in a block have the same Credible Full Names (CFN), regardless of which identified author groups he/she belongs to. This rule makes the dataset more challenging than other LN (Last Name)- or LNFI (Last Name First Initial)-based datasets.

In [5] the authors considered using the CFN instead of LNFI or FN to further aggregate the identified author group into blocks. This dataset has a similar structure to the existing block-based datasets. However, unlike them, blocks arranged in this way have two major advantages. First, it is more challenging to disambiguate CFN blocks than LNFI blocks. In LNFI blocks, the authors may largely be known by completely different full names; for example, in the "Freyman R" block, different authors "Richard Freyman" and "Robin Freyman" may exist, which would simplify the dataset and lead to unnecessary computations. For CFN blocks, e.g., "Freyman Richard", it is usually more difficult to disambiguate, since all authors in this block can be named "Richard Freyman". Second, unlike FN blocks, CFN is more authoritative for representing the block. In the ORCID system, CFN can only be maintained by the author, which is displayed directly in the ORCID interface without intermediate processes. Note that there seems to be no better way to accurately identify an actual name than to retrieve the author-maintained name (CFN) [5].

3.4 Similarity Indicators

The similarity indicator for each selected attribute is shown in Table 2.. The first six attributes are the same used in [5], and we add the normalized weight attribute related to the network of co-author relationships. An appropriate measure or model is used to determine a similarity value for each attribute. For the features full author name, publication year, venue, and affiliation, a common word-level measure, i.e., Jaccard, is used. For content-based features such as title and abstract, TFIDF and a representation learning model are used in addition to Jaccard, i.e., Doc2vec [27] to determine the similarities.

3.5 Deep Learning

Deep Neural Networks (DNNs) can represent higher complexity functions, and according to the results reported in [5], DNNs seem to be a great solution to the problem AND. In this work, Gated Recurrent Unit (GRU), Long Short-Term Memory (LSTM) and Multilayer Perceptron (MLP) models are compared to select the best performing algorithm to be included in the proposed framework. Since the LAGOS-AND dataset is huge and a solution with DNNs is computationally expensive, the simulations are performed with small datasets. We start by randomly forming four disjoint groups with 250, 500, 1000, and 1500 author names, respectively, from the "pairwise dataset". Note that the following steps are performed for each set so that a solution is given for each problem and then we compare the results. We construct the co-authorship graph according to the "block-based dataset", considering not only the previously selected authors,

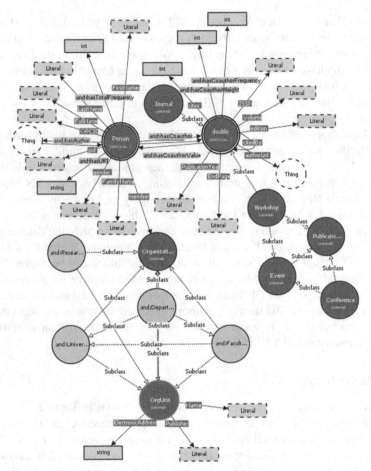

Fig. 2. Part of the AND Ontology Graph.

Table 2. Similarity Indicator of each attribute selected for AND. Based on [5].

Attribute	Data Type	Similarity Indicator/Dependent Model
Full author-name	String	Jaccard (2gram) char-level
Publication year	Integer	Normalized Absolute difference
Venue	String	Jaccard word-level
Affiliation	String	Jaccard word-level
Title	String	Jaccard word-level, TFIDF, Doc2vec, neural network
Abstract	String	Jaccard word-level, TFIDF, Doc2vec, neural network
Normalized weight	Double	Mean

but all co-authors who share at least one publication with them. We then calculate the exclusivity per publication, the co-authorship frequency and the total co-authorship frequency. These metrics are needed to calculate the normalized weight representing the co-authorship relationship between each pair of authors. Since the relationship is bidirectional, there are two values for each pair of authors. Therefore, a single value is used to represent the entire relationship, and this is added to the similarity indicators. For each pair of authors, the mean value between the normalized weights in both directions is selected as the representative value (see Table 2.). The inputs to the classifier come from the preprocessed features of the previous stage. The goal is to determine whether two authors are the same or not for each instance in the training dataset.

4 Results and Discussion

We implement the proposed framework in Python using our data, which is partially based on [5], but the proposed framework and the model used are different. Protégé is used to design the AND ontology and to test the semantic rules and queries implemented in a small part of the LAGOS-AND dataset. The Karma data integration tool is used to convert the LAGOS-AND dataset, previously in CSV format, into RDF format using the R2RML mapping language. Then, Stardog triple store is used to host the data portions of the LAGOS-AND dataset in RDF format. Later, the similarity indicators need to be computed to feed the deep learning model, which is trained with the ultimate goal of determining whether two presented authors are the same or not. All experiments were run on a cluster with two x Intel Xeon E5-2630 v3 (Haswell) 16 cores 2.4 GHz CPU and 128 GB RAM 4 x 1 Ethernet GB.

4.1 Experimental Results

We designed our experiments with four data partitions of 250, 500, 1000, and 1500 author names, extracting instances from the "pairwise dataset" that includes authors in the co-authorship graph. In each set of instances, we use 60% for training and 20% for validation. During training, we set some hyper-parameters for each model (GRU, LSTM and MLP). We tune the number of hidden layers (1, 2, 3), the number of input units for each layer (32, 64, 128), the learning rates (0.1, 0.01, 0.001), the activation functions for the hidden layers (Rectified Linear Unit (ReLU), Scaled Exponential Linear Unit (SELU), hyperbolic tangent), and the use or non-use of dropout (0, 1, 0.5). This tuning of hyperparameters is performed in the training phase, and the configuration with the best performance for each model is selected for re-training, where the training and validation sets (80% of all data) are merged. Finally, testing is performed on the remaining 20% of the data to select the best algorithm. It is worth noting that there are some aspects in the models that we do set as fixed values, such as the optimizer (Adam), the loss function (binary cross entropy), the metric (accuracy), the activation function in the last layer (softmax), the stack size (64), and the number of epochs (100).

According to the different combinations of these hyperparameters, the LSTM, GRU and MLP models are trained and their validation performance is compared to select the best representative of each model. It should be noted that this procedure is performed for each data partition. Table 3. shows the evaluation results of the four experiments and the three models (using the best hyperparameter setting for each), specifying the different partitions of the dataset LAGOS-AND. The metrics used are in percentages and include accuracy (Acc), precision (Pre), recognition (Rec), and F1 score (F1). The comparison table also uses the Name similarity results presented in [5], MAG author ID and the best model from LAGOS-AND as reference models.

- Name similarity: It is a basic method that uses only name differences to disambiguate authors [5].
- MAG Author ID: The ID system is disambiguated by the Microsoft Academic research team for its over 560 million authorships [5].
- LAGOS-AND: The best model identified in [5] is based on features and content characteristics (bf + cfnn)) that are used in the content similarity score by the neural network.

Table 3. Evaluation results of methods for AND. Based on [5].

Method-#Authors	Accuracy(%)	Precision(%)	Recall(%)	F1 score(%)
MLP-250	80.68	83.32	62.24	71.25
MLP-500	81.02	84.63	62.56	71.94
MLP-1000	81.25	85.05	64.12	73.12
MLP-1500	82.37	86.56	65.71	74.71
GRU-250	79.68	89.97	72.21	80.12
GRU-500	83.25	89.93	74.54	81.51
GRU-1000	85.54	91.72	77.56	84.05
GRU-1500	85.66	**92.86**	**82.21**	**87.21**
LSTM-250	83.69	87.54	78.02	82.51
LSTM-500	84.24	88.60	78.16	83.05
LSTM-1000	**86.35**	89.20	79.25	83.93
LSTM-1500	89.20	92.03	80.36	85.80
Name similarity	54.79	64.39	21.46	32.19
MAG Author ID	81.87	**98.49**	64.74	78.13
LAGOS-AND	**90.08**	93.23	**86.44**	**89.71**

As it is shown in Table 3., GRU wins on precision, recall, and F1 score, but LSTM has a higher value on average accuracy. GRU wins on precision, recall, and F1 score, but LSTM has a higher value on average accuracy. GRU wins slightly over LSTM, and MLP lags further behind, but LSTM's performance improves as the number of data increases. The best performers for MLP, GRU, and LSTM in the test phase are those with 1500 author names, as shown in Fig. 3. At first glance, we can see a relative proportionality function between the results: As the number of authors increases, the performance of the methods increases slightly.

4.2 Findings

Our method behaves similarly to the method in (LAGOS-AND), but we add semantic rules and queries in our framework to increase the semantic rigor of the weighted co-authorship network. Our contribution helps disambiguate author names in scientific papers, which was not considered in [5]. Although the comparison was performed on different parts of the LAGOS-AND dataset, the results presented are not significantly different from those in [5] (see Fig. 3). The LAGOS-AND dataset is more difficult to disambiguate than other datasets, not only because of its dimensionality, but also because it contains synonyms and homonyms of author names at the same time. In addition, the block technique used is more challenging. However, the results obtained with GRU and LSTM are very close to those presented in [5], with GRU achieving slightly better results.

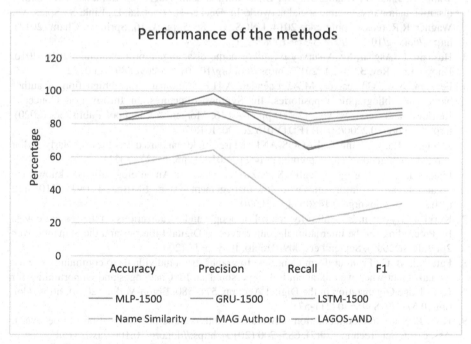

Fig. 3. Performance of the methods based on accuracy, precision, recall and F1.

5 Conclusions

In this work, we have presented a framework for author name disambiguation using a hybrid approach of an ontological model and a deep learning model. We have described the design of the ontology model, the process of automatic ontology generation, and the construction of a weighted co-author network through a set of semantic rules and queries. Then, we preprocessed the selected features during the attribute engineering process to measure the similarity indicator of each feature. The proposed framework was evaluated on four different data portions of the LAGOS-AND dataset using three different deep learning models, which show similar results to those presented in [5], with the GRU model performing slightly better in terms of precision, recall and F1 score. In future work, we will repeat the experiment with the entire LAGOS-AND dataset under better hardware conditions. We will also include other deep learning models in the comparison and apply the framework in other real-world scenarios.

References

1. Shoaib, M., Daud, A., Amjad, T.: Author name disambiguation in bibliographic databases: a survey. arXiv prepre arXiv:2004.06391, pp. 1–24 (2020)
2. Wang, P., Zhao, J., Huang, K., Xu, B.: A unified semi-supervised framework for author disambiguation in academic social network. In: Decker, H., Lhotská, L., Link, S., Spies, M., Wagner, R.R. (eds.) Conference 2014, LNCS, vol. 8645, pp. 1–16. Springer, Cham (2014). https://doi.org/10.1007/978-3-319-10085-2_1
3. Hussain, I., Asghar, S.: A survey of author name disambiguation techniques: 2010–2016. Knowl. Eng. Rev. **32**, 1–24 (2017). https://doi.org/10.1017/S0269888917000182
4. Ferreira, A.A., Gon,calves, M.A., Laender, A.H.F.: Automatic disambiguation of author names in bibliographic repositories. In: Synthesis Lectures on Information Concepts, Retrieval, and Services, vol. 12 (1), pp. 1—146. Morgan & Claypool Publishers (2020). https://doi.org/10.2200/S01011ED1V01Y202005ICR070
5. Zhang, L., Lu, W., Yang, J.: LAGOS-AND: a large, gold standard dataset for scholarly author name disambiguation. arXiv prepre arXiv:2104.01821, pp. 1—27 (2021)
6. Fiannaca, A., La Rosa, M., Gaglio, S., Rizzo, R., Urso, A.: An ontological-based knowledge organization for bioinformatics workflow management system. EMBnet. J. 18(B), 110—112 (2012). https://doi.org/10.14806/ej.18.B.570
7. Kurki, J., Hyvönen, E.: Authority control of people and organizations on the semantic web. In: Proceedings of the International Conferences on Digital Libraries and the Semantic Web 2009 (ICSD2009), September 2009, Trento, Italy, p. 15 (2009)
8. Pattuelli, M. C.: From uniform identifiers to graphs, from individuals to communities: what we talk about when we talk about linked person data. In: Challenges and Opportunities for Knowledge Organization in the Digital Age, pp. 571–580. Ergon-Verlag (2018). https://doi.org/10.5771/9783956504211-571
9. Kim, J.: Scale free collaboration networks: an author name disambiguation perspective. J. Assoc. Inf. Sci. Technol. **70**(7), 685–700 (2019). https://doi.org/10.1002/asi.24158
10. Thenmozhi, D., Aravindan, C.: Ontology-based Tamil-English cross-lingual information retrieval system. Sadhana **43**(157), 1–14 (2018). https://doi.org/10.1007/s12046-018-0942-7
11. Zaman, G., et al.: An ontological framework for information extraction from diverse scientific sources. IEEE Access **9**, 42111–42124 (2021). https://doi.org/10.1109/ACCESS.2021.306 3181
12. Hassell, J., Aleman-Meza, B., Arpinar, I.B.: Ontology-Driven Automatic Entity Disambiguation in Unstructured Text. In: Cruz, I., Decker, S., Allemang, D., Preist, C., Schwabe, D., Mika, P., Uschold, M., Aroyo, L.M. (eds.) ISWC 2006. LNCS, vol. 4273, pp. 44–57. Springer, Heidelberg (2006). https://doi.org/10.1007/11926078_4

13. Park, Y.-T., Kim, J.-M.: OnCU system: ontology-based category utility approach for author name disambiguation. In: 2nd International Conference on Ubiquitous Information Management and Communication Proceedings, pp. 63–68. New York, USA (2008). https://doi.org/10.1145/1352793.1352807

14. Lu, Z., Yan, Z., He, L.: OnPerDis: ontology-based personal name disambiguation on the web. In: 2013 IEEE/WIC/ACM International Joint Conferences on Web Intelligence (WI) and Intelligent Agent Technologies (IAT) Proceedings, vol. 1, pp. 185–192. IEEE (2013).https://doi.org/10.1109/WI-IAT.2013.28

15. Kurakawa, K., et al.: Researcher Name Resolver: identifier management system for Japanese researchers. Int. J. Digit. Libr. 14(1–2), 39–58 (2014). https://doi.org/10.1007/s00799-014-0109-z

16. Han, H., Yao, C., Fu, Y., Yu, Y., Zhang, Y., Xu, S.: Semantic fingerprints-based author name disambiguation in Chinese documents. Scientometrics 111(3), 1879–1896 (2017). https://doi.org/10.1007/s11192-017-2338-6

17. Bravo, M., Reyes-Ortiz, J.A., Cruz, I.: Researcher profile ontology for academic environment. Book Sect. Adv. Intell. Syst. Comput. 943, 799–817 (2019). https://doi.org/10.1007/978-3-030-17795-960

18. Färber, M., Ao, L.: The microsoft academic knowledge graph enhanced: author name disambiguation, publication classification, and embeddings. Quantitative Sci. Stud. 3(1), 51–98 (2022). https://doi.org/10.1162/qss_a_00183

19. Santini, C., Gesese, G.A., Peroni, S., Gangemi, A., Sack, H., Alam, M.: A knowledge graph embeddings based approach for author name disambiguation using literals. Scientometrics 127(8), 4887–4912 (2022). https://doi.org/10.1007/s11192-022-04426-2

20. Gnoyke, P., Matta, K.: Author name disambiguation by clustering based on deep learned pairwise similarities, pp. 0—12, May (2020)

21. Firdaus, F., et al.: Author identification in bibliographic data using deep neural networks. TELKOMNIKA (Telecommun. Comput. Electron. Control) 19(3), pp. 911–919 (2021). https://doi.org/10.12928/telkomnika.v19i3.18877

22. Ahmedi, L., Abazi-Bexheti, L., Kadriu, A.: A uniform semantic web framework for co-authorship networks. In: IEEE Ninth International Conference on Dependable, Autonomic and Secure Computing Proceedings, no. 2, pp. 958–965 (2011). https://doi.org/10.1109/DASC.2011.159

23. Gómez-Pérez, A., Suárez-Figueroa, M.C.: NeOn methodology for building ontology networks: a scenario-based methodology (2009)

24. Suárez-Figueroa, M.C., Gómez-Pérez, A., Mariano, F.-L.: The NeOn methodology framework: a scenario-based methodology for ontology development. Appl. Ontol. 10(2), 107–145 (2015). https://doi.org/10.3233/AO-150145

25. Leiva-Mederos, A., García-Duarte, D., Gálvez-Lio, D., Hidalgo-Delgado, Y., Senso-Ruíz, J.S: An ontological model for the failure detection in power electric systems. In: Iberoamerican Knowledge Graphs and Semantic Web Conference Proceedings, pp. 130–146 (2020). https://doi.org/10.1007/978-3-030-65384-2

26. Díaz-de-la-Paz, L., Riestra-Collado, F. N., García-Mendoza, J. L., GonzálezGonzalez, L. M., Leiva-Mederos, A. A., Taboada-Crispi, A.: Weights estimation in the completeness measurement of bibliographic metadata. Comput. Sist. 25(1), 117–128 (2021). https://doi.org/10.13053/cys-25-1-3355

27. Le, Q. V., Mikolov, T.: Distributed representations of sentences and documents. In: International Conference on Machine Learning Proceedings, arXiv Prepr. arXiv:1405.4053, vol. 32 (2), pp. 1188–1196 (2014)

On Contrasting YAGO with GPT-J:
An Experiment for Person-Related Attributes

David Martin-Moncunill[1], Miguel-Angel Sicilia[2(✉)], Lino González[1],
and Diego Rodríguez[1]

[1] Computing and Artificial Intelligence Lab (CAILab), School of Science and Technology,
Camilo José Cela University, C/Castillo de Alarcón, 49, Urb. Villafranca del Castillo 28692
Madrid, Spain
{david.martin,lgonzalezg,druben.rodriguez}@ucjc.edu
[2] Computer Science Department, University of Alcalá, Polytechnic Building. Ctra. Barcelona
km. 33.6, 28871 Alcalá de Henares (Madrid), Spain
msicilia@uah.es

Abstract. Language models (LMs) trained or large text corpora have demonstrated their superior performance in different language related tasks in the last years. These models automatically implicitly incorporate factual knowledge that can be used to complement existing Knowledge Graphs (KGs) that in most cases are structured from human curated databases. Here we report an experiment that attempts to gain insights about the extent to which LMs can generate factual information as that present in KGs. Concretely, we have tested such process using the English Wikipedia subset of YAGO and the GPT-J model for attributes related to individuals. Results show that the generation of correct factual information depends on the generation parameters of the model and are unevenly balanced across diverse individuals. Further, the LM can be used to populate further factual information, but it requires intermediate parsing to correctly map to KG attributes.

Keywords: YAGO · Ontology · Language models · GPT-J · Factual information

1 Introduction

Knowledge Graphs (KGs) are structured knowledge representations that attempt to encode relations among entities and associated information, aimed at supporting diverse tasks including reasoning, search or recommendation. Widely used KGs as Wikidata [10] are constructed from facts curated by humans, following different approaches and resulting in different levels of quality along a number of dimensions [5].

Recent progress in language models (LMs), such as BERT and GPT-2/3, has led to improved results even outperforming humans in a wide range of tasks, e.g., sentence classification or question answering. These LMs automatically acquire factual knowledge from large-scale corpora of text and build statistical models, in contrast with the approaches of KGs that rely on structuring data in some form of formal or semi-formal language, usually followed by curation workflows. Of special practical impact is the

B. Villazón-Terrazas et al. (Eds.): KGSWC 2022, CCIS 1686, pp. 234–245, 2022.
https://doi.org/10.1007/978-3-031-21422-6_17

availability of pre-trained LMs (PLMs) [9] that are used for numerous NLP tasks, used by researchers via fine-tuning for the task, directly using the PLMs, or reformulating the task as a text generation problem with application of PLMs to solve it accordingly.

Since LMs are apparently able to acquire factual knowledge without previous curation similar to that of KGs, the question of the extent to which they end up with representations that are equivalent, more comprehensive or less accurate than traditional KG approaches arises. In this direction, the ability of LMs in generating factual sentences has been subject to combination with KGs [8] which suggests KGs and LMs can effectively complement each other in some tasks. This is an important question to complement approaches that attempt to construct KGs from pre-trained language models without human supervision [15, 11] or to hybrid KG/LM approaches to tasks as question answering [16] or query answering [7]. However, it is still important to understand the current effectiveness and performance of existing, widely used PLMs to output factual information when prompted.

In this paper, we discuss the results of an experiment in contrasting a traditional KG with a large scale LM in the basic task of generating factual sentences about individuals. These are preliminary results to get insights towards a more comprehensive understanding of the extent to which structured KGs can be substituted by LMs or the extent to which they have different properties as *representations of factual information*. Concretely, we have extracted a subset of sentences from the YAGO ontology[1] (which is built from Wikidata) regarding persons, and used an openly available large PLM to generate text from prompts. This gives a lower bound of the capabilities of the PLM in capturing and subsequently generating text containing facts. The observation that various facts are produced for a single template/predicate suggests that a strategy that generates longer sequences of text and then processes those further searching different patterns may be more effective.

The rest of this paper is structured as follows. Section 2 briefly discusses background information on KGs and LMs relevant to the rest of the sections. Then, in Sect. 3 the methods, software, models and data used for the experiment is detailed. Section 4 provides the discussion on the evaluation of the experiment. Finally, conclusions and outlook are provided in Sect. 5.

2 Background

2.1 Knowledge Graphs and Text

Huaman and Fensel [5] list 20 quality dimensions for KGs that may have different importance for different scenarios. Here we are mostly concerned with openly accessible, and pragmatically considers their accuracy by recurring to collaborative curation as done in Wikipedia. Since we are focusing on factual data around persons, the degree of interpretation is limited, and this has guided our selection of KG for experimentation.

Wikidata has evolved into one of the world's largest publicly available KG. It is a community effort where anybody can contribute facts. YAGO is thus a simplified, cleaned, and "reasonable" version of Wikidata [12] that uses some filtering mechanisms.

---

[1] https://yago-knowledge.org/.

It contains more than 50 million entities and 2 billion facts. YAGO is built automatically which makes studies using it more reproducible or comparable, and provides a degree of homogeneity to the knowledge extracted from its sources, Wikidata and Schema.org.

While KGs are structured representations of data, aimed at queries with logics-based or general purpose querying languages, these can be transformed to text sentences to be used for several tasks. Agarwal et al. [1] converted an entire KG (Wikidata) to natural language text and used the results to improve other language modeling tasks.

2.2 Pre-trained Language Models and KGs

Pre-trained language models can be categorized in three classes [9]: autoregressive language models (e.g. GPT), masked language models (e.g. BERT), and encoder-decoder models (e.g. BART, T5). Here we have focused on the first category, since they are trained to predict the next word x_i given all previous words $x_1, x_2,... x_{i-1}$.

Petroni et al. [13] use the cloze-style task for relation/fact probing in PLMs, which is similar to the methods used here, further described in the next section. Jiang et al. [6] apply a similar method to a variety of languages other than English, demonstrating the difficulty of this task, and that knowledge contained in LMs varies across languages. However, here we approach the task of extracting factual information in the context of populating KGs, so that no cloze style tasks are prepared and instead we rely on templates or prompts related to certain types of facts, and inspect the initial few words obtained from the PLM.

PLMs have been used to extract KGs. Hao et al. [4] use BERT-like models to extract knowledge of arbitrary new relation types and entities, without being restricted by pre-existing knowledge or corpora, evaluating the outcomes using human effort. Wang et al. [15] proceed similarly but compare the quality with existing KGs, which is an approach closer to the experiment reported here.

3 Methods

3.1 Extracting Sentences from YAGO

We use here the KG selected as a gold-standard relational data that enables queries such as (Dante, born-in, X). These factual relational queries have a representation specific to YAGO that required pre-processing.

YAGO 4[2] is openly available as a set of.ntx files that contain RDF* (a.k.a. RDF star) N-Triples syntax. We have used the N3.js[3] library to extract the facts in the current study. The library reads the data and makes available a stream of quads that can be processed, which can be used to filter the facts that we have selected. That form of processing was found to be more practical than setting up a triplestore first with all the contents and then using SPARQL for the queries.

YAGO is distributed in three variants, full (all data in Wikidata), Wikipedia (only those items that have a Wikipedia article) and English Wikipedia (subset of the previous

---

[2] https://yago-knowledge.org/.
[3] https://github.com/rdfjs/N3.js/.

with a Wikipedia article in English). Since we are not dealing with the multilingual impact on the performance of LMs, we use the last, smaller version only in English.

Concretely, the extraction was made in two steps, described in what follows.

Extracting a List of Persons from YAGO. This was done via processing the yago-wd-full-types.nt file (that contains all rdf:type relations), and collecting the subjects of the triples for which the object was Human[4]. This resulted in 1.691.133 identifiers.

It should be noted that the item Human in Wikidata is defined as "common name of Homo sapiens, unique extant species of the genus Homo". While this restricts the domain of the task to non fictional humans, it is still shaped by the inclusion criteria in Wikipedia, concretely to the concept of *notability* used by editors to decide whether a given topic warrants its own article. Notability is related to verifiability (existence of reliable, independent sources about the topic), which provides a degree of quality control. However, the results of studies like the one presented here are subject to biases in Wikipedia contents that have been subject to study in the literature for elements as gender [14] or culture [2]. We do not deal with these potential biases in the filtering of persons, since we operate in this study under the assumption that those biases will also be present in LMs that have overlapping or similar sources.

Removing Ambiguous Identifiers. The list of person ids from the previous step was used in the selection of samples described below. However, many of the names were ambiguous, e.g. refer to different persons. This is encoded in Wiki-data with different item identifiers, but in YAGO this can be processed since the URIs retain the disambiguation pattern in Wikipedia. For example, Frank Barnes was found in 5 different ids with the pattern Frank Barnes (profession) were the profession in that case was one of "politician", "left handed pitcher", "actor", "right handed pitcher" and "gunsmith". We decided to remove all identifiers including a disambiguation pattern. The rationale is that our study provides no or little context to LMs to be able to correctly disambiguate, and these ambiguous names will introduce an error that is subject to the lack of context and not the capturing of factual information, and we did not consider disambiguation as part of the task evaluated. The resulting list of identifiers contained 1.492.517 items, about 88% of the initial list.

Extracting Facts About the Identified Items. The last step involved the extraction of triples related to the identifiers selected in the previous step. This involved processing the complete and large yago-wd-facts.nt file containing all the facts, and selecting those having as subject one of the human identifiers. This resulted in 9.734.498 triples.

3.2 Generation of Questions and Factual Sentences

The analysis of the actual predicates found in the facts dataset revealed 29 predicates only, and the following were discarded as having low occurrence rates (lower than 1000): faxNumber, callSign, weight, netWorth, owns, telephone, height, worksFor, gender and affiliation. Table 1 describes the kinds of predicates not included in relation with their suitability for the task.

---

[4] http://yago-knowledge.org/resource/Human.

Table 1. Predicates excluded from the experiment

Group	Predicates	Inclusion
Redundant	givenName, familyName, nationality	Already appears in the prompt to the LM or partially redundant with other predicates
Minor relevance	honorificPrefix	Considered unimportant as factual information
Non textual	url, image	Not included as not appropriate for the text generation task

The distribution of predicates across subjects is shown in Fig. 1. In both cases, it is skewed to the right, and in the case of retained predicates, it is important to notice that most subjects have 5 or less of these predicates attached. The probing method relies on manually crafted templates. As recognized by Petroni et al. [13] it can be expected that the choice of templates has an impact on the results, i.e. for some relations we find both worse and better ways to query for the same information by using an alternate template. This can be interpreted that as that we are measuring a lower bound for what language models know. In any case, the type of factual information used corresponds to simple templates that are provided in Table 2. Note that in the actual examples provided in some occasions as in the second, the template is producing more than one fact (in that case, both place and date of death). Also, in some cases, the results did not produce a complete sentence as in the example for birth date. Another important observation is that matches appear from prompts that were not intended for that particular fact type. An example in the Table is that of the match of parent that comes from the "daughter" form that was not used as template.

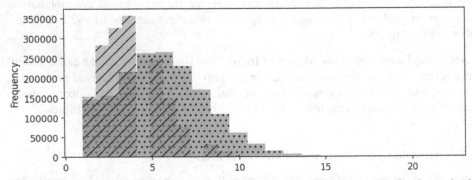

Fig. 1. Distribution of the predicates across subjects: overall (points), and only for the retained predicates (lines)

Predicting the token after a sequence of tokens is known as causal language modeling, and that is the procedure use to generate factual sentences.

3.3 Probing Procedure

The GPT-J model[5] is a GPT-2-like causal language model trained on the Pile dataset [3]. It is important to remark that the Pile includes in its sources the English Wikipedia, so that it has some indirect relation with YAGO through it.

The probing consisted on using all the templates with all the person identifiers in a given sample and recording the responses of the GPT-J model with some given configuration. Cloze templates have been used to probe PLMs starting from the work of Petroni et al. [13]. In that approach, they evaluate LMs based on how highly it ranks the ground truth token against every other word in a fixed candidate vocabulary. However, we are here interested in the broader (arguably more difficult) task of suggesting the next term, and then comparing it to the standard considered, in our case, facts directly extracted from the KG for very particular types of factual sentences.

We used the default configuration of GPT-J available in the transformers Python library, testing different values for the temperature parameter. Since we are interested in short, factual information, we set the parameter max length to produce just a few words typical of completing a sentence. That may have an influence in the experimental results since in some cases the facts are produced in subsequent text generated, but we took this decision since we are evaluating the ability of PLMs to produce factual information first.

Table 2. YAGO relations, templates used with GPT-J and example outputs produced for each

Relation	Template	Example
birthPlace	—was born in—	Luna Bijl was born in in the Netherlands on 20 January 1967
deathPlace	—died in—	Sergey Reformatsky died in Moscow in 1981
birthDate	—was born in the date of—	Georg von Peuerbach was born in 1816 in—
deathDate	—died in the date of—	Vitale da Bologna died in 1643 in Bologna
knowsLanguage	—spoke—	Anke Brockmann spoke in the German language
award	—was awarded—	Enzo Calzaghe was the father of Stefan Calzaghe, a former world champion boxer
parent	—father/mother was—	Manuela Zu¨rcher was born in Switzerland. She is the daughter of the Swiss composer and organist Hans Zu¨rcher

(*continued*)

---

[5] https://huggingface.co/docs/transformers/modeldoc/gptj.

Table 2. (*continued*)

Relation	Template	Example
spouse	—married with—	Paul Alfons von Metternich-Winneburg was married to Princess Marie of Saxe-Coburg and Gotha
children	—was the father/mother *same as for parent* of—	Same as for parent
homeLocation	—lived at—	Martine Chartrand was born in Montr´eal, Qu´ebec on September 17, 1984 and grew up there and in Ontario
alumniOf	—studied at—	Jesu´s Olalde was born in Mexico City, Mexico in 1973. He studied at the Uni versidad Iberoamericana,
memberOf	—was member of—	Baldur von Schirach father was a member of the Nazi Party and the leader of the Hitler Youth
hasOccupation	—worked as—	Jason Klein was born in New York City, New York, United States. He is an American actor, writer, producer, director

3.4 Contrasting the Outputs Generated

One problem with this approach is that the outputs of the PLM can be accurate but do not literally match due to differences in spelling. We have used an approximate solution to this matching by using the Python thefuzz library[6]. Concretely, that library provides a function partial ratio that ows us to perform substring matching. It works by taking the shortest string and matching it with all substrings that are of the same length. This allows for an approximation of full matches and a metric for partial matches, accounting for small spelling differences. An alternative approach may be that of using a word embedding to compute similarities, but given the relatively short and concrete nature of the factual information produced, we have discarded it in this experiment.

In addition, we have approached the probing by random sampling of the KG. This can be considered to be closer to a situation of using the PLM "in the wild". It is to be expected that the performance of the PLM widely varies with the "popularity" or degree of influence of the triggering instance. For example, in our case, popular historical personalities or maybe recent popular characters may have better responses than other less known, simply due to the fact that PLMs are trained on text sources that are not required to be balanced in coverage.

---

[6] https://github.com/seatgeek/thefuzz.

4 Results and Discussion

Since the generation of the outputs requires processing a large amount of sentences, we approached the testing by selecting random samples of 1000 persons, and repeated the sampling. Since no relevant differences were found across samples, here we report the results as a single experiment.

4.1 Experiment Results

The experiment consisted in taking random samples of the sentences extracted from YAGO and probing them according to the methods described in Sect. 3. Concretely, samples of size 1000 persons were taken.

The temperature parameter in GPT-J controls how much randomness is in the output. Overall, the lower the temperature, the more likely it will choose words with a higher probability of occurrence, and will produce the same results more consistently. However, the scores for different levels of partial match shown in Table 3 do not show a clear pattern or correlation. Further, the Table shows that the amount of sentences matched is relatively low, with around 17% in the higher cases. Since we found that the text generated for a particular template and YAGO predicated (see examples in Table 2) often produces facts for others, in the table X/Y pairs of percentages of matches are shown, with X matching only with the intended predicate and Y matching all the predicates and selecting the maximum score. Note that Y is possibly overcounting samples, since in many cases it counts the same match for more than one of the templates. The higher values of the Y figures account for typical patterns in which facts are generated together, for example, the case of "was born in [date] in [place]", so that these higher numbers in this case are arguably of relative importance.

It should be noted however that lower match thresholds as M 50 would likely include a significant amount of false positive matches that may not be recognized for example when using a NER trained model.

Table 3. Partial matches for different temperature configurations. Figures X/Y in the cells account for matching only the intended predicate (X) or matching all the predicates in the selected text (Y)

Temperature	M100	M80	M60	M50
0.1	.0194/.0350	.0266/.0586	.0717/.1725	.1447/.3443
0.25	.0198/.0379	.0253/.0586	.0620/.1645	.1337/.3346
0.5	.0202/.0350	.02666/.0569	.0667/.1632	.1413/3443
0.75	.0143/.0342	.0202/.0552	.0612/.1628	.1354/.3388

An important observation is that exact partial matches only appear systematically in the different samples for some predicates. For example, with temperature 0.01 only birthPlace, deathPlace and knowsLanguage, with a larger proportion for the first. If we relax the matching threshold for example to M60, some other appear as children, parent and spouse. This may be related to the effectiveness of the templates for different predicates, so that we have looked in detail at the matches independent of the prompt, to assess the overall effectiveness for all the templates together (Table 4).

Table 4. Partial matches for temperature 0.5. For the different predicates

Predicate M60	M100 M50	M80
birthPlace	.0206	.0244
birthDate	.0050	.0092
knowsLanguage	.0016	.0042
deathPlace	.0046	.0063
deathDate	.0029	.0059
spouse	–	.0016
parent	–	.0029
children	–	.00211

These results may be limited by the parameter max length that was set to 30, and limits the amount of text produced by GPT-J. Setting higher values of this parameter may give higher levels of partial matches, among the "less likely" words that may appear in longer texts.

Another interesting observation is the distribution of matches per YAGO entity. Figure 2 shows the distribution of the sum of partial ratings per entity (person) in log scale, for temperature 0.5. In addition to a significant amount of zero matches, it shows that most entities have the equivalent of one or two matches, with only a few about the equivalent of three. This is in contrast with the 13 potential relations to be identified.

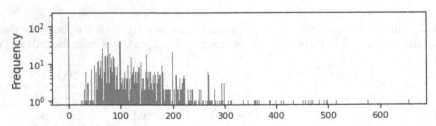

Fig. 2. Distribution of the sum of patial match scores across entities (log scale)

4.2 Discussion and Observations

The results presented point out to an unbalanced capability of eliciting facts of different templates, and the potential of eliciting factual information from a template or prompt that is more generic or usual and results in GPT-J text that contains more facts intermingled. The strategy to design templates specific to certain predicates in the KG is thus questionable, and NLP tasks oriented to extract facts from text in general may represent a more effective alternative. This appears to be consistent with the fact that PLMs are trained in large corpora and factual information often comes from descriptions of entities as persons that are found as is in the corpora (especially this is so in our case since Wikipedia is one of the sources for GPT-J [3]). However, the results presented here raise the question of the effectiveness of PLMs to aid in population of KGs given that there were no matches for many of the entities (persons), and a limited number of entities for which the matches were 4 or more.

In addition to the overall quantitative results presented, we have manually inspected the responses generated and identified a number of limitations and problems of the methods used in the experiment.

Issues Related to Levels of Aggregation. Some of the factual terms probed entail some form of inclusion or part-of relation. For example, in some cases the match of a date could be specified as a year, month and year o concrete day. The same occurs with places, which can be expressed as countries, cities or other geopolitical aggregates. This could be mitigated by matching with knowledge representations associated to each of the domains involved.

Issues Related to Entities in Time. When speaking of places, countries or cities may be referred to contemporary geopolitics or in the case of historical personalities, to names of places that have changed over the course of history. This aspect has some interaction with the previous one, and the use of supplementary representations for disambiguation would require diachronic models of places.

5 Conclusions and Outlook

KGs provide human-curated resources including factual information that require significant effort to maintain and complete. In contrast, pre-trained LMs leverage recent advances in machine learning to provide large models encoding factual information acquired automatically. This points to the possibility of using PLMs to populate KGs. However, this requires an understanding of how prompting tasks typically used with PLMs for different purposes as auto-completing text should be adapted to that particular objective. The experiments reported here are early work aimed at understanding the potential problems of that approach and getting evidence on lower bounds for the capabilities of PLMs for the task. We have reported our first experiment using YAGO, a mature and curated ontology built from Wikidata and GPT-J, one of the latest PLMs available openly. The experiment has focused on facts about persons, which appear as a task adequate for eliciting factual information. The setting takes purposefully short templates since in a typical KG semi-automated population setting, further additional text that could be used as context would likely not be available or be scarce. The results can be considered a lower threshold of the capabilities of the PLM to aid in populating KGs. Overall, reliable matches are around 2% and relaxing the potential of the quality of the matches, it can go up to 17%, which points to a significant but not high capability. It should be noted that there are many limitations that may make these initial results, including the fact that GPT-J is trained on a set of resources that include Wikidata, the source of YAGO. Also, the templates selected, the parameters chosen and the use of a concrete partial matcher based on edit distance may have an influence that requires further systematic experimentation.

Future work should extend the experiments both in broadening the samples and the KG/PLM pairs considered, but also in the aspects identified as relevant for the task, including the templates used, the methods of matching generated text to entities and parameters of the generation. In a different direction, studies should be adapted to different contexts of use for the task of curating KGs, in which the assumption made here that there is no context should be relaxed.

References

1. Agarwal, O., Ge, H., Shakeri, S., Al-Rfou, R.: Knowledge graph based synthetic corpus generation for knowledge-enhanced language model pre-training (2020). arXiv preprint arXiv: 2010.12688
2. Callahan, E.S., Herring, S.C.: Cultural bias in Wikipedia content on famous persons. J. Am. Soc. Inf. Sci. Technol. **62**(10), 1899–1915 (2011)
3. Gao, L., et al.: The pile: An 800GB dataset of diverse text for language modeling (2020). arXiv preprint arXiv:2101.00027
4. Hao, S., Tan, B., Tang, K., Zhang, H., Xing, E.P., Hu, Z.: BertNet: Harvesting Knowledge Graphs from Pretrained Language Models (2022). arXiv preprint arXiv:2206.14268
5. Huaman, E., Fensel, D.: Knowledge graph curation: a practical frame-work. In: The 10th International Joint Conference on Knowledge Graphs, pp. 166–171 (2021)
6. Jiang, Z., Anastasopoulos, A., Araki, J., Ding, H., Neubig, G.: X-FACTR: Multilingual factual knowledge retrieval from pretrained language models (2020). arXiv preprint arXiv:2010. 06189
7. Kalo, J.-C., Fichtel, L., Ehler, P., Balke, W.-T.: KnowlyBERT - hybrid query answering over language models and knowledge graphs. In: Pan, J.Z., et al. (eds.) The Semantic Web – ISWC 2020. LNCS, vol. 12506, pp. 294–310. Springer, Cham (2020). https://doi.org/10.1007/978-3-030-62419-4_17
8. Logan IV, R.L., Liu, N.F., Peters, M.E., Gardner, M., Singh, S.: Barack's wife hillary: Using knowledge-graphs for fact-aware language modeling (2019). arXiv preprint arXiv: 1906.07241
9. Min, B., et al.: Recent advances in natural language processing via large pre-trained language models: a survey (2021). arXiv preprint arXiv:2111.01243
10. Mora-Cantallops, M., Sánchez-Alonso, S., García-Barriocanal, E.: A systematic literature review on Wikidata. Data Technol. Appl. **53**(3), 250–268 (2019)
11. Omeliyanenko, J., Zehe, A., Hettinger, L., Hotho, A.: LM4KG: improving common sense knowledge graphs with language models. In: Pan, J.Z., et al. (eds.) The Semantic Web – ISWC 2020. LNCS, vol. 12506, pp. 456–473. Springer, Cham (2020). https://doi.org/10. 1007/978-3-030-62419-4_26
12. Tanon, T.P., Weikum, G., Suchanek, F.: YAGO 4: a reasonable knowledge base. In: Harth, A., et al. (eds.) The Semantic Web. LNCS, vol. 12123, pp. 583–596. Springer, Cham (2020). https://doi.org/10.1007/978-3-030-49461-2_34

13. Petroni, F., et al.: Language models as knowledge bases? (2019). arXiv preprint arXiv:1909.01066
14. Tripodi, F.: Ms. Categorized: Gender, notability, and inequality on Wikipedia. New Media Soc. 14614448211023772 (2021)
15. Wang, C., Liu, X., Song, D.: Language models are open knowledge graphs (2020). arXiv preprint arXiv:2010.11967
16. Yasunaga, M., Ren, H., Bosselut, A., Liang, P., Leskovec, J.: QA-GNN: reasoning with language models and knowledge graphs for question answering (2021). arXiv preprint arXiv:2104.06378

Easy and Complex: New Perspectives for Metadata Modeling Using RDF-Star and Named Graphs

Florian Rupp[✉], Benjamin Schnabel, and Kai Eckert

Stuttgart Media University, Stuttgart, Germany
rupp@hdm-stuttgart.de
http://wiss.iuk.hdm-stuttgart.de/

Abstract. The Resource Description Framework is well-established as a lingua franca for data modeling and is designed to integrate heterogeneous data at instance and schema level using statements. While RDF is conceptually simple, data models nevertheless get complex, when complex data needs to be represented. Additional levels of indirection with intermediate resources instead of simple properties lead to higher barriers for prospective users of the data. Based on three patterns, we argue that shifting information to a meta-level can not only be used to (1) provide provenance information, but can also help to (2) maintain backwards compatibility for existing models, and to (3) reduce the complexity of a data model. There are, however, multiple ways in RDF to use a meta-level, i.e., to provide additional statements about statements. With Named Graphs, there exists a well-established mechanism to describe groups of statements. Since its inception, however, it has been hard to make statements about single statements. With the introduction of RDF-star, a new way to provide data about single statements is now available. We show that the combination of RDF-star and Named Graphs is a viable solution to express data on a meta-level and propose that this meta-level should be used as first class citizen in data modeling.

Keywords: Data modeling · RDF-star · Named graphs · Meta-level

1 Introduction

Data can come in many different forms and there are nearly infinite ways to model data describing the same information. Depending on the use case, different forms might be preferable: a simple view on data is the easiest to work with and facilitates data reuse. On the other hand, more complex applications need advanced data models providing the full complexity of the data with all available details. The Resource Description Framework (RDF) [1] provides a common basis for the publication of data in form of statements on the Web but makes no assumptions how these statements actually look like and how entities and their relations are used to represent the desired information. Thus, various modeling approaches can represent the same information. The Semantic Web aims at

B. Villazón-Terrazas et al. (Eds.): KGSWC 2022, CCIS 1686, pp. 246–262, 2022.
https://doi.org/10.1007/978-3-031-21422-6_18

providing interlinked data based on the "follow your nose" principle [4], i.e.: users and applications can get data about a resource by dereferencing its URI.

There are two possible solutions, when data is to be provided both in simple and complex[1] forms for different applications:

1. Different data representations can be provided; in this case, a mechanism is required that allows the application to request one of these representations (see Sect. 2).
2. The data is provided in a simple data model and additional information is provided on a meta-level in the form of statements that provide further explanations.

In this article, we discuss the latter approach. We argue that there are benefits of using such a meta-level approach versus the provision of different data formats: there is only one representation required, it is straight-forward to extend existing data representations while maintaining backwards compatibility and it fits the "follow your nose" principle of the Semantic Web where applications simply can ask if there is further information about a statement.

A meta-level can be created by providing a further statement describing a single or multiple statements. To provide statements about statements in RDF, a mechanism is needed to identify statements. The identification can happen in two ways: either a single statement is identified (statement-level) or a group of statements – usually called a graph – is identified and meta-statements are used to further describe all statements in this graph (graph-level).

For statement-level data there always has been the optional RDF reification, for graph-level statements, Named Graphs have been introduced. Particularly RDF reification, however, has two major shortcomings: it is inefficient for exchanging RDF data and writing queries to access statement-level metadata is cumbersome [7].

Now there is a new way to provide statement-level metadata within RDF itself with the recent specification of RDF-star[2]. With increasing support in RDF databases and tools and the specification of various serializations including TriG-star[3], it provides all that is needed to use both statement-level metadata with RDF-star and graph-level metadata with Named Graphs.

---

[1] While it depends on the perspective, we consider data complex that uses several layers of properties and dependent classes to describe a resource.
[2] https://www.w3.org/2021/12/rdf-star.html.
[3] Throughout this paper, we use TriG-star syntax in the examples. For reference this is a list of the implied namespaces, ex is the default or empty namespace: rdf: http://www.w3.org/1999/02/22-rdf-syntax-ns#; owl: http://www.w3.org/2002/07/owl#; ex: http://example.org/; dcat: http://www.w3.org/ns/dcat#; dct: http://purl.org/dc/terms/; foaf: http://xmlns.com/foaf/0.1/; prov: http://www.w3.org/ns/prov#; dbp: http://dbpedia.org/property/; dbo: http://dbpedia.org/ontology/; gn: http://www.geonames.org/ontology#; gndo: http://d-nb.info/standards/elementset/gnd#.

In this article, we explore the possibilities and limitations of RDF-star and Named Graphs with respect to the creation of data models which are simple and easy to use, but nevertheless contain the full complexity of the data on a meta-level so that it can be used when needed. We start with an overview on related work focusing on providing data to applications with different requirements. In Sect. 3, we identify three abstract patterns that show how common use cases can be modeled with metadata. In Sect. 4, further examples about querying and working with the meta-level are given. The paper concludes with a discussion of the advantages and disadvantages of data modeling with a meta-level as first-class citizen.

2   Related Work

Different applications have different requirements towards data. Consider a personal blog that wants to provide a thumbnail view of a book versus a library that needs data about a book to include in its catalog. It is common practice in many open data portals that data can be accessed in different formats, e.g. Dublin Core for simple data vs. MARC for all the details. In the Semantic Web, however, this poses a problem. If a URI for a book is resolved automatically by an application, what data should be provided? In our paper this is not directly solved, but arguably our meta-level approach allows applications to decide if they want to access the additional information. For each resource there can only be one data description. The provider of the data needs to decide, how this data is modeled and which vocabularies are used. To address this shortcoming, the Dublin Core community developed the concept of *Application Profiles* [8] to describe the data model for a specific application. Based on the notion of Application Profiles, there is a recent RFC draft to use HTTP for the negotiation of different data representations on content level, similar to the well-known content negotiation for different RDF serializations [18]. Using such mechanisms, however, requires proper support in applications and it is not yet clear if this will actually be used in practice.

Another aspect is the provision of metadata to describe the actual data, mostly to provide information on the *provenance* of the data [5,6,13,14,16,17]. There are generally two aspects to consider: (1) how can the subject, i.e., the data, be identified to allow statements about it, and (2) how can the provenance of data be described. As we do not focus specifically on the description of provenance, we are mainly concerned with the former.

Besides RDF-star [12] and Named Graphs [3], other approaches have been proposed. The oldest way to do this is *RDF reification*, which is available since the very first versions of RDF, cf. [15]. However, making statements about statements using reification is hard, mainly due to its verbose syntax and the fact that every statement needs to be reified, even when many statements share the same meta information. This might be a reason why reification is rarely used in real world applications[4]

Singleton statements [11] use unique identifiers which are added to the predicate statement. Following this method, the meta-level can be expressed by using the predicate in combination with the assigned identifier. Singleton properties lead to compact representations of the meta-level, but obviously affect the querying of the data. There is support for singleton properties in some triple stores, which makes the application more feasible. Even then it is not recommended, for instance see the documentation of GraphDB[5]).

These approaches all address statement-level metadata. There are many cases where an identification of single statements is not needed and even not helpful. Probably the best example is provenance information. Usually, many statements share the same provenance when they are created together by a single process.

For the separation of different descriptions of the same resource, as an alternative, *Proxy entities* [9] or resource maps in OAI-ORE [10] have been proposed, that use proxies as distinct placeholder entities that all represent the same actual entity but with different URIs to be able to distinguish statements about the entity from different sources. This raises the complexity of the data and requires applications to correctly interpret the proxy entities.

A problem with Named Graphs is that organizing structures, for example nested graphs or subgraphs can only be represented via additional data describing *graph relationships*. This can be used to identify provenance information in hierarchical graphs [5]. A proposal for the structured use of several named graphs are *nano publications* [2] that also explicitly support provenance.

3   Meta Modeling Patterns

Based on our experience with data applications, we can identify three abstract modeling patterns that illustrate the benefits of a meta-level. For each pattern, we formulate a problem statement, the solving approach, an example and a conclusion. The three patterns are:

---

[4] To get evidence of reification usage in real world application we checked a given list of SPARQL endpoints. In favor we use the Wikidata SPARQL endpoint list at https://www.wikidata.org/wiki/Wikidata:Lists/SPARQL_endpoints. This examination was done automatically with a suitable SPARQL query. The list, however, contains 139 entries and can be, according to its small size, considered as a sample only. Merely 78 endpoints were reachable via GET or POST requests (on July 20th 2022). 7 out of 78 remaining endpoints were not suitable for this study, being no actual SPARQL endpoint or the domain has been sold. Out of 72 endpoints 5 are using reification. At just 7,7% that's a small percentage. This supports the thesis that reification is rarely used.

[5] https://graphdb.ontotext.com/documentation/free/devhub/rdf-sparql-star.html, Singleton Properties.

- Pattern 1: RDF provenance modeling in the meta-level.
- Pattern 2: Extending an existing data model.
- Pattern 3: Shifting proxy entities to the meta-level.

We will not distinguish statement-based and graph-based metadata. With RDF-star and Named Graphs, there are now viable solutions for both levels and it is best decided by the data modeler or the data provider which one is more suitable. This of course requires that consuming applications understand both approaches, i.e., TriG-star needs to be supported as serialization.

3.1   Pattern 1: RDF Provenance Modeling in the Meta-level

Problem Statement: The provision of provenance data *in RDF* is straight forward, for example using PROV or Dublin Core [6]. Nevertheless provenance data is not often provided on a meta-level *for RDF* data. Instead, entities are introduced that either represent the data (e.g. a `dcat:CatalogRecord`) or that act as a placeholder for the actual entity for the description from one specific source (e.g. a `ore:Proxy`). Both approaches are problematic when data from different sources is to be merged due to both approaches must now fit into the same data model.

Even unfavorable is the mixing of data about a resource and about its description. Let's say, we want to model the provenance of a statement that provides a title for an entity `<E>`. In the following example, the information is added directly to the instance of `<E>`.

```
<E> rdf:type Entity_E ;
    dct:title "title" ;
    prov:wasDerivedFrom <source> .
```

But what is derived here? Is it the title of the resource or the resource itself? As all statements describe their subject, it is clearly the resource `<E>` and not the title or other descriptive data about the resource.

Solving Approach: Provenance is a prime example where Named Graphs should be used as usually many statements share the same provenance:

```
<entity> ex:ID <ID> .

:data {
    <entity>
      ex:data "data" ;
}

:data
    prov:wasDerivedFrom <entity> ;
    ...
```

Example DBpedia: The DBpedia project[6] is one of the largest Linked Open Datasets on the Web. It converts Wikipedia articles into RDF. The data is expressed using the DBpedia ontology. The information from which Wikipedia article data was derived is expressed using the `prov:wasDerivedFrom` predicate. Additional provenance information is given by further statements such as `dbo:wikiPageID` giving the ID of the Wikipedia page, `dbo:wikiPageRevisionID` stating the revision ID or the number of characters of the original article in `dbo:wikiPageLength`. Indeed, the provenance information is related to the article, but is mixed with the description of the entity the article is about:

```
<entity>
   dbp:size "63" ;
   dbp:built "1889" ;
   prov:wasDerivedFrom <article> ;
   dbo:wikiPageID "123" ;
   dct:date "2022-05-21" .
```

Here the statement level is not sufficient thus multiple statements share the same provenance. A cleaner solution is a model with a meta-level. It can also be argued that the provenance information of the data is only important for a small subset of applications, for example when ensuring the data quality such as the relevancy and timeliness of an article. Here is the example data using Named Graphs:

```
:data {
   <entity>
     dbp:size "63" ;
     dbp:built "1889" ;
}

<article> dbo:wikiPageID "123" .

:data
   prov:wasDerivedFrom <article> ;
   dct:date "2022-05-21" ;
```

Conclusion: The meta-level is the best and most suitable way how to model provenance data in RDF: This avoids levels of indirections, i.e. the usage of proxy resources or specific constructs such as the notion of records. The provenance information can be kept separate from the data and may only be queried if needed, which makes the data model lightweight. While Named Graphs are usually the best fit for provenance data, it has to be noted that RDF-star can also be used – even additionally – to provide the provenance for a single statement. This first pattern addresses the most common use case for metadata. With the following two patterns, we would like to extend its use to new applications.

---

[6] https://www.dbpedia.org/.

3.2   Pattern 2: Extending an Existing Data Model

Problem Statement: Data models usually evolve over time and changes to the model are inevitable. This poses a problem when these changes are not backwards compatible and break existing applications. We therefore propose to use the meta-level to extend and improve an existing data model.

For example, an existing knowledge graph could be enriched to provide additional confidence values. Sometimes an existing data model is wrong, unfitting or simply corrupted resulting in loss of context.

For illustration, we assume we have a simple data model where a relation of two entities <A> and <B> can be expressed:

```
<A> :conformsTo <B> .
```

Now this model is to be enriched with additional information such as a confidence score. To achieve this, however, various approaches are possible without using the meta-level, for example n-ary entities. This would require a remodeling which is a high risk of breaking existing applications, for instance:

```
<A> :hasConformanceStatement <C> .
<C> a :ConformanceStatement;
    :conformingTo <B>;
    :confidence 0.8 .
```

Solving Approach: We propose a modeling of this new information in the meta-level without touching the existing model to ensure backwards compatibility. The additional information is integrated by using RDF-star.

```
<A> :conformsTo <B> .
<< <A> :conformsTo <B> >> :confidence 0.8 .
```

Example 1 DBpedia: All data within DBpedia is derived from Wikipedia articles. Among other information, many articles have a thumbnail image related to its article. In DBpedia a thumbnail is attached to an article with the predicate `dbo:thumbnail`. However, Wikipedia is a Web page where the thumbnail is an HTML image tag. In DBpedia the src-attribute of the tag is converted to the URI describing the thumbnail. Furthermore, an HTML image tag is described by its caption attribute and an alternative text (alt attribute). Both are included in DBpedia via `dbp:caption` and `dbp:alt`. However, the relation to which thumbnail a caption or alternative text conforms is gone.[7] Due to this simplification the relation is lost. This becomes really disadvantageous if articles have multiple thumbnails where it is impossible to resolve these relationships. While this obviously should be fixed in the data, it might be that there are applications using the data that rely on the current data representation. With RDF-star, however,

---

[7] According to the DBpedia ontology, this should actually be modeled with an n-ary entity, but at least in the current version of DBpedia, the described problem exists.

we can provide additional information to deliver the relationship information to applications that need it, without changing the asserted statements in the RDF data. One way would be to add the caption to the relationship:

```
<entity> dbo:thumbnail <thumbnail> ;
  dbp:caption "Portrait of X" .

<< <entity> dbo:thumbnail <thumbnail> >>
  dbp:caption "Portrait of X" .
```

This is arguably again not ideal as a caption should probably refer to a thumbnail, not a thumbnail assignment. On the other hand, if the same thumbnail is used in different contexts with different captions, the above solution is good. For the sake of the argument however, let us assume that the caption should actually be assigned to the thumbnail. So the data should look like this:

```
<entity> dbo:thumbnail <thumbnail> .
<thumbnail> dbp:caption "Portrait of X" .
```

To actually fix the data, in this case the subject of the original statement needs to be changed. We could provide a vocabulary for such cases that would be universally understandable by applications supporting it, i.e. the replaceSubjectBy-statement:

```
<entity> dbo:thumbnail <thumbnail> ;
  dbp:caption "Portrait of X" .

<< <entity> dbp:caption "Portrait of X" >>
  ex:replaceSubjectBy <thumbnail> .
```

This means that an application should replace the subject of the original statement (`<entity>`) with the new subject `<thumbnail>`. As can be seen here, RDF-star opens many interesting ways to provide a history or change requests for statements. Consider the following example where statements are related to each other:

```
<entity> dbo:thumbnail <thumbnail> .
<thumbnail> dbp:caption "Portrait of X" .

<< <thumbnail> dbp:caption "Portrait of X" >>
  ex:replaced << <entity> dbp:caption "Portrait of X" >> .
```

In this case, the actually wrong statement `<entity> dbp:caption "Portrait of X"`. would only be available as part of the RDF-star triple stating that it has been replaced. Nevertheless, it is still part of the graph and old applications could still use it.

Example 2 GeoNames: GeoNames[8] is a graph for geographical data. It includes alternative names, but not information on historical names. This example shows how historical names of a city, or a place can be added to the current data and extended including more specific information. Currently in GeoNames there is only a distinction of `gn:name` and `gn:alternateName`. The alternate name can contain a language tag, but it does not indicate when a name was used in case of historical names. For example: The German city of Chemnitz used to be called "Karl-Marx-Stadt" between May 10th, 1953 and July 1st, 1990.

The listing below shows an excerpt of the RDF entry in GeoNames[9]:

```
<https://sws.geonames.org/2940132/>
  gn:name "Chemnitz" ;
  gn:alternateName "Chemnitz"@de ;
  gn:alternateName "Chemnitz"@en ;
  gn:alternateName "Karl-Marx-Stadt"@de .
```

We could add the missing information about the time span, for example by using the Common Authority File (Gemeinsame Normdatei, GND) of the German National Library, which provides the following data:

```
<https://d-nb.info/gnd/2015221-8>
  gndo:preferredNameForThePlaceOrGeographicName
    "Karl-Marx-Stadt";
  gndo:dateOfEstablishment "10.05.1953" .
  gndo:dateOfTermination "31.05.1990" ;
```

We can take this data from the GND and apply it to GeoNames to add the additional data to the entities, such as date of establishment and date of termination:

```
<https://sws.geonames.org/2940132/>
  gn:alternateName "Karl-Marx-Stadt"@de .

<< <https://sws.geonames.org/2940132/>
  gn:alternateName "Karl-Marx-Stadt"@de >>
  ex:valid_from "09.05.1953"^^xsd:date ;
  ex:valid_to "01.06.1990"^^xsd:date .
```

This way, the data is still provided in a very simple manner that nevertheless is useful for many applications that do not need the additional information, for example for named entity resolution. Nevertheless, the additional information can be provided in a modular way if it is needed.

---

[8] http://www.geonames.org/.
[9] https://sws.geonames.org/2940132.

Conclusion: RDF-star is a great fit when additional information about a statement needs to be provided. In particular when a data model is already used in applications it is an adaption to provide more complex data avoiding incompatible changes. By doing so, a simple model retains still simple by providing an additional context in the meta-level. The backwards compatibility is obtained as well (see also Sect. 4).

3.3   Pattern 3: Shifting Complex Relations to the Meta-level

Problem Statement: Whenever additional data about a relationship between two entities is needed, an additional *n-ary* entity can be created for representation. As entities (subjects and objects in statements) are substantives, this results in graphs with many nominalized relationships, for example:

```
<E> :hasSubject <SubjectAssignment1> .
<SubjectAssignment1>
    :hasHeading "Data Modeling" ;
    :fromVocabulary <TopicsVocabulary> .
```

Here, a complex entity of the class `SubjectAssignment` is used to represent the subject of entity `<E>`. The problem is that this structure feels cumbersome for many applications that might only be interested in the subject heading and that do not care about the vocabulary the subject heading is coming from. Even if uncontrolled subject headings (free tags) are used, the intermediate subject assignment is still required as the data model is created this way.

Solving Approach: Similar to pattern 2, we propose to use a meta-level, but this time from the beginning, so that the additional and perhaps even optional information can be pushed to the meta-level:

```
<E> dc:subject "Data Modeling" .
<< <E> dc:subject "Data Modeling" >>
    :fromVocabulary <TopicsVocabulary> .
```

Example DCAT: In this example we demonstrate how such n-ary entities can be avoided directly at the time of designing ontologies by shifting entities to the meta-level. DCAT is an ontology enabling publishers to describe datasets and its properties. The datasets can be listed in a `dcat:Catalog` entity. The `dcat:Dataset` entity holds meta information of the actual data, which is linked to as a `dcat:Distribution` entity. When listing a dataset in the catalog, the optional entity `dcat:CatalogRecord` may be used to express metadata about the listing such as the issue date.

As of the current DCAT 2 ontology specification[10], the `dcat:CatalogRecord` is provided for the purpose of adding additional metadata for the description of the listing of a resource, e.g., datasets in the catalog. The following example lists

---

[10] https://www.w3.org/TR/vocab-dcat-2/.

a dataset in a catalog. To add metadata such as the issued date or the title, the `CatalogRecord` entity is applied providing these information.

```
ex:catalog dcat:record ex:catalogRecord .

ex:dataset a dcat:Dataset .

ex:catalogRecord a dcat:CatalogRecord ;
  dct:issued "05.04.2022" ;
  dct:title "record title" ;
  dct:description "record description" ;
  foaf:primaryTopic ex:dataset .
```

However, the `dcat:CatalogRecord` entity adds metadata only to the actual listing of the catalog and the dataset. This can be shortened by rewriting it in the following RDF-star syntax to shift the `dcat:CatalogRecord` entity to the meta-level. The attributes of this entity can be used to describe the provenance of the relation directly:

```
ex:catalog dcat:dataset ex:dataset .

<< ex:catalog dcat:dataset ex:dataset >>
  a dcat:CatalogRecord ;
  dct:issued "05.04.2022" ;
  dct:title "record title" ;
  dct:description "record description" .
```

This information can be queried easily from the meta-level using SPARQL-star. Here we give an example on how to query the issued date of the CatalogRecord denoted before:

```
SELECT ?date WHERE {
  << ex:catalog dcat:dataset ex:dataset >> dct:issued ?date .
}
```

As RDF data can easily be split in several junks of statements, the metadata could also be separated from the core data, containing only the RDF-star statements.

Conclusion: On the one hand, complexity can be shifted to the meta-level and the data model is improved towards simpleness. In the course of this, the data may only be queried when needed reducing the size of query results. On the other hand, the simple data model might seem to be oversimplified to a data modeler. If the vocabulary scheme belongs to the subject heading and is not optional, it might seem arbitrary to put this information to the meta-level. And if the majority of applications need the complete data, the querying and using of the meta-level can feel more cumbersome than a more complex data model. To conclude, there are situations where the data modeled in an entity really should be on the same level as the rest of the actual data. In other situations, using the meta-level might just be the ideal solution where a simple model can be provided for simple applications and additional information is available if needed.

4 Querying and Constructing the Meta-level with SPARQL-Star

In the last section we have shown how data using the meta-level in its data model could look like. For a broad adoption of the proposed patterns, it is important that working with the data is effortless and data on the meta-level can easily be found and processed when needed. For this, two aspects should be considered:

1. How can an application find out if there is data on the meta-level?
2. How can data using the meta-level be transformed to other data models that potentially do not use the meta-level?

How can an application find out if there is data on the meta-level? For this question, it is important to adjust some expectations that an application might have when dealing with RDF and Linked Data. A modern RDF application that supports data on the meta-level is required

1. to support Named Graphs and RDF-star as well as at least TriG-star[11] as standard format for any responses from a Linked Data server as well as for data dumps,
2. to expect Named Graphs to be used for the actual data, so the internal organisation of data, e.g. for provenance tracking, must support graph hierarchies, such as a provenance context [5],
3. to check for the existence of a meta-level in form of Named Graphs and/or RDF-star triples.[12]

The following SPARQL-star query returns all RDF-star triples:

```
SELECT DISTINCT ?s WHERE {
  ?s ?p ?o .
  FILTER( isTRIPLE(?s) )
}
```

The function isTRIPLE is new in SPARQL-star and returns TRUE if the parameter is an RDF-star triple. Together with a query for the existence of Named Graphs, this enables an application to check if a meta-level is available.

---

[11] https://w3c.github.io/rdf-star/cg-spec/2021-04-13.html#trig-star.
[12] https://w3c.github.io/rdf-star/cg-spec/2021-04-13.html#dfn-triple.

How can data using the meta-level be transformed to other data models that potentially do not use the meta-level? The ability to transform data from one data model to another by means of CONSTRUCT queries is one of the most powerful features of RDF. This is not restricted by the use of meta-level data, as the following queries demonstrate:

In Pattern 3 (Sect. 3.3) we have shown how to shift the meta information encapsulated in an n-ary entity to the meta-level. Using SPARQL-star and CON-STRUCT queries, it is possible to rebuild the former data model, e.g. for the dcat:CatalogRecord. To extract the subject or object of a triple (annotated with RDF-star), the SPARQL-star functions SUBJECT and OBJECT are used:

```
CONSTRUCT {
  ?catalog :record ?record .
  ?record :primaryTopic ?dataset .
  ?record ?pred ?obj .
} WHERE {
  ?star ?pred ?obj
  FILTER(isTRIPLE(?star)) .
  BIND(SUBJECT(?star) as ?catalog) .
  BIND(OBJECT(?star) as ?dataset) .
  BIND(IRI(CONCAT(STR(?dataset), "/record")) as ?record) .
}
```

Here, string functions are used to coin an IRI for the new n-ary entity ?record, by appending /record to the IRI of ?dataset. This is to avoid a blank node, but it depends on the application if and how this should done.

The other way round is also possible, i.e., to create a simple data model with meta-level data from a complex data model:

```
CONSTRUCT {
  ?star ?pred ?obj .
} WHERE {
  ?catalog :record ?record .
  ?record :primaryTopic ?dataset .
  ?record ?pred ?obj .
  BIND(TRIPLE(?catalog, :dataset, ?dataset) as ?star).
}
```

This time, the SPARQL-star function TRIPLE is used to create an RDF-star triple from a subject, a predicate, and an object.

5   Discussion

With this paper, we ask the following question: can and should meta-level concepts be used for data modeling in RDF? Besides Named Graphs and RDF-star, several concepts have been introduced to construct a meta-level as well.

However, they all have weaknesses regarding e.g., feasibility or performance. Named Graphs have been around for a long time now and are well-supported in SPARQL. With RDF-star, we now see a serious contender for an RDF extension that allows statements about statements in a concise and straight-forward way, also with a SPARQL extension. So yes, we can use it for data modeling. But should we?

We identified three patterns that are suitable for a more detailed discussion of the potential benefits of meta-level modeling: (1) using the meta-level for actual metadata, i.e. data about RDF data like provenance data; (2) using the meta-level for the extension of existing data models; and (3) using the meta-level from the get-go to get simpler and more modular data models.

The first pattern is arguably the most trivial one, but addresses the reasons why Named Graphs and RDF-star have been developed in the first place. While at least Named Graphs exist for quite some while now, we still see only rarely data models that make use of them. Instead, mostly workarounds within vanilla RDF are used to artificially create a meta-level when it is needed. This should change and data about RDF data should be where it belongs: in the meta-level.

With the second pattern, we showed that existing simple data models can be extended by additional meta statements, with two main scenarios in mind: First, additional data can be added to provide more complex information to applications that need it. Second, this can be used to provide corrections and missing information to a data model without compromising existing applications. Adding the additional information in the meta-level does not only preserve backwards compatibility of the data model. It also keeps the core model simple and allows a modular distribution of additional data for the applications that need it. This observation leads to the third pattern where we proposed to use the meta-level to create simple and modular data models from the beginning.

We have further shown in several examples from real world data (including DCAT, DBpedia, and GeoNames), how the application of these patterns could look like and how they would affect the provided data. And finally, we have shown that working with the meta-level is straight-forward, as long as applications expect a meta-level and support Named Graphs and RDF-star, with TriG-star as preferred serialization. Particularly also the transformation of data into and out of the meta-level using `CONSTRUCT` queries is possible.

For the question, if meta-level modeling should be used, we conclude with the following thoughts:

Is the Provision and Management of Meta-level Data Possible? This depends on the technology stack that is used. Named Graphs are usually not a problem for triple stores but might lead to problems when the meta-level is to be preserved in other systems. For data serializations, a compatible file format such as TRIG must be used. File-based representations of graphs (i.e., one RDF file per graph) are also possible, but require additional infrastructure and/or tooling to make sure all meta information is properly preserved. RDF-star is very new, but at least a transparent serialization via RDF reification is possible.

Is There a Clear Decision for Named Graphs and/or RDF-star? As stated in the paper, it depends on the use case and sometimes even on the data. It might be worthwhile to just use one of the mechanisms to reduce complexity within a project. Alternatively, both can be used, but for different parts of the data model, for instance Named Graphs to provide provenance information and RDF-star to provide additional information on entity links. And finally, this could be left to the implementation and applications would have to expect such information being provided with either of the mechanisms.

Can we Still "Follow the Nose"? One of the main advantages of RDF is that it is easy to get data about a resource and use subsequent queries to find out more or even just to check if there are any more information. This is in our opinion the main reason why meta-level information is not used as often as it should. Applications often expect Turtle or even still RDF/XML as RDF serializations. They might not be able to deal with TriG. With SPARQL, Named Graphs can be queried. However, the application needs to know that there is useful information attached to a Named Graph. This is similar for RDF-star. The application needs to ask (or understand from the serialization) if there are additional statements about a statement. On the other hand, checking for the existence of Named Graphs or RDF-star triples is possible in SPARQL and SPARQL-star. So this is only a matter of getting used to it.

6   Conclusion and Future Work

Using the meta-level actively for data modeling opens interesting new perspectives on the task. In this paper we formulated three abstract patterns on how and when to use the meta-level in the modeling process actively. Therefore we gave several examples on real world applications.

Usually, the data modeler has to decide if the model should be optimized for data consumption, i.e. as easy to digest as possible and only as complex as necessary. However, more often than not, it is rather created with the most complex possible data in mind. Often data that does not even exist (yet) in reality - but certainly would be great to have at some point so the data model should better be ready. Meta-level modeling actually provides a means to define a simple core data model - particularly with simple, direct relations. Additional data can be provided in a modular way, with additional RDF data that can contain additional statements further describing resources.

For future work, we aim at using the meta-level in our own projects on production level to gain further experience with it and to support a wider adoption of Named Graphs and RDF-star in public data models. We are interested in the potential of meta-level modeling for the representation of changes in data to support data transparency and better provenance tracking. Currently we are exploring the creation of an ontology to capture the abstract entities appearing in meta-modeling, based on the patterns we identified so far and potentially more patterns that will arise with further applications.

Acknowledgements. This research was partially supported by the Volkswagen Foundation (Project: Consequences of Artificial Intelligence on Urban Societies, Grant 98555) and the German Research Foundation (Project: Specialized Subject Service for Jewish Studies, Grant 286004564). We thank Magnus Pfeffer for his valuable feedback.

References

1. Resource Description Framework (RDF): Concepts and Abstract Syntax. https://www.w3.org/TR/rdf-concepts/   .
2. Bucur, C.-I., Kuhn, T., Ceolin, D.: A unified nanopublication model for effective and user-friendly access to the elements of scientific publishing. In: Keet, C.M., Dumontier, M. (eds.) EKAW 2020. LNCS (LNAI), vol. 12387, pp. 104–119. Springer, Cham (2020). https://doi.org/10.1007/978-3-030-61244-3_7
3. Carroll, J.J., Bizer, C., Hayes, P., Stickler, P.: Named graphs, provenance and trust. In: Proceedings of the 14th International Conference on World Wide Web - WWW 2005, pp. 613–622. ACM Press, Chiba (2005)
4. Dodds, L., Davis, I.: Follow your nose. In: Linked Data Patterns - A pattern catalogue for modelling, publishing, and consuming Linked Data (2012). https://patterns.dataincubator.org/book/follow-your-nose.html
5. Eckert, K.: Provenance and annotations for linked data. In: Linking to the Future: 2013 Proceedings of the International Conference on Dublin Core and Metadata Applications, pp. 9–18 (2013)
6. Garijo, D., Eckert, K.: Dublin core to PROV mapping (2013). https://www.w3.org/TR/prov-dc/
7. Hartig, O.: Foundations of RDF* and SPARQL*. In: Proceedings of the 11th Alberto Mendelzon International Workshop on Foundations of Data Management and the Web 2017. CEUR Workshop Proceedings, vol. 1912, pp. 1–11 (2017)
8. Heery, R., Patel, M.: Application profiles: mixing and matching metadata schemas. Ariadne (25) (2000). http://www.ariadne.ac.uk/issue/25/app-profiles/
9. Isaac, A.: Europeana Data Model Primer (2013). https://pro.europeana.eu/files/Europeana_Professional/Share_your_data/Technical_requirements/EDM_Documentation/EDM_Primer_130714.pdf
10. Lagoze, C., Van de Sompel, H., Johnston, P., Nelson, M., Sanderson, R., Warner, S.: ORE User Guide - Primer (2008). http://openarchives.org/ore/1.0/primer
11. Nguyen, V., Bodenreider, O., Sheth, A.: Don't like RDF reification? Making statements about statements using singleton property. In: Proceedings of the ... International World-Wide Web Conference. International WWW Conference 2014, pp. 759–770, April 2014
12. Olaf Hartig, P.-A.C., Kellogg, G., Seaborn, A.: RDF-star and SPARQL-star (2021). https://www.w3.org/2021/12/rdf-star.html
13. Orlandi, F., Graux, D., O'Sullivan, D.: Benchmarking RDF metadata representations: reification, singleton property and RDF. In: 2021 IEEE 15th International Conference on Semantic Computing (ICSC), pp. 233–240, January 2021. ISSN 2325-6516
14. Orlandi, F., Passant, A.: Modelling provenance of DBpedia resources using Wikipedia contributions. J. Web Semant. **9**(2), 149–164 (2011)
15. Patrick J. Hayes, P.F.P.S.: RDF 1.1 Semantics (2014). https://www.w3.org/TR/rdf11-mt/
16. Pérez, B., Rubio, J., Sáenz-Adán, C.: A systematic review of provenance systems. Knowl. Inf. Syst. **57**(3), 495–543 (2018)

17. Sikos, L.F., Philp, D.: Provenance-aware knowledge representation: a survey of data models and contextualized knowledge graphs. Data Sci. Eng. **5**(3), 293–316 (2020). https://doi.org/10.1007/s41019-020-00118-0
18. Svensson, L.G., Verborgh, R., Sompel, H.V.d.: Indicating, discovering, negotiating, and writing profiled representations. Internet Draft draft-svensson-profiled-representations-01, Internet Engineering Task Force, March 2021. https://datatracker.ietf.org/doc/draft-svensson-profiled-representations-01

Multi-aspect Sentiment Analysis Using Domain Ontologies

Srishti Sharma[1], Mala Saraswat[2](✉), and Anil Kumar Dubey[3]

[1] The NorthCap University, Gurugram, India
[2] School of Computing Science and Engineering, Galgotias University, Greater Noida, India
malasaraswat@gmail.com
[3] ABES Engineering College, Ghaziabad, U.P, India

Abstract. Various aspects or characteristic features of an entity come into interplay to create an underlying fabric upon which sentiments blossom. In multi aspect Sentiment Analysis (SA), potentially related aspects of an entity under review are discussed in a single piece of text such as an online review. In this work, we use domain ontologies for enabling multi-aspect Sentiment Analysis. Since, domain ontologies contain the entire domain knowledge, they assist in enhanced aspect identification and detection of the latent or hidden aspects in a review document. We illustrate our approach by developing a system named Ontology driven Multi Aspect Sentiment Analysis (OMASA) system. We provide hotel reviews as input to this system and identify the panorama of explicitly expressed and latent aspects in a review using hotel domain ontology. After detecting the aspects, we link them with the corresponding opinions to gauge the sentiment pertaining to the aspects extracted. OMASA first computes sentiment scores for every aspect of the hotel. It then evaluates the overall sentiment score. On comparing with the baseline, the experimental results of OMASA show a marked improvement in the aspect level evaluation metrics Δ_{aspect}^2 and ρ_{aspect} after detecting the hidden aspects. This shows that OMASA has the potential to identify the latent aspects in text thereby improving the quality of SA.

Keywords: Sentiment analysis · Domain ontologies · Machine learning · Sarcasm · TripAdvisor · Amazon · Twitter

1 Introduction

There exists abundant literature on successful approaches for document-level sentiment classification [1, 2]. Today, there exist a number of online portals wherein users can pen down their reviews of the merchandise purchased, facilities availed, such as Zomato, Makemytrip. Most of the reviews are self-contained and comprehensive utilized to forecast rankings of merchandise or facilities. E-commerce sites incorporate the feature for reviews and comments as a complementary method of enhancing customer relationship management. Currently, Machine Learning (ML) and neural network driven methods are the contemporary approaches for SA of long documents [3, 4].

B. Villazón-Terrazas et al. (Eds.): KGSWC 2022, CCIS 1686, pp. 263–276, 2022.
https://doi.org/10.1007/978-3-031-21422-6_19

However, predicting one sentiment score for each review text is not sufficient, because a review might include different aspects or features of the artefact or amenity under discussion. In the instance of a hotel, different aspects such as "food", "location" and "service" may be discussed in the document, each having its own significance and bearing its own sentiment.

For example, Fig. 1 shows a sample hotel review. The author reviews the different aspects of a hotel such as hotel location, service, room, etc. These aspects help one to understand the major highlights and shortcomings of the hotel. Compared to an overall rating, users may feel encumbered to explicitly rate each and every aspect separately. However, during free flowing writing of reviews and comments, such multi-aspect sentiments are expressed naturally. Therefore, it is practical to perform multi- aspect SA to forecast distinct scores for every feature rather than ascribe total document score.

Fig. 1. A sample hotel review

2   Related Work

Some of the earliest works in the area of SA can be traced back to the efforts of Turney et al. [5] in discovering sentiments in product reviews. Around the same time, Pang et al. [6] focused on extracting sentiments from movie reviews. In the beginning, SA was mostly limited to document analysis to determine their sentiment polarity to be either positive or negative. Currently, SA can be carried out at the document or review level [5–7] at the sentence level [8] and at the feature level [9]. At the document level SA, the complete document's opinion is divided in different sentiment like positive or negative or neutral. In this, we presume that the entire document conveys only a single opinion only, rather than a combination of sentiments. In sentence level SA, we analyze if the given sentence conveys a positive or a negative or a neutral sentiment. At sentence level, we distinguish the objective sentences from the subjective sentences. In this, our main goal is to identify and extract the object feature/aspect about which the sentiment has been expressed and to categorize those sentiments as positive or negative or neutral.

Chakraverty et al. [10] in their paper developed a systematic approach to extract and analyse in real time, the macro-level emotional transitions of a city's inhabitants from their Twitter postings [10]. Saraswat et al. in their research harness reviews as the content generated from user to exploit, emotion as a basis for generating recommendations [11].

Sharma et al. developed a novel SA algorithm utilising the contextual information of the words comprising a document. Using this as a foundation we were able to measure the degree to which the words in a document influence each other and the impact of the dynamics of this relationship on the overall document sentiment [12]

Ontologies are being progressively used to capture the semantics of information from various sources. Saraswat et al. [13] focused on designing and implementing an efficient algorithm that will find similar objects in a given ontology tree using Earth Movers distance using MESH (Medical Subject Heading) ontology and Ohsumed Test Collection.

Xianghua et al. in their paper propose an unsupervised approach to automatically discover the aspects discussed in Chinese social reviews and also the sentiments expressed in different aspects. They apply the Latent Dirichlet Allocation (LDA) model to discover multi-aspect global topics of social reviews, and then extract the local topic and associated sentiment classified by HowNet lexicon [14].

Since domain ontologies contain the entire domain knowledge, they assist in enhanced aspect identification and detection of the latent or hidden aspects in a review document. Different aspects may often be linked to one another and occur together naturally. For instance, the expression, "the chamber was messy" describes the explicitly mentioned feature "chamber". In addition, it gives an opinion about the independent aspect "cleanliness" which is not explicitly mentioned in the phrase. OMASA has the capability to identify such latent and hidden aspects and mine sentiments from them appropriately, which is not possible by aspect based sentiment analysis approaches proposed earlier in literature.

Proposed approach also incorporates appropriate mechanisms to identify and handle sarcastic text, which may otherwise affect the results of Sentiment Analysis. This is often ignored in SA approaches.

3 Proposed Work

According to [2], an opinion is a quintuple $(o_j, f_{jk}, so_{ijkl}, h_i, t_l)$ where o_j is the target object, f_{jk} is an aspect k of the object j, so_{ijkl} is the sentiment value of the opinion of the opinion holder h_i on aspect f_{jk} of object o_j at time t_l. The sentiment factor so_{ijkl} may be positive, negative, neutral or it may express an extra granulated assessment. In terms of this definition, SA is the problem of estimating correctly the sentiment so_{ijkl} at time l as expressed by the author i on an aspect k of the entity j which is discussed in any opinionated document d. It indicates the opinion of the reviewer regarding a corresponding feature or aspect of the item. We explain the steps of the OMASA approach for Aspect based SA in the following sub-sections.

3.1 Pre-processing

The OMASA system first converts all the words in the text to lower case to maintain a consistent casing. Next, co-reference resolution is performed. Co-reference resolution is the problem of identifying all the terms that address the same real world entity in a writing [15]. For instance, consider the sentence, *"Eve said Jack was her brother"*. In this sentence, the proper noun *Eve* and the pronoun *her*, both denote the same entity, *Eve*. Similarly, the nouns *Jack* and *brother* refer to the same entity, Jack. In this example, *Eve* and *her* represents a pronominal co-reference whereas *Jack* and *brother* represents a nominal co-reference.

In order to arrive at the precise meaning and polarity from a writing, pronouns and other such referring terms need to be linked accurately with the entities that they actually denote. Set of rules aimed at determining the co-references usually look for the previously occurring, bordering entities attuned to the referring term. We perform co-reference resolution using Stanford CoreNLP [16, 17]. Stanford CoreNLP implements pronominal as well as nominal co-reference resolution. The output of the co-reference resolution engine of Stanford CoreNLP is a co-reference graph. We save this co-reference graph and use it later when the system processes the sentiments associated with all the entities.

3.2 Sarcasm Detection

Content posted by users on the internet such as blogs, microblogs, online reviews, etc., is often loaded with sarcasm. One of the core challenges of SA is to evaluate phrases loaded with sarcasm. Understanding the underlying context of a text is important to understand whether sarcasm is present as a hidden sentiment or not. Hence, it is a challenging assignment in the domain of Natural Language Processing (NLP).

The authors in [18] explored the presence of sarcastic expressions in tweets. In particular, they calculated the effect of sarcasm on SA. Their experiments show that if sarcasm is identified and accounted for appropriately, SA improves by approximately 50% as compared to SA without detecting sarcasm.

3.2.1 Training Corpus

Our training dataset consists of a pre-compiled sarcasm dataset from Twitter [19] and a corpus of Amazon product reviews. The Twitter dataset has a total of 1000 tweets. It is composed of tweets that are author annotated for sarcasm, bearing hashtags #sarcasm or #irony, similar to the approach described in [20]. The corpus of Amazon product reviews contains a total of 1995 Amazon product reviews, taken from the Sarcasm Corpus v1 [21].

3.2.2 Aspect Term Extraction Using Domain Ontology

Domain ontologies present domain knowledge in a hierarchical fashion. A domain ontology shows the top level concepts in the domain and illustrates their comprehensive descriptions and interrelationships with other domain concepts. It represents the entities (classes) in a domain, objects, object properties and the links that exist among them. Domain ontologies are a tool that enable collaboration of domain knowledge. Mapping with domain ontologies is carried out to identify the aspects on which a reviewer has commented. Domain ontologies allow for enhanced aspect identification and detection.

Since we are working on the hotel domain, we map with a domain ontology for the hotel domain. However, this is just to illustrate the utility of this method. This may be done for other domains as well like movies or any other products like camera, etc. We map with a hotel domain ontology namely Hontology [22]. Hontology contains overall 282 concepts organized into 16 top-level concepts and sub-concepts. All the concepts and sub-concepts in the ontology have some relevance to the hotel domain, and the occurrence level of a concept in the domain hierarchy is indicative of the concept's relevance to the domain. The depth of Hontology is 5. The top level concepts are Accommodation, Facility, Service, Staff, Room and Guest Type, Design, Meal, Points of Interest, Price, Rating and Staff, etc. Fig. 2 gives a representation of Room types, Service and Staff in Hontology. As we can clearly see, for the top level concept service, there exists a number of sub-concepts like Housekeeping, Airport Shuttle, Babysitting, etc. These sub-concepts also provide us valuable information about how an aspect is discussed by a reviewer. Hence, we also take the sub-concept words of every top level concept into account for aspect identification and detection.

Hontology contains information about 16 concepts corresponding to the hotel domain, however in our hotel review dataset, we have hotel reviews, overall hotel ratings and ratings for 7 aspects of the hotel domain per review. Hence, in this work we evaluate a hotel only on 7 aspects for which we have ground truth ratings. However, using Hontology we could evaluate a hotel on total 16 aspects. We map all the aspects that we need to evaluate on with Hontology concepts. For example, the aspect Rooms is mapped to concepts Room and Accommodation in Hontology, aspect Location is mapped to concepts Location and Points of Interest, aspect Service is mapped to Hontology concepts Service and Staff, aspect Check in/Front Desk is mapped to Hontology concept Reception-Check in service-Check out service, etc. Once an aspect is mapped to a Hontology concept, we extract all the concept and sub-concept words of that particular Hontology concept and add them to the set of aspect keywords corresponding to that particular aspect.

Fig. 2. Representation of room types, service and staff in hontology

Aspect Term Extraction using Hontology is carried out as explained in process *GetAspectAnnotations(.)* in Fig. 3. In line 1, the input review *R* to be analyzed is split into its constituent sentences. Starting from line 2, the review is processed sentence-wise. In line 3, $Count(j,i)$ that records the matching hits between sentences and aspects is initialized to zero for each sentence S_j and aspect K_i. The value of i varies from 1 to 7 as there are seven aspects to be evaluated. In line 4, sentence S_j is split into its corresponding token set T_j. In lines 5–7, first of all we compare each token t_z in sentence S_j with the set of aspect keywords K_i corresponding to every aspect a_i. If t_z is in K_i, this is a hit for sentence S_j and aspect a_i. In this case, $Count(j,i)$ is incremented by 1. Otherwise, we compute the semantic similarity between any token t_z in Sj and the all the aspect keywords for each of the aspects a_i. If this similarity for some aspect keyword K_{ni} of aspect $a_i$comes out to be more than a threshold θ, we record this as a hit for sentence S_j and aspect a_i and increment $Count(j,i)$ by 1 (lines 8–10). This helps us discover the latent or hidden aspects in a review that a reviewer may have commented on without explicitly using any aspect keyword. We explain the semantic similarity computation in the next subsection. We determine the threshold θ experimentally. In line 11, we label a sentence S_j by aspect a_i if $Count(j,i) > 0$. A sentence may be labeled by multiple aspects, indicating that one sentence contains information pertaining to multiple aspects. Finally, in line 13 we output all the sentences in the review annotated with their aspect assignments.

Semantic Similarity Computation
In the *GetAspectAnnotations(.)* algorithm discussed in Fig. 3, we compute the semantic similarity between all the aspect keywords of every aspect and all the words in each sentence of a review. This step helps in discovering the latent or hidden aspects in a review that the reviewer may have commented on without explicitly using any aspect keyword. For example, a hotel review may contain a sentence, "The room was filthy". In this sentence, there is explicit mention of the aspect room and implicit mention of the aspect cleanliness. The method proposed for Aspect Term Extraction discovers such relationships. For semantic similarity computation, we want a measure reflective of the semantics of a term, rather than its lexical category. For instance, the verb "marry" should be semantically similar to the noun "wife". Hence, we use Word vector- based similarity or discordance to estimate the semantic similarity. We use two different types of word embedding to capture this similarity, namely:

– GloVe: We utilize the pre-trained vectors of the GloVe project. The lexicon magnitude is 2,195,904 [23].
– Word2Vec: We utilize pre-trained Google word vectors. These were trained using Word2Vec tool on the Google News corpus. The lexis magnitude for Word2Vec is 3,000,000 [24].

3.3 Target Opinion Extraction

After identifying the multiple aspects that a reviewer has commented on in a review, the next task is Target Opinion Extraction. This task identifies the opinion corresponding to each target, that is, aspect and estimates its polarity.

Features having both syntactic and semantic content, like dependency trees, frequently exhibit better performance in SA systems [25]. The dependency parser is capable of handling effectively conjunctions, negations and bigram aspect terms. As such, it exhibits the best performance as per [26].

GetAspectAnnotations(.)

Input: A review R and a set of aspect keywords $K_i = \{K_{1i}, K_{2i}, \ldots \ldots K_{Nij}\}$ corresponding to each aspect a_i in A for $i=1$ to 7

Output: R fragmented into sentences annotated with feature labels.

Begin

1. Fragment R into sentences, $R = \{S_1; S_2; \ldots \ldots S_J\}$
2. **For** each sentence S_j in R:
3. Initialize $Count(j,i)=0$ for $i=1$ to 7
4. Split the sentence into a set of unigram tokens $T_j = \{t_1, t_2, \ldots t_Z\}$
5. **For** each token t_z in S_j:
6. **If** t_z in K_i:
7. $Count(j,i) = Count(j,i)+1$
8. **Else If** $sim_{vn}(t_z, K_{ni}) > \Theta$
9. $Count(j,i) = Count(j,i)+1$
10. **End for**
11. Assign the sentence S_j an aspect label a_i if $Count(j,i)>0$. A sentence may be assigned with multiple aspects.
12. **End for**
13. Output the annotated sentences with aspect assignments.

End

Fig. 3. The aspect segmentation algorithm

3.3.1 Neighborhood Relation

Any two consecutive words words w_i, w_{i+1} S, if w_i, $w_{i+1} \in$ Stopwords in a sentence S in a review R are related by the neighborhood relation. *Stopwords* is a list of words like 'he', 'she', 'the', etc. We ignore these words and do not take them into consideration because they do not contribute to sentiments.

3.3.2 Significant Dependency Relations

Any two words w_i and w_j in a sentence S of review R are related by a dependency relation if $(w_i, w_j) \in$ Significant_*Dependency_Relation*. We extract Significant Dependency Relations from the dependency parse tree of S. The set of significant dependency relations listed above is not minimal because not all the relations in the set are equally significant for Target Opinion Extraction. Taking cue from past work as in [27], we illustrate pruning the relation set and obtaining a minimal set of significant relations, by leave-one-relation out test or ablation test on a seed set of data.

3.4 Classification

We represent every sentence S in a review R as a set of vectors V where each vector consists of an aspect a_i in A and its associated opinion words w obtained from SRS and Neighborhood Relation set. These set of vectors then form the input for any supervised classification system.

4 Experiments and Results

We now describe the dataset used, Review level and Aspect level SA results.

4.1 Dataset Used

We use the hotel review dataset of Wang et al. 2010 [28] for our valuation. This dataset encompasses approximately 250,000 periodical evaluations of hotels from the website Tripadvisor that mention about 1850 hotels covering over 60 different sites. For each review there exists text, associated mathematical evaluations and metadata. In this corpus, reviews have overall ratings, and ground truth aspect ratings on 7 aspects: Value, Room, Location, Cleanliness, Check in/front desk, Service, and Business service. All the rankings in this corpus, vary from 1 star to 5 stars.

4.2 Review Level SA Results

We evaluate the reviews in the dataset for overall and aspect wise predictions. For overall review evaluation, we divide all the reviews in this dataset into three classes, that is, Positive, Negative and Neutral. We consider Reviews rated 4-5 as positive, reviews rated 3 as neutral and those rated 1–2 as negative. After this, total 178,487 reviews are positive, 36316 negative and 25262 neutral. Since this data is highly skewed, while training our classifiers we perform random under sampling. We train on 50–80% of the reviews after under sampling in each of the positive, negative and neutral classes. We test on the remaining samples. We illustrate the experimental results by three classifiers Multiclass Support Vector Machine (SVM), Naïve Bayesian (NB) and Maximum Entropy (MaxEnt) in Tables 1, 2, 3 and 4. After classification, we reverse the polarity of reviews labeled as sarcastic in the test set.

In Table 1 that depicts the results for 80% training and 20% test data, SVM performs the best with 82.18% accuracy. The accuracies obtained by NB and MaxEnt are almost comparable. The TPR values in all cases are significantly higher than the corresponding FPR values. SVM outperforms the other two classifiers in case of all positive, negative and neutral samples in terms of accuracy, FPR and TPR barring FPR for negative samples. FPR for negative samples is least by NB classifier, yielding only 5.94%

Table 1. Review level quantitative evaluation of the performance of different classifiers

Classifier	TPR (in percent)			FPR (in percent)			Accuracy
	Positive	Negative	Neutral	Positive	Negative	Neutral	
Naïve Bayes	80.17	81.32	84.56	8.49	**5.94**	12.54	80.40
SVM	**81.92**	**83.77**	**85.21**	**7.33**	6.74	**10.47**	**82.18**
Max Entropy	79.73	81.16	83.27	7.82	7.58	12.51	79.96

In Table 2 that depicts the results for 70% training and 30% test data, SVM performs the best with 78.79% accuracy. The TPR values in all cases are significantly higher than the corresponding FPR values. SVM outperforms the other two classifiers in case of all positive, negative and neutral samples in terms of accuracy, FPR as well as TPR.

Table 2. Review level quantitative evaluation of the performance of different classifiers

Classifier	TPR (in percent)			FPR (in percent)			Accuracy
	Positive	Negative	Neutral	Positive	Negative	Neutral	
Naïve Bayes	76.99	79.10	80.01	9.32	9.20	13.42	77.33
SVM	**78.22**	**81.54**	**84.18**	**8.10**	7.32	**12.61**	**78.79**
Max Entropy	76.11	78.73	81.35	10.68	8.38	12.84	76.58

In Table 3 that depicts the results for 60% training and 40% test data, NB classifier performs the best with 76.65% accuracy. SVM is a close second with 76.56% accuracy. The TPR values in all cases are significantly higher than FPR values. SVM outperforms the other two classifiers in case of all positive, negative and neutral samples in terms of accuracy and TPR. FPR values are best for NB classifier for all positive, negative and neutral samples.

Table 3. Review level quantitative evaluation of the performance of different classifiers

Classifier	TPR (in percent)			FPR (in percent)			Accuracy
	Positive	Negative	Neutral	Positive	Negative	Neutral	
Naïve Bayes	74.31	76.02	77.26	**0.11**	**0.10**	**0.15**	**76.65**
SVM	**75.92**	**79.03**	**82.26**	9.51	9.14	12.75	76.56
Max Entropy	73.37	75.26	78.69	12.43	9.75	14.16	73.85

In Table 4 that depicts the results for 50% training and 50% test data, SVM performs the best with 72.74% accuracy. NB is a close second with 72.17% accuracy. The TPR values in all cases are significantly higher than FPR values. SVM outperforms the other two classifiers in case of all positive, negative and neutral samples in terms of accuracy, FPR as well as TPR barring FPR for negative samples which is least by NB classifier.

Table 4. Review level quantitative evaluation of the performance of different classifiers

Classifier	TPR (in percent)			FPR (in percent)			Accuracy
	Positive	Negative	Neutral	Positive	Negative	Neutral	
Naïve Bayes	69.90	70.27	72.48	13.76	9.94	16.48	72.17
SVM	72.08	75.11	76.94	10.93	13.42	13.58	72.74
Max Entropy	68.68	70.85	73.41	13.25	12.12	18.16	69.23

4.3 Aspect Level SA results

Let $D = \{R_1; R_2; \ldots\ldots R_j\}$ be a set of review text documents. Given the complete set of review documents D where each review R is associated with I aspects $\{A_1; A_2; \ldots\ldots A_I\}$ to be analyzed. We use two measures to quantitatively evaluate the performance of different classifiers aspect wise. We explain them as under:

(1) *Mean square error on aspect rating prediction* (Δ^2_{aspect}) Suppose f, R, is the ground-truth or the actual rating obtained from the training data for aspect A in review R. Δ^2_{aspect} is indicative of the difference amid the predicted aspect rating fR_i and f, R_i. It is represented as explained in Eq. 1.

$$\Delta^2_{aspect} = \sum_{R=1}^{|D|} \sum_{i=1}^{I} \left(f_{Ri} - f^*_{Ri}\right)^2 / I * |D| \tag{1}$$

where $|D|$ is the total amount of review documents.

(2) *Aspect correlation inside reviews* (ρ_{aspect}): ρ_{aspect} measures how accurately the expected aspect scores can preserve the comparative order of aspects in a review as indicated by the ground-truth ratings. For instance, in a review, the author may have enjoyed the locality more than the sanitation. The factor ρ_{aspect} measures whether the projected scores retain this inclination. ρ_{aspect} is defined as given by Eq. 2.

$$\rho_{aspect} = \sum_{R=1}^{|D|} \rho_{f_R, f_{R*}} / |D| \tag{2}$$

where $\rho_{f_R, f_{R*}}$ is the Pearson correlation coefficient between two vectors f_R and f^*_R.

Pearson correlation coefficient is used because it is indicative of the link amid two variables. So, this helps to measure the association between the ground truth and predicted ratings. The value of this coefficient varies from -1 to 1. If the assessment is close to 1, it indicates perfect correlation: as one variable surges, the other variable also inclines to surge. If the assessment is close to -1, as one variable surges, the other variable inclines to shrink. If the assessment is zero, it indicates no correlation. The Pearson correlation coefficient amid any two variables x and y is given by r as shown in Eq. 3.

$$r_{xy} = \frac{n \sum x_i y_i - \sum x_i \sum y_i}{\sqrt{n \sum x_i^2 - \left(\sum x_i\right)^2} \sqrt{\sum y_i^2 - \left(\sum y_i\right)^2}} \tag{3}$$

where n represents sample size and x_i, y_i are the distinct sample points indexed with i.

We summarize the aspect level results in Table 8. Though SVM performs much better than NB and MaxEnt classifiers in review level analysis as well as on ρ_{aspect}, it does not perform the best on $\Delta^2{}_{aspect}$. NB has the best value of 0.575. A high value of ρ_{aspect} indicates that SVM is more superior in differentiating the ratings of diverse aspects inside a review. Such a hypothesis concerning the relative preferences on different aspects cannot be acquired with a single overall rating.

The factor $\Delta^2{}_{aspect}$ indicates the nonconformity of each predicted aspect rating and ground-truth rating independently. Thus, the lesser this value is, the better the classifier. However, $\Delta^2{}_{aspect}$ does not reflect how well the relative order of aspects in a review is preserved. For example, take a case where there are five aspects and a review having an overall rating of 4 and ground-truth aspect ratings of (3; 4; 4; 5; 4). An aspect-unaware prediction of (4; 4; 4; 4; 4), cannot differentiate the different aspects. It has $\Delta^2{}_{aspect} = 0.67$. It has a ρ_{aspect} of 0. This indicates no correlation between the ground truth and predicted aspect ratings.

Now take another prediction (2; 3; 3; 5; 4), which can tell the real difference between aspects. It has a $\Delta^2{}_{aspect} = 1$, which is higher, thus showing more difference between predicted and gold standard ratings. This is worse than the aspect unaware prediction. However, in this case the ρ_{aspect} is 0.93 which shows that the aspect ratings order is preserved. We clearly see that even an aspect unaware prediction result can have a lesser individual deviation in aspect ratings from gold standard. We conclude that even though $\Delta^2{}_{aspect}$ indicates the difference between the predicted and gold standard ratings, it is of little significance without ρ_{aspect} alues to correlate along with the relative rating order.

In Table 5 we present the results of the aspect-level quantitative evaluation of the performance of different classifiers by OMASA and baseline. For OMASA, the NB classifier achieves the best $\Delta^2{}_{aspect}$ of 0.575, but since it under-performs on ρ_{aspect} having the lowest value of 0.139, it cannot be considered the best. On the other hand, SVM performs better by having a $\Delta^2{}_{aspect}$ of 0.684 and ρ_{aspect} of 0.472. In the baseline evaluation, we do not carry out hidden aspect detection and significant relation set extraction. Taking a look at Table 8, we can see that ρ_{aspect} is lesser for baseline by each of the three classifiers that we applied. This shows that OMASA is better at preserving the relative order of aspects in the ground-truth ratings than the baseline. On the other hand, the baseline $\Delta^2{}_{aspect}$ values are higher than OMASA in each case which indicates more deviation from ground truth ratings. This clearly establishes the superiority of OMASA over the baseline.

Table 5. Aspect-level quantitative evaluation of the performance of different classifiers

Classifier	$\Delta^2 aspect$		$\rho aspect$	
	OMASA	Baseline	OMASA	Baseline
NB	**0.575**	0.693	0.139	0.120
SVM	0.684	0.701	**0.472**	**0.411**
Max Ent	0.672	**0.686**	0.294	0.267

5   Conclusions and Future Work

In this work, we used domain ontologies for performing multi-aspect SA of hotel reviews by developing our system OMASA. Since the domain ontology on hotels contain the entire domain knowledge, we were able to improve aspect identification in the corpus of hotel reviews. Specifically, we were able to detect the latent or hidden aspects in the review. We employed dependency parsing and neighborhood relations to extract opinion expressions pertaining to the identified aspects. The system learnt the minimal set of significant relations used by the dependency parser. We performed both document (review) level SA and aspect level SA on these reviews using three different ML algorithms- SVM, NB and MaxEnt. SVM classifier reported the best results amongst the three.

We compared our approach with a baseline that does not have hidden aspect detection and does not learn the significant relations of the dependency parser. The aspect level SA results reported by OMASA were better than the baseline in terms of the evaluation metrics ρ_{aspect} as well as Δ^2_{aspect}. The best value of ρ_{aspect} reported by OMASA is 6.1% more than the best value reported by the baseline, indicating better association between the ground truth and predicted aspect ratings by OMASA. The best value of Δ^2_{aspect} by the baseline is 11.1% higher than OMASA, indicating lesser deviation between ground truth and predicted aspect ratings by OMASA than by the baseline. This establishes the superiority of OMASA over the baseline. This re-affirms our belief that indeed there are certain latent or hidden aspects present in online reviews and detecting these helps improve the accuracy of SA. In the next chapter, we summarize all our works and present the conclusions.

References

1. Pang, B., Lee, L.: Opinion mining and sentiment analysis. In: Foundation and Trends in Information Retrieval, p. 2(1–2), 1–135 (2007)
2. Liu, B.: Sentiment Analysis and Subjectivity-Handbook of Natural Language Processing (2010)
3. Tang, D., Qin, B., Liu, T.: Document modeling with gated recurrent neural network for sentiment classification. In: EMNLP, pp. 1422–1432 (2015)
4. Yang, Z., Yang, D., Dyer, C., He, X., Smola, A., Hovy, E.: Hierarchical attention networks for document classification. In: Proceedings of NAACL-HLT, pp. 1480–1489 (2016)
5. Turney, P.: Thumbs up or thumbs down? Semantic orientation applied to unsupervised classification of reviews. In: Proceedings of the Association for Computational Linguistics, pp. 417–424 (2002)

6. Pang, B., Lee, L., Vaithyanathan, S.: Thumbs up? Sentiment classification using machine learning techniques. In: Proceedings of the Conference on Empirical Methods in Natural Language Processing (EMNLP), pp. 79–86 (2002)
7. Gräbner, D., Zanker, M., Fliedl, G., Fuchs, M.: Classification of customer reviews based on sentiment analysis. In: Fuchs, M., Ricci, F., Cantoni, L. (eds.) Information and Communication Technologies in Tourism 2012, pp. 460–470. Springer Vienna, Vienna (2012). https://doi.org/10.1007/978-3-7091-1142-0_40
8. Celikyilmaz, A., Hakkani-Tur, D., Feng, J.: Probabilistic model-based sentiment analysis of Twitter messages. In: The 2010 IEEE International Conference on Spoken Language Technology, Workshop, pp. 79–84 (2010)
9. Singh, V.K., Piryani, R., Uddin, A., Waila, P.: Sentiment analysis of movie reviews. In: Proceedings of International Multi Conference on Automation, Computing, Control, Communication, and Compressed Sensing, pp. 712–717. Kerala, India (2013)
10. Chakraverty, S., Sharma, S., Bhalla, I.: Emotion–location mapping and analysis using twitter. J. Inf. Knowl. Manag. **14**(03), 1550022 (2015)
11. Saraswat, M., Chakraverty, S.: Emotion distribution profile for movies recommender systems. In: Sharma, H., Gupta, M.K., Tomar, G.S., Lipo, W. (eds.) Communication and Intelligent Systems. LNNS, vol. 204, pp. 365–373. Springer, Singapore (2021). https://doi.org/10.1007/978-981-16-1089-9_30
12. Sharma, S., Chakraverty, S., Sharma, A., Kaur, J.: A context-based algorithm for sentiment analysis. Int. J. Comput. Vis. Robotics **7**(5), 558–573 (2017)
13. Saraswat, M.: Efficiently finding similar objects on ontologies using earth mover's distance. In: Bringas, P.G., Hameurlain, A., Quirchmayr, G. (eds.) DEXA 2010. LNCS, vol. 6262, pp. 360–374. Springer, Heidelberg (2010). https://doi.org/10.1007/978-3-642-15251-1_29
14. Xianghua, F., Guo, L., Yanyan, G., Zhiqiang, W.: Multi-aspect sentiment analysis for Chinese online social reviews based on topic modeling and HowNet lexicon. Knowl. Based Syst. **37**, 186–195 (2013)
15. Manning, C.D., Surdeanu, M., Bauer, J., Finkel, J., Bethard, S.J., McClosky, D.: The stanford CoreNLP natural language processing toolkit. In: Proceedings of the 52nd Annual Meeting of the Association for Computational Linguistics: System Demonstrations, pp. 55–60 (2014)
16. Clark, K., Manning, C.D.: Deep reinforcement learning for mention-ranking Coreference Models. In: Proceedings of EMNLP (2016)
17. Maynard, D., Greenwood, M.A.: Who cares about sarcastic tweets? Investigating the impact of sarcasm on sentiment analysis. In: Lrec, 4238–4243 (2014)
18. Sharma, S., Chakraverty, S.: Sarcasm detection in online review text. ICTACT J. Soft Comput **08**(03) (2018)
19. Data Mining Project focused on Twitter Sarcasm Measurement. https://github.com/dmitry vinn/twitter-sarcasm-measurement
20. Liebrecht, C., Kunneman, F., Van den Bosch, A.: The perfect solution for detecting sarcasm in tweets #not. In: Proceedings of the 4th Workshop on Computational Approaches to Subjectivity, Sentiment and Social Media Analysis (2013)
21. Lukin, S., Walker, M.: Really? Well. Apparently bootstrapping improves the performance of sarcasm and nastiness classifiers for online dialogue. In: The Workshop on Language Analysis in Social Media (LASM), at The Conference of the North American Chapter of the Association for Computational Linguistics (NAACL), Atlanta, Georgia, USA (2013)
22. Chaves, M.S., Freitas, L., Vieira, R.: Hontology: a multilingual ontology for the accommodation sector in the tourism industry. In: KEOD2012- International Conference on Knowledge Engineering and Ontology Development, pp. 149–154 (2012)
23. Pennington, J., Socher, R., Manning, C.D.: GloVe: Global Vectors for Word Representation (2014)

24. Mikolov, T., Chen, K., Corrado, G., Dean, J.: Efficient Estimation of Word Representations in Vector Space. CA, USA (2013)
25. Nakagawa, T., Inui, K., Kurohashi, S.: Dependency tree-based sentiment classification using CRFs with hidden variables. In: Human Language Technologies: The 2010 Annual Conference of the North American Chapter of the ACL, pp. 786–794, Los Angeles (2010)
26. Moghaddam, S., Ester, M.: On the design of LDA models for aspect-based opinion mining. In: CIKM'12, Maui, HI, USA (2012)
27. Mukherjee, S., Bhattacharyya, P.: feature specific sentiment analysis for product reviews. In: CiCLing 2012
28. Wang, H., Lu, Y., Zhai, C.: Latent aspect rating analysis on review text data: a rating regression approach. In: The 16th ACM SIGKDD Conference on Knowledge Discovery and Data Miining (KDD 2010), pp. 783–792 (2010)

Popularity Driven Data Integration

Fausto Giunchiglia[1]([✉]) [ID], Simone Bocca[1] [ID], Mattia Fumagalli[2] [ID],
Mayukh Bagchi[1] [ID], and Alessio Zamboni[1] [ID]

[1] Department of Information Engineering and Computer Science (DISI), University of Trento,
Trento, Italy
{fausto.giunchiglia,simone.bocca,mayukh.bagchi,
alessio.zamboni}@unitn.it
[2] Conceptual and Cognitive Modeling Research Group (CORE), Free University of
Bozen-Bolzano, Bolzano, Italy
mattia.fumagalli@unitn.it

Abstract. More and more, with the growing focus on large scale analytics, we are confronted with the need of integrating data from multiple sources. The problem is that these data are impossible to reuse *as-is*. The net result is high cost, with the further drawback that the resulting integrated data will again be hardly reusable *as-is*. *iTelos* (Not to be confused with Telos [11]) a general purpose methodology aiming at minimizing the effects of this process. The intuition is that data will be treated differently based on their *popularity*: the more a certain set of data have been reused, the more they will be reused and the less they will be changed across reuses, thus decreasing the overall data preprocessing costs, while increasing backward compatibility and future sharing.

Keywords: Data reuse · Data sharing · Knowledge graphs

1 Introduction

More and more, with the growing focus on large scale analytics, we are confronted with the need of integrating data from multiple sources. The key issue is how to handle the *semantic heterogeneity* which is intrinsic in such data. Two main approaches have been proposed. The first approach consists of using *ontologies*, where the goal is to agree on a fixed language and/or schema towards facilitating future sharing [3]. The second consists of exploiting the flexibility of *Knowledge Graphs (KGs)* [9], as the means for facilitating the adaptation and integration of heterogeneous data. However the problem is far from being solved. No matter the approach, it is just impossible to reuse data *as-is*. The net result is usually a lot of data preprocessing (e.g., cleaning, normalization) which results in high cost, with the further drawback that the resulting integrated data will again be hardly reusable *as-is*. It is a negative loop which consistently reinforces itself.

The research by F. Giunchiglia, M. Bagchi and S. Bocca has received funding from the *"DELPhi - DiscovEring Life Patterns"* project funded by the MIUR (PRIN) 2017. The research by A. Zamboni was supported by the *InteropEHRate* project, EC Horizon 2020 programme under grant number 826106.

B. Villazón-Terrazas et al. (Eds.): KGSWC 2022, CCIS 1686, pp. 277–284, 2022.
https://doi.org/10.1007/978-3-031-21422-6_20

In this paper, we propose *iTelos*, a general purpose methodology whose main goal is to minimize the high costs of this loop of reuse. *iTelos* is crucially based on the use of KGs and ontologies, i.e., *reference schemas*, as the schemas of KGs. However, its novel underlying intuition is to treat data differently, depending on their *popularity*. In particular, the idea is to select first and minimize changes on those data which are more (re)-used thus decreasing the preprocessing costs, while increasing backward compatibility and future sharing. To this extent, *iTelos* distinguishes among three categories of data. That is: *Common*, which are used across domains (e.g., data about space, time, transportation), *Core*, which, while being more vertical than common, are extensively used in the domain under consideration (e.g., in tourism, all the data that can be found in Open Data portals), and *Contextual*, namely, data specific to the application at hand (e.g., the data extracted on purpose from legacy systems). Popularity also drives the selection of the reference schemas, based on the intuition that, also at the schema level, more reuse is a good motivation for further reuse. Thus we have *Common* reference schemas (e.g., standards about space, time, transportation, *schema.org*[1]), *Core* reference schemas, (e.g., in Health, FHIR[2]), and *Contextual* reference schemas (e.g., as they apply the current application).

iTelos implements the above idea based on a precisely articulated data integration process, based on three key assumptions, as follows:

– *data* and *reference schemas* should be integrated under the overall guidance of the needs to be satisfied, formalized as *competence queries* (*CQs*) [8];
– The requirements, including those on data and reference schema reuse, as well as CQs, should be known a priori as part of an application *purpose*;
– A difficulty is that the schemas of the data to be reused usually do not map to reference schemas, the mapping being usually arbitrarily complex. The idea is to build the integrated KG via a sequence of *middle-out* iterations where, first, CQs are used to drive the selection and preprocessing of the data, largely independently from the reference schemas, and where, in a second step, reference schemas, suitably and independently integrated among them, are adapted to fit best the integrated data minimizing the negative effects on sharability.

This paper is organized as follows. In Sect. 2 we describe the *iTelos* process. In Sect. 3 we describe how *iTelos* enhances the reusability, with minimal changes, of the available data. In Sect. 4 we describe how *iTelos* enhances the future sharability of the resulting KG. Finally, Sect. 5 syntethically describes the case studies to which *iTelos* has been applied.

2   The Process

The *iTelos* process is depicted in Fig. 1. The *User* provides in input the specification of the problem, the *Purpose*, and receives in output an integrated KG, i.e., the *Entity Graph*. The purpose contains three main elements, as follows:

---

[1] https://schema.org/.
[2] http://hl7.org/fhir/.

- The functional requirements of the KG to be generated, that we assume to be ultimately formalized as CQs;
- The *datasets* to be reused. We assume that these datasets consist of *Entity Graphs (EGs)*, namely graphs where nodes are *entities* (e.g., my cat *Garfield*), decorated with data property values and linked among them via object property links;[3]
- The *Ontologies* to be reused. We assume that these ontologies, i.e., reference schemas, consist of *entity type (etype) Graphs (ETGs)*, namely KGs which define the schema of EGs. In ETGs nodes are *etypes*, namely classes of entities (e.g., the class *cat*), decorated by the data and object properties which define the EG structure.[4]

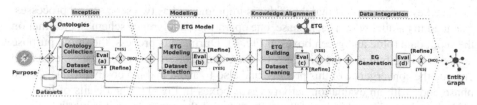

Fig. 1. The *iTelos* process.

The overall *iTelos* process is articulated in four phases. The *Inception* phase takes as input the purpose and collects the ontologies and datasets needed to build the target EG. During this phase, the functional requirements are encoded into a list of CQs which are then matched to the input datasets and ontologies thus implementing the activities of *Ontology Collection* and *Dataset Collection*, as from Fig. 1. The main objective of the *Modeling* phase is to build the most suitable model of the ETG to be used as the schema of the final EG, which in Fig. 1 is called the *ETG model*. In practice, the ETG model includes all the etypes and properties needed to represent the information required by each CQ, possibly extended by extra etypes and properties suggested by the datasets. The *Dataset Selection* activity finalizes the selection of the datasets selecting whenever possible, the ones which are most popular. The goal of *Knowledge Alignment* is to enhance the sharability of the final EG, building in turn a shareable ETG that fits in the best possible way the datasets to be integrated. The input ETG model is itself a possible solution. However the set of reference schemas provides more possibilities, in terms of etypes and properties available, that can be adopted to implement a final ETG (called ETG in Fig. 1) easier to share and reuse. Here again preference is given to those reference schemas which are most popular. The *Dataset Cleaning* activity performs the final cleaning of the datasets consistently with the ETG, trying to minimizing it and concentrating the preprocessing on those datasets which are less popular. As described in detail in Sect. 4, this may require building an ETG which is an adaptation of the input reference schemas, e.g., in the selected etypes, properties, data types and formats.

---

[3] Many Open data portals provide such datasets, at different levels of formalization in the 5STAR Open Data Model.

[4] Many such schema repositories are already available; see for instance: LOV (https://lov.lin keddata.es/), LOV4IoT (http://lov4iot.appspot.com/), DATAHUB (https://old.datahub.io/) and *LiveSchema* (http://liveschema.eu/).

The last phase is *Data Integration*. The objective is to build the EG, what we call *EG Generation*, integrating the ETG with the data resources. To do that, the ETG and datasets are provided in input to a specific data mapping tool, called *KarmaLinker*, which consists of the *Karma* data integration tool [10] extended to perform Natural Language Processing on short sentences (i.e., what we usually call *the language of data*) [1]. The process is described in some detail in [7]. The first activity in this phase maps the data to the etypes and properties of the ETG. The following step is the generation of the entities that are then matched and, whenever they are discovered to be different representations of the same real world entity, merged. These activities are fully supported by Karmalinker. The above process is iteratively executed over the list of selected datasets. The process concludes with the export of the EG into an RDF file.

The key observation is that the *iTelos* is implemented as two separate subprocesses, executed in parallel within each phase, one operating on reference schemas, the other on the input data (blue and green boxes in Fig. 1). During this process, the initial purpose keeps evolving building the bridge between CQs, datasets and reference schemas. To enforce the convergence of this process, and also to avoid making costly mistakes, each phase ends with an evaluation activity (*Eval* boxes in Fig. 1). The details of how the evaluation is performed is out of the main scope of the paper. Here it is worth noticing that, within each phase, the evaluation aims to verify that the target of that phase is met, namely: aligning CQs with datasets and ontologies in phase 1, thus maximizing reusability; aligning the ETG model with the datasets in phase 2, thus guaranteeing the success of the project; and aligning ETG and ontologies in phase 3, thus maximizing sharability. The evaluation in phase 4 has the goal of checking that the final EG satisfies the requirements specified by the purpose.

3   Enhancing Data Reuse

Reusability is enhanced during the phases of inception and modeling, whose main goal is to progressively transform the specifications from the purpose into the ETG Model. This process happens according to the following steps:

1. generation of a list of natural language sentences, each informally defining a CQ, as implicitly or explicitly implied by the purpose;
2. generation of a list of relevant etypes and corresponding properties, which formalize the informal content of CQs, as from the previous step;
3. selection of the datasets whose schema informally matches the CQs, as from the previous step;
4. generation of a list of etypes with associated properties, from the selected datasets, which match the etypes and properties from the CQs;
5. construction of the ETG model;

Steps 1–4 happen during inception, while step 5 happens during the modeling phase. The key observation is that, starting from an analysis of the etypes and properties inside CQs, the two types of resources involved (i.e., ontologies, datasets) are handled through a series of three iterative executions, each corresponding to a specific category, following a decreasing level of reusability. The categories are defined as follows:

Common: this category involves resources associated with aspects that are common to all domains, also outside the domain of interest. Usually, these resources correspond to abstract etypes specified in *upper level ontologies* [4], e.g., *person, organization, event, location*, and/or to etypes from very common domains, usually needed in most applications, e.g., *Space* and *Time*. The data that are found in Open Data sites as well the ontologies which can be found in the repositories mentioned above are examples of common resources.

Core: this category involves resources associated with the more core aspects of the domain under consideration. They carry information about the most important aspects considered by the purpose, information without which it would be impossible to develop the EG. Consider for instance the following purpose:

"There is a need to monitor Covid-19 data in the Trentino region (Italy), to understand the diffusion of the virus and the social restriction caused by the virus, with the possibility to identify new outbreaks".[5]

In this example, core resources could be those data values reporting the number of Covid-19 infections in the specified region. Examples of common resources are the data of certain domains, e.g., public sector facilities (e.g., hospitals, transportation, education), domain specific ontologies that can be found again in the repositories above, as well as domain specific standards (e.g., Health, interoperability standards of various types). In general, data are harder to find than ontologies, in particular when they are about the private sector, where they carry economic value and are often collected by private bodies.

Contextual: this last category involves resources that carry specific, possibly unique, information from the domain of interest. These are the resources whose main goal is to create added value. If core resources are necessary for a meaningful application, contextual resources are the ones which can make the difference with respect to the competitors. In the above example, examples of contextual resources can be those data describing the type of social restrictions adopted to contrast the virus. At the schema level, contextual etypes and properties are.

Those which differentiate the ontologies which, while covering the same domain, actually present major differences. Data level contextual resources are usually not trivial to find, given their specificity and intrinsic. In various applications we have developed in the past, this type of data have turned out to be a new set of resources those had to be generated on purpose for the application under development, in some cases while in production.

The overall conclusive observation is that the availability of resources, and of data in particular, decreases from common to contextual. On top of this and because of this, as part of the *iTelos* strategy, a decreasing effort is made, moving from common to core to contextual data, in maintaining high the level of reusability and sharability, thus concentrating the preprocessing costs in the latter categories.

---

[5] This is a small example extracted from the project which built a KG, following the *iTelos* methodology, on *"Integration of medical data on Covid-19"* developed by Antony, N., Gotca, D., Jyate, M., Donini, L. The complete material and description can be found at the URL https://github.com/UNITN-KDI-2020/ COVID-data-integration.

4   Enhancing Data Sharability

The knowledge alignment phase aims to enhance sharability, by aligning and possibly modifying the ETG model to take into account the etypes and properties coming from the reference ontologies. The key observation is that the alignment mainly concerns the common and, possibly, the core types with much smaller expectations on contextual etypes. Notice how the alignment with the most suitable ontology will enable the reuse of data, at least for what concerns common and sometimes core etypes. As an example, the selection of Google GTFS or FOAF as reference ontologies ensures the availability of a huge amount of data, a lot of which are open data. This type of decision should be made during inception; if delayed up to here, it might generate backtracking.

We align the ETG model with the reference ontologies, by adapting the *Entity Type Recognition (ETR)* process proposed in [6]. This process happens in three steps as follows:

Step 1: ontologies selection. This step aims at selecting the set of reference ontologies that best fit the ETG model. As from above the first step is to rank the ontologies, as selected during the Inception phase, based on their popularity. Then, moving from top to down in the list, as from [6], this selection step occurs by measuring each reference ontology according to three metrics, which allow:

- to identify how many etypes of the reference ontologies are in common with those defined in the ETG model, and
- to measure a property sharability value for each ontology etype, indicating how many properties are shared with the ETG model etypes.

  The output of this first step is a set of selected ontologies, which best cover the ETG model, and that have been verified fitting the dataset's schema, at both etypes and etype properties levels.

Step 2: Entity Type Recognition(ETR). The main goal here is to predict, for each etype of the ETG model, which etype of the input ontologies, analyzed one at the time, best fits the ETG. In practice, the ETG model's etypes are used as labels of the classification task and, as mentioned in [6], the execution exploits techniques that are very similar to those used in ontology matching (see, e.g., [5]). The final result is a vector of prediction values, returning a similarity score between the ETG model's etypes and the selected ontology etypes.

Step 3: ETG generation. This step identifies, by using the prediction vector produced in the previous step, those etypes and properties from the ontologies which will be added to the final version of the ETG. Notice how this must be done while preserving the mapping with the datasets' schemas. The distinction among common, core and contextual etypes and properties plays an important role in this phase and can be articulated as follows:

- The common etypes should be adopted from the reference ontology, in percentage as close as possible to 100%. This usually results in an enrichment of the top level of the ETG model by adding those top level etypes (e.g., *thing, product, event, location*)

that usually no developer considers, because too abstract, but which are fundamental for building a highly shareable ETG where all properties are positioned in the right place. This also allows for an alignment of those *common isolates* (see Sect. 3) for which usually a lot of (open) data are publicly available (e.g., *street*);
- The core etypes are tentatively treated in the same way as common etypes, but the results highly depend on the ontologies available. Think for instance of the GTFS example above;
- Contextual etypes and, in particular, contextual properties are mainly used to select among ontologies, the reason being that, they allow distinguishing the most suitable among a set of ontologies about the same domain [6].

5   Case Studies

The specification of *iTelos* is in its early phases, in particular in terms of tool support. However, a lot of work has been dedicated to the refinement of the single steps of the overall approach and on their extensive evaluation. In particular, *iTelos* has been validated during the past four Academic Years as part of the.

Knowledge and Data Integration (KDI) class, a six credit course of the Master Degree in Computer Science of the University of Trento.[6] During this class, 2–5 students per group, must generate an EG using the pipeline above starting from.

a high level problem specification. The overall project has an elapsed time of fourteen weeks during which students have to work intensely. We estimate the overall effort each group puts into building an EG in around 4–8 person-months, depending on the case.

As of today we have piloted around 30 projects and 90 evaluations of the *iTelos* methodology as a whole. The details of this work cannot be reported here for lack of space. However the results, restricted to the first three years.

are described in some detail in [2]. We report below the most relevant answers provided by the students, which were asked to fill a very detailed questionnaire containing a set of qualitative questions about the methodology.

- *(Strength)* the step by step, precisely articulated, *iTelos* process is easy to follow;
- *(Strength)* the stepwise iterative evaluation process supports well the refinement of the Entity Graph;
- *(Weakness)* A wrong decision made in the early phases is quite difficult to remedy, with this possibility being very high during the inception phase;
- *(Weakness)* The work between the schema and the data layer is unbalanced in favour of the second, in particular during the *informal modeling* phase. This complicates the synchronization of the work with the possibility of misalignments, mainly because of misunderstandings, that have to be handled very carefully by the project manager.

---

[6] http://knowdive.disi.unitn.it/teaching/kdi/ contains the material used during the last two editions of the course. This material consists of theoretical and practical lectures, as well as demos of the tools to be used, some of which have been mentioned above.

6 Conclusions

In this paper we have introduced *iTelos*, a novel methodology whose ultimate goal is to implement a *circular* development process. By this we mean that the goal of *iTelos* is to enable the development of EGs via the *reuse* of already existing EGs and ETGs, while being simultaneously developed to be later easily *reused* by other applications to come.

References

1. Bella, G., Gremes, L., Giunchiglia, F.: Exploring the language of data. In: Proceedings of the 28th International Conference on Computational Linguistics, pp. 6638–6648 (2020)
2. Bocca, S., Dragoni, M., Giunchiglia, F.: iTelos - case studies in building domain specific knowledge graphs. In: International Semantic Intelligence Conference (2022)
3. Ekaputra, F., et al.: Ontology-based data integration in multi-disciplinary engineering environments: a review. Open J. Inf. Syst. 4(1), 1–26 (2017)
4. Gangemi, A., Guarino, N., Masolo, C., Oltramari, A., Schneider, L.: Sweetening Ontologies with DOLCE. In: Gómez-Pérez, A., Benjamins, V.R. (eds.) EKAW 2002. LNCS (LNAI), vol. 2473, pp. 166–181. Springer, Heidelberg (2002). https://doi.org/10.1007/3-540-45810-7_18
5. Giunchiglia, F., Autayeu, A., Pane, J.: S-match: an open source framework for matching lightweight ontologies. Semantic Web 3(3), 307–317 (2012)
6. Giunchiglia, F., Fumagalli, M.: Entity type recognition – dealing with the diversity of knowledge. In: Knowledge Representation Conference (KR) (2020)
7. Giunchiglia, F., Zamboni, A., Bagchi, M., Bocca, S.: Stratified data integration. In: 2nd International Workshop On Knowledge Graph Construction (KGCW) (2021)
8. Grüninger, M., Fox, M.S.: The Role of Competency Questions in Enterprise Engineering. In: Rolstadås, A. (ed.) Benchmarking — Theory and Practice. IAICT, pp. 22–31. Springer, Boston, MA (1995). https://doi.org/10.1007/978-0-387-34847-6_3
9. Kerjriwal, M.: Domain-specific knowledge graph construction. SCS, Springer, Cham (2019). https://doi.org/10.1007/978-3-030-12375-8
10. Knoblock, C.A., Szekely, P.: Exploiting semantics for big data integration. AI Mag. 36(1), 25–38 (2015)
11. Mylopoulos, J., Borgida, A., Jarke, M., Koubarakis, M.: Telos: Representing knowledge about information systems. ACM Trans. Inf. Syst. (TOIS) (1990)

Methodology for Creating a Community Corpus Using a Wikibase Knowledge Graph

Sara Assefa Alemayehu[1], Kushagra Singh Bisen[1], Pierre Maret[1,4(✉)],
Alexandra Creighton[2], Rachel Gorman[2], Bushra Kundi[2], Thumeka Mgwgwi[3],
Fabrice Muhlenbach[1], Serban Dinca-Panaitescu[2], and Christo El Morr[2]

[1] Université Jean Monnet Saint Etienne, Saint-Étienne, France
pierre.maret@univ-st-etienne.fr
[2] York University, School of Health Policy and Management, Toronto, Canada
[3] School of Gender, Sexuality and Women's Studies,
York University, Toronto, Canada
[4] The QA Company, Saint-Étienne, France

Abstract. The Wikibase environment is provided by the Wikimedia foundation. It provides a suite of tools and a collaborative environment and implements Semantic Web techniques to model and store data. It has been shown that the Wikibase environment is well adapted to store and manage documents with their meta-data and that it can be associated with a search mechanism that facilitates the accessibility to this set of information. Several implementations of such a web platform have been proposed for various communities, but no methodology has been yet proposed. A methodology would help to identify tasks and easily reproduce them for the new implementation of community Wikibases, together with the search function that provides access to the information stored. In this paper, we propose a methodology for domain-specific Wikibase instantiation, for documents and their meta-data, and we identify some tools related to this methodology.

Keywords: Wikibase · Methodology · Semantic web · Domain-specific corpus · Knowledge graph

1 Introduction

With the rise of the Semantic Web [2], technologies assisting better representation of data and entities have emerged. Knowledge Graphs [10] aims to serve as an ever-evolving shared substrate of knowledge within an organisation or community [14]. Searching for domain-specific information on the web is rather tough. Communities employ domain-specific vocabulary and context while accessing information which is often overlooked while searching for information on the web. Knowledge Graph infrastructures such as Wikibase can be employed to

© The Author(s), under exclusive license to Springer Nature Switzerland AG 2022
B. Villazón-Terrazas et al. (Eds.): KGSWC 2022, CCIS 1686, pp. 285–297, 2022.
https://doi.org/10.1007/978-3-031-21422-6_21

build a community-specific knowledge graph to improve domain-specific representation and access. The Wikibase environment is a set of software provided by the Wikimedia Foundation and is the environment behind Wikidata [1]. Wikidata supports the Wikimedia family of projects including Wikipedia, Wikivoyage, Wiktionary, Wikisource, and others. Due to the success of Wikidata, the Wikibase environment has received increasing interest in storing textual data. It is well suited for setting up data archives that interoperate with the semantic web via open standards.

Large amounts of textual data are continuously communicated around the world. Textual data can be unstructured or structured. Unstructured textual data (also called free text) doesn't have a predefined data model, as opposed to structured data which is commonly modelled in a relational or graph data model [5]. A relational or graph data model is a robust data structure used for retrieving, organizing and managing data.

The Wikibase environment implements structured data, and more specifically a relational data model internally, and a graph data model for its interface. Several implementations of community-specific Wikibases have been created. Web sites built on top of these Wikibases facilitate the management and access to the information they contain [4,6]. In [17] (manuscript in evaluation) it has been shown that the Wikibase environment is well adapted for the management of documents and their meta-data. In this implementation, the search function enables free text queries to uniformly access the document meta-data as well as the document contents.

Alternatives to Wikibase are Semantic MediaWiki (SMW) and Cargo [11]. SMW and Cargo are not developed for collaborative creation and maintenance of a knowledge base containing data that can be used by third parties systems. This is enabled in Wikimedia projects, and therefore, in the Wikibase environment. Thus, organizations can curate data for third-party consumption, for instance, GLAM (galleries, libraries, archives, and museums) institutions [3].

Therefore, the Wikibase environment seems well adapted for organizing data and serving as a data store to a web platform for efficiently accessing documents and their meta-data. Despite this, to the best of our knowledge, there is no methodology described to implement this kind of architecture. Thus, in this paper, we propose a methodology to organise the design and the implementation process of domain-specific Wikibases and a search function which substantially helps in the access to community information and documents.

The rest of this paper is organized as follows: In Sect. 2, we list some related work done on methods and tools for building and querying domain-specific corpora and for structuring and implementing community Wikibase for domain-specific documents. In Sect. 3, we describe the methodology we propose, we describe the tools that can be used in each task of the process, and we give an illustration of each task using examples from the Disabilitywiki project. This project aims at collecting information and improving access to disability and human-right related data and documents. It implements a Wikibase and a Natural Language Processing (NLP) enabled multilingual search engine [8]. We conclude and highlight future work in Sect. 4.

2 Related Work

Many studies have been recently conducted on the implementation of document repositories for accessing the information on domain-specific corpora. However unstructured data affect querying process performance and give the difficulty of the user to manage or retrieve it. Many attempts have been made to reorganize or directly process this data. Yafooz et al. discussed in [18] methods of managing unstructured data in the relational database management system and this study showed the significance of managing this data. Furthermore, he highlights the differences in managing such data with relational or NoSQL databases. He presents the methods that are often used to manage unstructured data in relational databases, however, how this structured data is queried by end-users, and what method and tools are used, are not specified.

Pampari et al. proposed in [15] a novel methodology to generate domain-specific large-scale question answering (QA) datasets by re-purposing existing annotations. This study demonstrates an instance of this methodology in generating a large-scale QA dataset for unstructured electronic medical records by leveraging existing expert annotations on clinical notes for various NLP tasks from the community-shared i2b2 datasets. The resulting corpus (emrQA) has 1 million question-logical forms and 400,000+ question-answer evidence pairs. The authors finally characterize the dataset and explore its learning potential by training baseline models for the question to logical form and question-to-answer mapping. Therefore, this paper addresses the lack of publicly available EMR (unstructured Electronic Medical Records) by creating a large-scale dataset emrQA. In this study, they show how to structure and access textual data but no methodology and tools are proposed.

The Enslaved project has developed a platform with linked open data (LOD) i.e. structured data available under an open license under the Wikibase environment, to interconnect individual projects and databases. A LOD-based approach facilitates federated searching and browsing across all linked project data on Enslaved.org [6]. They build a robust, open-source architecture to discover, connect, and visualize 600,000+ people's records and 5 million data points from archival fragments and spreadsheet entries that show the lives of the enslaved in richer detail. The enslaved project doesn't have any automatic data upload workflow and it always requires computer experts to feed information directly to the data store.

Zhou et al. proposed in [19] a real-world dataset from the Enslaved project as a potential complex alignment benchmark. The benchmark consists of two resources, the Enslaved Ontology along with a Wikibase repository holding a large number of instance data from the Enslaved project, as well as a manually created reference alignment between them. The alignment was developed in consultation with domain experts in the digital humanities. The two knowledge graphs and the reference alignment were designed and created by ontologists and historians to support data representation, sharing, integration, and discovery. Additionally, they take advantage of Wikibase as a tool to represent the data, which is convenient for users with any level of expertise in its use. Zhou et al. didn't introduce any methodology and tools suite for their process.

Diefenbach et al. present in [4] how Wikibase is used as the infrastructure behind the "EU Knowledge Graph", which is deployed at the European Commission. This graph mainly integrates projects funded by the European Union and is used to make these projects visible to and easily accessible by citizens with no technical background. Moreover, the authors explain this deployment compared to a more classical approach to building RDF knowledge graphs, and they point to other projects that are using Wikibase as the underlying infrastructure. This paper presents how Wikibase can be used as an infrastructure for knowledge graphs and shows that while Wikibase is not as flexible as a traditional RDF deployment, it offers many out-of-the-box services that are either necessary or convenient for deploying a knowledge graph infrastructure. No methodology is described to build such a system.

Li et al. in [12] propose AliMe KG, a domain knowledge graph in E-commerce that captures user problems, points of interest (POI), item information and relations thereof. It helps to understand user needs, answer pre-sales questions and generate explanation texts. They applied AliMe KG to several online business scenarios such as shopping guides, question answering over properties and selling point generation. In the paper, they systematically introduce how they construct a domain knowledge graph from free text, and demonstrate its business value with several applications.

3   Proposed Methodology

Our target to propose methodology for the design and implementation of systems that enable the implementation of Wikibases and facilitate access to community documents and their meta-data. These systems shall be composed of three main components: a Wikibase instance used to store the meta-data of the documents; File repositories used to store the documents; and a User interface (UI) used to upload, query, and access data and documents.

We propose the description of this methodology in three parts as illustrated in Fig. 1. Each part is divided into several tasks. We use event-driven process chain (EPC) diagrams to describe the control-flow structure of each part as a chain of events and tasks. Part 1 concerns the *domain knowledge modelling and Wikibase setup*. Part 2 relates to the process of *uploading documents*. It is started when the knowledge environment is ready and when documents have to be loaded into the system. And part 3 relates to the implementation of *search function* over the documents and meta-data. It is started once the knowledge environment is ready.

3.1   Domain Knowledge Modeling

After the identification of the domain of interest for the community, this part aims at preparing the knowledge environment in which the targeted system will operate. The tasks of this part are presented in Fig. 2.

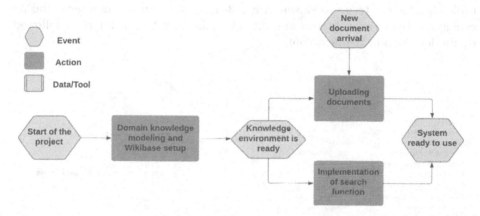

Fig. 1. General description of the methodology

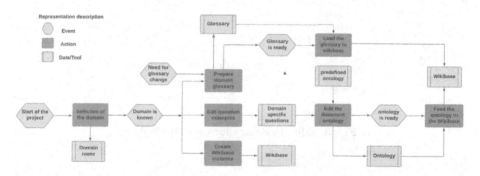

Fig. 2. Part 1 of the methodology: Domain knowledge modelling and Wikibase setup

Prepare Domain Glossary. This task consists of the preparation of domain-specific vocabulary and the definition of terms. The output of this process consists of a list of terms, synonyms and definitions. The glossary content will be added to the Wikibase in a further task.

The glossary will serve as an input of the task *Annotate document* that will associate documents or sub-parts of documents (called Content box in the ontology) to glossary terms.

Preparing the glossary can be done manually by domain experts or with the help of tools reading automatically documents and extracting terms (i.e. MonkeyLearn) [7]. The automatic glossary term extraction shall be seen as a bootstrapping technique that generates an initial list of terms that domain experts will revise [16]. Some of examples of tools used to automatically read documents and extract terms are IBM Watson and MonkeyLearn . The output format of this task should be a CSV file that will serve as an input to the tasks *Load glossary to Wikibase* and *Annotate Document*.

In the Disability wiki project, domain experts have manually prepared in a table the glossary terms as shown in Fig. 3. Each line represents a term and its synonyms. Definitions of terms are edited in a different file, each term is followed by its definition in the next column.

	A	B	C	D	E	F
1	Label	alias1	alias2	alias3	alias4	alias5
2	access to information					
3	access to justice					
4	access to medical treatment					
5	accessibility	access	access audit			
6	accommodations	accommodation				
7	adult guardianship act					
8	advancement	promotion	rise	upgrade	upgrading	advancements
9	age of majority					
10	alternative display format					
11	amendments	amendment				
12	American sign language					
13	analyze	analyzing				
14	assistance	white cane				
15	assistive technologies					
16	auditory handicap					
17	authentic texts					
18	authenticity					
19	autonomy	free will	self-determination	volition	will	
20	awareness					
21	barrier	barricade	environmental barrier			
22	basic human rights					
23	behavior	actions				
24	behavioral					

Fig. 3. Edited glossary for disabilityWiki

Edit Question Examples. In this task, domain experts prepare prototype questions, which will be used to illustrate the types of knowledge expected by end-users and to test and train the search functionality. End-users can also be involved in this task. Experts can be guided by question types such as retrieval definitions, search for document meta-data, search for a unique answer from the documents, search for multiple answers from the documents, and search terms from the text.

As example, domain experts in the illustrative project prepared questions such as:

- What is the definition of health?
- Subject of crpd article 11
- When starts prison uprising? (grammatically wrong questions is supported by the search engine)

Edit the Document Ontology. In this task, one has to create a formal description of the documents and their meta-data used for the targeted domain. The questions previously prepared by domain experts are used as illustrations of the need for knowledge to be represented in the documents. We propose a generic document ontology (Fig. 4) that should be specifically adapted to the types and content of the documents used in the chosen domain.

Editing the ontology can be done manually or utilizing a graphical tool such as Protégé, which is the most used one [13] and supports many input and output formats. The output format of the ontology should be a file in CSV format (generated with a Protégé dedicated plugin). This file will be consumed later in the methodology by the automated task *Feed the ontology into the Wikibase*.

As an illustration from the Disabilitywiki project, we can mention that the generic ontology has been enriched with the concept *Article*. It is related to the concept *Paragraph* with a *Kind of* relation. The concept *Article* is necessary for this domain because the corpus will contain also legal texts which will be better represented by using this new concept.

3.2 Wikibase Setup

Wikibase can be described as a tool that offers a connected infrastructure to implement and maintain a large-scale knowledge graph. Launching the Wikibase environment for collecting and accessing information for a specific domain requires two main processes:

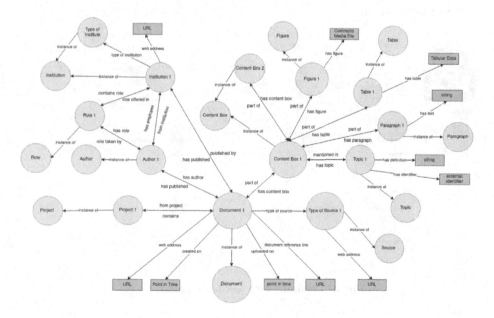

Fig. 4. Edited ontology for DisabilityWiki

Creating Wikibase Instance. The Wikibase proposes a Docker image for the Wikibase stack. After installing Docker, the user should clone locally this Wikibase Docker. image[1]. Wikibase requires also the installation of Docker images for the Blazegraph database, MySQL database and some other components. A Docker-compose file organizes these images. A bot account is necessary for launching bot programs for populating the wikibase with terms: items and properties. Terms can be also added by hand using special Wikibase pages.

Figure 5 shows the fresh look of the Disability wiki main page after installation of the Wikibase.

Load the Glossary to the Wikibase. Using the bot account, bulk edits can be launched using the pywikibot tool provided with the platform. Pywikibot is a Python library that interacts with the MediaWiki API. The terms from the domain glossary shall be loaded with this tool.

Load the Ontology to the Wikibase. As in the previous task, the prepared document ontology is loaded to the Wikibase by using the same bot program.

3.3   Upload Document Process

The uploading process contains several tasks to upload and manage documents and manage the meta-data. It is summarized in Fig. 6. A repository is used to store the documents, and Wikibase is used to store the meta-data as well as the links to the documents.

Fig. 5. Disabilitywiki wikibase main page

---

[1] https://medium.com/@thisismattmiller/wikibase-for-research-infrastructure-part-1-d3f640dfad34.

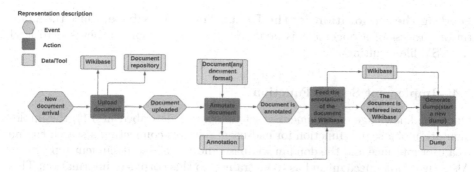

Fig. 6. Part 2 of the methodology: Document upload process

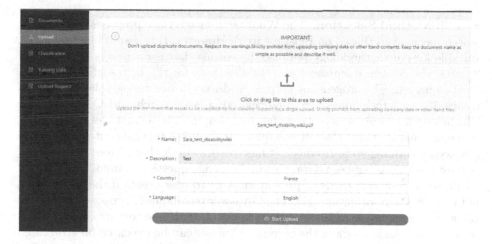

Fig. 7. Document uploading process in disabilitywiki

Upload Documents. Figure 7 shows the user interface for the upload of documents and the edition of the first set of metadata related to them. Some of the meta-data defined in the document ontology (such as author, title, URL, country, institution, etc. in the Disabilitywiki project) is edited or automatically extracted by document analysis. Once checked (and possibly modified), this information is loaded to the Wikibase.

Annotation Process. The annotation process consists of labelling the content of the uploaded documents with glossary terms (meta-data). The document ontology is used to organize structured data from these annotations. The result of the annotation task is stored in a structured way(CSV, RDB).

In the Disabilitywiki project, each paragraph from uploaded documents is analyzed by an integrated machine learning algorithm to be tagged with glossary terms. The algorithm is using Latent Dirichlet Allocation method, which is an unsupervised learning that identifies hidden relationships in data. Figure 8 shows the user interface where a tagged paragraph has to be associated with glossary terms. The terms can be approved, deleted or extended by the user.

Feeding the Annotations of the Document to Wikibase. Once the annotation process of a document is done, the Pywikibot tool can be used to load the CSV file resulting.

3.4 Implement Search Function

Once the knowledge environment is set up (part 1 described in 3.1), the implementation of a search function for end-users requires connecting a search engine to a user interface and the domain wikibase and its related document repository. Also, the search mechanism has to be trained to this corpus of information. This 3rd part of the methodology is summarized in Fig. 9 and is described hereafter.

Search Method Selection. The Wikibase user interface proposes a by-default search mechanism (based on Elasticsearch). However, this search function is not suitable for our methodology because the corpus that is created is composed of 2 parts: the domain documents and the Wikibase for the meta-data.

To query the heterogeneous corpus produced in our methodology, two customized search techniques are implemented: Elasticsearch and QAnswer. The Elasticsearch technique is a classic full-text search. The advantages of this technique are that it is real-time, it can query any kind of textual content, and does not need a training step. However, it has the drawback that it does not consider semantic relations between data, natural language questions cannot be answered with precision, and it tends to present answers to user events if the corpus does not contain the information [17]. The QAnswer technique ?? proposes to analyse the semantic meaning of the question expressed with free-text or keywords and to search for the answers in the corpus. QAnswer can be executed on structured data (for example: knowledge graphs) or unstructured data (documents). The

Fig. 8. Tagged paragraph of documents with glossary terms

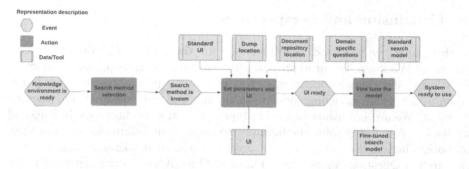

Fig. 9. Part 3 of the methodology: search process implementation

combination of the two techniques (heterogeneous corpus composed of structured and unstructured) has been proposed in [9]. The QAnswer search technique is therefore adapted to our methodology.

Parameters and User Interface Setup. In this task, one has to set the parameters of the selected search method and implement a user interface (UI) for the domain-specific user searches. A standard user interface (if available) can be used as an input and can be customized, or a newly created UI shall be created. The document repository location and Wikibase dump location are examples of parameters that should be entered to parameterize the system.

Fine Tuning of the Search Model. This step is only valid in the case of the selection of a model that has to be trained. The questions previously prepared by domain experts in the *Edit question examples* task (See 3.1) are used as inputs for domain experts to train the system. A standard selected search model is used and trained with these questions to produce a fine-tuned search model. Additional domain questions can be added to possibly train the search model for better results.

An example of the search result on disabilitywiki UI is shown in Fig. 10.

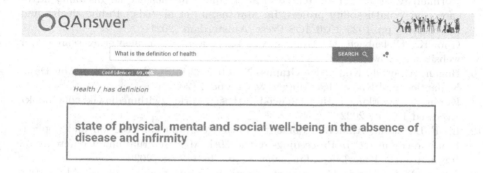

Fig. 10. Search result on Disabilitywiki UI

4   Conclusion and Perspectives

In this paper, we proposed a methodology for the design of a community corpus using Wikibase. In our approach, the community corpus consists of domain knowledge and domain documents. The methodology consists of three parts: domain knowledge modelling, document upload, and implementation of a search function. We are not aware of the existence of such a methodology before, and we hope it could help some communities to implement Wikibases to share their corpora. The search function over the created corpus can take advantage of Elasticsearch or Question-Answering techniques. The QAnswer search engine has the advantage of accepting free text and keyword questions, and it can query uniformly the wikibase information as well as the document content.

The methodology could be integrated into a framework that would incorporate automated or semi-automated tools taking advantage of various techniques such as topic extraction, document summarization, etc.

References

1. Wikidata:wikibase, May (2022). https://cutt.ly/aLF2R6Y Accessed 16 July 2022
2. Berners-Lee, T., Hendler, J., Lassila, O.: The semantic web. Sci. Am. **284**(5), 34–43 (2001)
3. Jeroen De Dauw. Semantic mediawiki vs wikibase vs cargo, May (2022). https://cutt.ly/5LnHLQP Accessed 20 Jun 2022
4. Diefenbach, D., Wilde, M.D., Alipio, S.: Wikibase as an infrastructure for knowledge graphs: the EU knowledge graph. In: Hotho, A., et al. (eds.) ISWC 2021. LNCS, vol. 12922, pp. 631–647. Springer, Cham (2021). https://doi.org/10.1007/978-3-030-88361-4_37
5. Cecilia Eberendu, A., et al.: Unstructured data: an overview of the data of big data. Int. J. Comput. Trends Technol. **38**(1), 46–50 (2016)
6. Enslave.org. People of the historical slave trade. https://enslaved.org/ Accessed 20 Jun 2022
7. Pascual Espada, J., Solís Martínez, J., Cid Rico, I., Emilio Velasco Sánchez, L.: Extracting keywords of educational texts using a novel mechanism based on linguistic approaches and evolutive graphs. Expert Syst. Appl. 213, 118842 (2022)
8. Gorman, R., et al.: The potential of an artificial intelligence for disability advocacy: the WikiDisability project. In: Mantas, J., et al. (eds.) Public Health and Informatics, pp. 1025–1026. IOS Press, Amsterdam (2021)
9. Guo, K., Diefenbach, D., Gravier, C.: Towards question answering search over websites, Qanswer (2022)
10. Hogan, A., et al.: Knowledge Graphs. Number 22 in Synthesis Lectures on Data, Semantics, and Knowledge. Morgan & Claypool (2021)
11. Koren, Y.: Working with mediawiki. https://workingwithmediawiki.com/book/ Accessed 19 July 2022
12. Li, F.-L., et al.: AliMeKG: domain knowledge graph construction and application in e-commerce. In Proceedings of the 29th ACM International Conference on Information & Knowledge Management, pp. 2581–2588 (2020)
13. Musen, M.A.: The protégé project: a look back and a look forward. AI Matters **1**(4), 4–12 (2015)

14. Noy, N., Gao, Y., Jain, A., Narayanan, A., Patterson, A., Taylor, J.: Industry-scale knowledge graphs: lessons and challenges. ACM Queue, 17(2), April (2019)
15. Pampari, A., Raghavan, P., Liang, J., Peng, J.: emrQA: a large corpus for question answering on electronic medical records. arXiv preprint arXiv:1809.00732 (2018)
16. Qawasmeh, O., Lefranois, M., Zimmermann, A., Maret, P.: Computer-assisted ontology construction system: focus on bootstrapping capabilities. In: Gangemi, A., et al. (eds.) ESWC 2018. LNCS, vol. 11155, pp. 60–65. Springer, Cham (2018). https://doi.org/10.1007/978-3-319-98192-5_12
17. Bisen, K.S., et al.: Wikibase as an infrastructure for community documents: the example of the disability wiki platform. In Semantics 2022–18th International Conference on Semantics Systems, Vienna, Austria, September (2022)
18. MS Yafooz, W., ZZ Abidin, S., Omar, N., Drus, Z.: Managing unstructured data in relational databases. In 2013 IEEE Conference on Systems, Process & Control (ICSPC), pp. 198–203. IEEE (2013)
19. Zhou, L., et al.: The enslaved dataset: a real-world complex ontology alignment benchmark using wikibase. In Proceedings of the 29th ACM International Conference on Information & Knowledge Management, pp. 3197–3204 (2020)

Understanding Patient Activity Patterns in Smart Homes with Process Mining

Onur Dogan[1(✉)], Ekin Akkol[2], and Muge Olucoglu[3]

[1] Department of Industrial Engineering, Izmir Bakircay University, 35665 Izmir, Turkey
onur.dogan@bakircay.edu.tr
[2] Department of Management Information Systems, Izmir Bakircay University, 35665 Izmir, Turkey
[3] Department of Computer Engineering, Izmir Bakircay University, 35665 Izmir, Turkey

Abstract. Especially in people over 50 years of age, sedentary lifestyle can cause muscle loss called sarcopenia. Inactivity causes undesirable outcomes such as excessive weight gain and muscle loss. Weight gain can lead to a variety of problems, including deteriorating of the musculoskeletal system, joint problems, and sleep problems. In order to provide better service, it can be beneficial to understand human behavior in terms of health services. Process mining, which can be considered a part of knowledge graphs, is a crucial methodology for process improvement since it offers a model of the process that can be analyzed and optimized. This study uses process mining approaches to examine data from three patient that were collected using indoor location sensors, allowing the collection of flows of human behavior in the home. The analyses indicated how much time was spent by the patients of the house in each room during the day as well as how frequently they occurred. The movement of patients from room to room was observed daily and subjected to a variety of analyses. With the help of user pathways, lengths of stay in the rooms, and frequency of presence, it has been possible to reveal the details of daily human behavior. Inferences about the habits of the participants were revealed day by day.

Keywords: Process mining · Indoor location system · Smart homes · Sensors

1 Introduction

Every human being is born in an environment of different systems. These systems include the family, education, belief, and economic systems [10]. All these systems directly affect human behavior. The creation of human behavior models is so important. Human behavior is determined by biological, psychological, and sociocultural factors that are dependent on multiple variables, making the development of general behavior models complex [36]. Human behavior, on the other hand, evolves in a non-homogeneous manner over time [2]. For instance, the rate of muscle and bone loss increases with age and might approach 50%. To prevent undesirable situations, elderly individuals should be protected against muscle and bone loss while staying at home. Because especially after the age of 50, muscle loss called sarcopenia begins. People over 50 should move frequently to prevent muscle loss from occurring while they are sedentary at home. In addition, there is a risk of gaining more weight in a lifestyle with little movement. Over time, this weight gain brings with it joint disorders, sleep disorders, musculoskeletal disorders.

New opportunities have emerged in the industrial environment as a result of the continuous development of human society and the advancement of scientific and technological innovation and information technology [23]. Innovation in science and technology has significantly altered human society [27]. The Internet of Things (IoT) has emerged as a disruptive technology, with applications in smart cities, retail, agriculture, industrial automation, electronic-health (e-health), mobile-health (m-health), and a variety of other fields [26]. Interconnected wearable devices for physiological activity tracking are rapidly emerging, forming a new market segment known as Wearable IoT [22]. For improving life expectancy and healthcare access, behavior recognition using motion sensors is gaining traction over other systems such as e-healthcare and life-log analysis systems, especially in the healthcare domain [24].

Digital technology and data analytics advancements have created unparalleled potential to evaluate and change health behavior, accelerating science's ability to comprehend and contribute to better health behavior and health outcomes [14]. The complexity and detail of human behavior is captured in digital health data, as well as the confluence of variables that influence behavior at any given time and the internal evolution of behavior over time. These data may contribute to translational science by supplying individualized and timely models of intervention delivery and discovery science by exposing digital signals of health/risk behavior [30]. Initiatives such as Obama's 4P [35] (personalized, predictive, preventive, and participatory) are, in his words, a pioneer in a new model of patient-powered research that promises to accelerate biomedical discoveries and provide clinicians with new tools, knowledge, and therapies to select which treatments will work best for which patients.

Information systems supporting healthcare processes have significant difficulties in terms of design, implementation, and diagnosis since these processes are preeminently heterogeneous and multidisciplinary [11,41]. Machine learning and pattern recognition techniques enable the development of models that represent human behaviors [15]. Although machine learning methods have contributed to individual tracking, they require complex iterations and have difficulty producing understandable visual results [12]. Process mining is a machine learning discipline that infers models from event logs and provides understandable human models by supplying significant human behavior details, which are typically in the form of workflows [13,38]. Workflows are a simple representation of processes that can help human behavior analysis not only detect behavioral changes but also provide an understandable view of a person's patterns and insights [1,18].

Process mining is a new technique in this field that uses data from multiple sources to give detailed models and information on the processes that are actually being executed. Process mining has been used in various studies to identify human activities by treating human routes like a commercial process. Using significant human behavior details, typically in the form of workflows, it is a machine learning discipline that infers models from event logs and produces intelligible human models [12,38]. Workflows are a straightforward depiction of processes that can help with human behavior analysis by not only spotting behavioral changes but also by providing a clear picture of a person's habits and insights [18].

The movements of three retired men over the age of 50 who need sarcopenia treatment in their houses are observed in this study. Being active is very important in the treatment of sarcopenia. Therefore, the behavior of the patients at home (the state of being still) should be monitored.In order to better interpret these behaviors, process mining method was applied by using human activity data collected through sensors. Thus, the time spent in the rooms and the transitions between the rooms were examined. In the rest of the article, firstly, the studies in the literature are explained and the links between them are revealed in Sect. 2. The methodology of the study for understanding patient behavior is described in Sect. 3. The application is thoroughly explained in Sect. 4. Section 5 contains a discussion of the findings and their limitations.

2 Literature Review

Recently, there has been an increase in interest in employing smart home technology to find patterns of human behavior for applications that monitor health. The main goal is to study the behavioral traits of the residents in order to comprehend and foresee their actions that can signify health problems. In this part, we examine previous research in the literature that uses information from smart homes to study user behavior. Table 1 summarizes advantages and disadvantage of the mentioned studies.

[8] focused on the detection of human activity in smart homes using data from smart meters. The paper proposes two approaches to analyze and detect user's routines. The first technique employs a Semi-Markov-Model (SMM) for data training and habit detection, while the second provides an impulse-based method to identify activity in daily living (ADL) that focuses on the temporal analysis of concurrent activities. Similar to this purpose, [33] suggested using sensors classified according to the primary functions of the smart home to identify human activity for elderly people's wellness monitoring. In order to establish the patterns of electrical appliance usage, the study collects preprocessed data from homes. A machine learning-based algorithm is then used to extract the main activities taking place inside the home. The problem is that to completely isolate the primary activities, the work needs to do two phases on the data.

Table 1. Studies based on human activity patterns

Study	Technique	Advantage	Disadvantage
[8]	Semi Markov Model	Analyze and detect user's routines	Not given
[33]	Machine Learning	Human activity detection for wellness monitoring of elderly people	The study has to perform two steps on the data to completely isolate the main activities
[4]	Decision Tree Correlations	A model that can forecast how often appliances will be used in homes so that the system may plan energy production and consumption and determine which appliance will be used each hour	The study takes only the last 24 hour window along with appliance sequential relationships
[5]	Decision Tree Correlations	The approach is both a knowledge driven and data driven one	The study takes only the last 24 hour window along with appliance sequential relationships
[21]	Bayesian Networks	A general method to predict users requests for services in energy consumption is proposed	Not suitable for real world scenarios
[19]	C-Means Clustering	They show that revealing household characteristics from smart meter data is feasible and provides an appealing visualization of general data patterns	Changes during device use are not taken into account
[18]	Process Mining	A tool and a process mining-based methodology that, through the use of indoor location systems, allows health staff to not only represent the process, but also to know precise information about its deployment in an unobtrusive and transparent way	Not given
[28]	Process Mining	A graphical insight about the human activity on daily basis	Only most frequent activity sequences are examined
[32]	Process Mining	The relationship between workload and service time is investigated with regression analysis	Simulation models, which is the method used in the study, are often based on incorrect assumptions
[31]	Process Mining	Support the redesigning and personalization of decision support systems	The study needs detailed navigation behavior of different target groups
[17]	Process Mining	The algorithm allows for the inference of parallel activities and sequences	The study needs to investigate data with more information about the user's daily actions
This study	Process Mining	Process management-based analysis of household activities of people in need of sarcopenia treatment	More participants can be included in the study to produce more results that can be validated

Studies [4] and [5], used time series multi-label classifier to forecast appliance usage based on decision tree correlations, however, the study only considers the most recent 24 h and the sequential relationships between appliances. Hawarah et al. [21] used Bayesian networks to forecast occupant behavior from collected smart meters data. It suggests behavior as a service based on a single appliance, but does not offer a model that can be used in scenarios that occur in the real world. Gajowniczek and Zabkowski [19] applied hierarchical and c-means clustering to discover utilization trends while taking into account the ON and OFF status of the appliances. The study, however, does not take into account the length of appliance usage or the predicted variations in the order of appliance utilization.

Previous studies using process mining to understand the trajectories and individual behavior of people have been performed in nursing homes [18], in a shopping mall [14] and in the daily living of individuals [37]. Maarif [28] presented human daily activity patterns in a graphical presentation using process mining. By taking into account the connection between workload and service time, Nakatumba and van der Aalst [32] looked into how workload affects service times. Additionally, it was used to study the habits of people in operating rooms using 25-week data from nine individuals that was gathered using RFID technology [17]. When Maruster et al. [31] used process mining to connect insights with decision-making processes, they created a user behavior model for farmers' behavioral patterns.

These studies show process mining can support to determine the personal behavior changes in specific day to day processes. However, human activities are not static and change in a not homogeneous way. The discovered trajectories by process mining algorithms are hard to interpret for experts. It creates the undesired spaghetti effect with high variability in human behaviors. One solution to overcome the spaghetti effect is to group similar behaviors [14]. Therefore, various grouping approaches including similar paths in process mining have been developed to decrease the spaghetti effect [6,40].

3   Methodology

As can be seen in Fig. 1, where all the stages are given together, the solution presented in the article consists of four main components to track patients in healthcare domain . These stages are data collection, preprocessing of the data collected by process mining, extracting the behavior of the user and finally creating the behavior map.

Technological infrastructures (hardware and software) have been used in different methods that enable patient to be monitored in their home environments. Among these technologies used are a wide variety of technologies such as RFID [25], Bluetooth [34] and ZigBee [16]. Such systems have previously been used to support people's activity recognition in smart homes [3], AAL solutions [7], and the design of activity primitives [29].

Phase 1 is the monitoring of human activities in all rooms of a house through a sensor located in each room. This study is based on a standard house model. It is thought that each house has five main rooms. These rooms; Bedroom, Living room, Kitchen, Bathroom and Hall. Sensors are installed and labeled in each room of individual residences. Thus, each position provided by the sensor corresponds to a single room and a single user. The sensors work with bluetooth technology. And a signal exchange is provided from these sensors every three seconds. If there is a signal from which sensor, it indicates that this person is in that room. Depending on the location of the wristband on the person's arm, if data is also received from another sensor in the house, the measurement is made by taking the data of the closest signal.

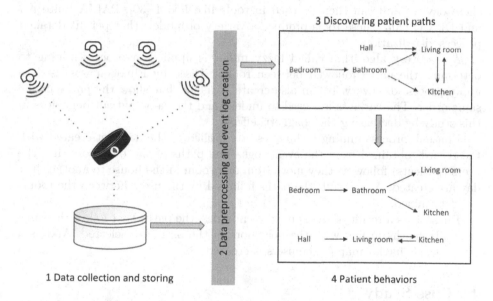

Fig. 1. Methodology for understanding patient behavior

Phase 2, detailed models of human behavior are created using the data received from the System. And process mining techniques are used to extract meaningful information from these data. Event logs containing the information of the records of each user's own ID are created. And it is saved in the dataset. An information table is created on the relevant day and time, when the person entered the room for the first time and how long they stayed in that room.

PatientID, a unique identifier (ID), was created to track every patient. The location shows the rooms in the smart home and is used to build patients' paths. The dataset also contains timestamp data, including start and end times for each localization data. Sessions are separated by day by day.

We applied some filters to extract more information. The room duration filter alters that the occurrence duration must be more than one minute. Otherwise, it is assumed that the patient data is captured while walking instead of representing human behavior. After the room duration filter, we fuse successive patient data to bypass the disappearance. This filter adjusts the ending time of occurrence of a room by subtracting one second before the starting of the consequent event. Only the signals from the nearest location detected by iBeacon devices are assumed to be where the customer is. The other signals are ignored and can be considered missing data.

Utilizing data from the system, process mining techniques are utilized to provide detailed models and knowledge about human behaviour. An event log containing the pertinent information is created using the data extracted from these devices and can then be used in tools like ProM [39], PALIA Suite [9], and DISCO [20]. These tools produce a variety of models that permit detailed process visualizations.

A discovery algorithm called fuzzy miner is applied in process mining to determine the paths followed between rooms. Disco by Fluxicon was used to apply process discovery. It can also create variants that show the paths in the same order. The variants are used to understand the same patient behaviors in this study by decreasing the spaghetti effect.

Phase 3, process mining techniques are applied to the data set created with the records obtained from the event logs. The paths of the behaviors that the users (patients) follow as they move from one room in the house to another in a day are created. As a result, the paths followed by the users between the rooms are determined.

Phase 4, as a result of the applied techniques, the paths that follow the same or similar paths in line with the behaviors of the users are selected. And as a result, the behavior map of the users is created.

4   Case Study

This study used data collected from three different participants. When the time spent in each of the five rooms in the houses is examined, it is discovered that the room in which the most time is spent varies from day to day. According to the number of incidents that occurred in the rooms, the participants were mostly seen in the living room. The place where they were seen the least was the hall. While the number of occurrence in the bedroom and kitchen are nearly equal, the number of occurrence in the bathroom is the second lowest after the hall. Although the participants occurrence in the hall the least, they spent the most time in the hall with an average of 292 min. The average time spent in the bedroom is 213 min per week, the average time spent in the hall is 164 min per week, the average time spent in the living room is 131 min, and the average time spent in the kitchen is 36 min. Figure 2 shows the daily average amount of time spent and the number of occurrence in rooms.

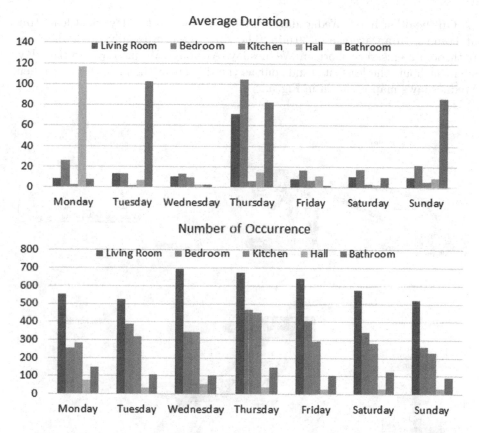

Fig. 2. Daily comparison for average duration and number of occurrence

Furthermore, many different results were revealed based on analyses conducted on weekdays and weekends within the scope of the study. For all week, while the bedroom was the room where the most time was spent, the living room was the room with the most activity. The time spent at home and the activity in the rooms are higher on weekdays than on weekends. In addition to all these, while the average time spent in all rooms on weekdays is 77.2 min, the average time spent in all rooms on weekends is 45.90 min. The average number of movements seen in all rooms on weekdays is 7466 and on weekends the number of movements in all rooms is 2490. The map showing the movements during weekdays and weekends was given in Fig. 3.

It has been observed in Fig. 4 that on Thursdays, much more time is spent at home than on other days. The time spent by the participants in the living room and bedroom is much longer than the other days. In addition, the number of occurrence in the bedroom and kitchen is higher than the other days.

On the other hand, Wednesdays were determined as the day spent least time at home with an average of 39 min. The time spent especially in the hall and bathroom seems very short on Wednesday. Although Wednesdays are the days spent at home the least, it stands out as the day spent the most in the kitchen. Wednesday's map is shown in Fig. 5.

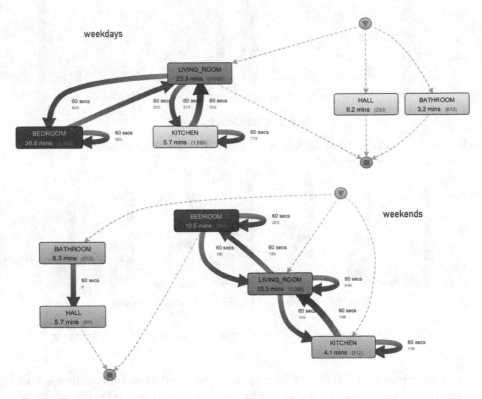

Fig. 3. Weekly comparison for average duration and number of occurrence

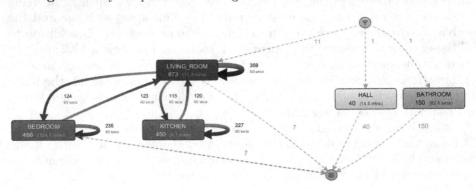

Fig. 4. Map showing movements on Thursday

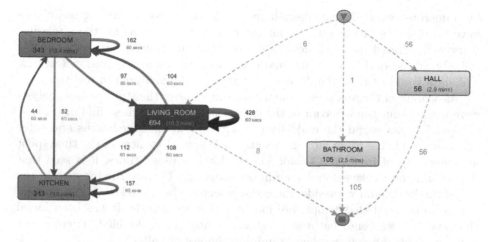

Fig. 5. Map showing movements on Wednesday

After Wednesday, Friday and Saturday are the days where the least time is spent at home. It is understood that people spend more time outside, especially on holidays and Fridays, the day before the holiday.

Participants who were tracked by sensors as part of the study were identified as P1, P2, and P3. When the activities of the participants are examined, the house of participant P1 has the most activity but the least amount of time spent. The P1 participant spends the most time in the kitchen. The P2 participant is the participant who spends the most time at home. P2 spent most of his time in the bedroom. The participant who spends the least amount of time at home is the P3 participant. The most of P3's time was spent in the living room.

5   Conclusion and Discussions

In this study, it was aimed to understand and analyze the individual behavior of users (patients) in the home. Human behavior models were discussed with process mining techniques. Because of the sensors located in all the rooms of the users' houses, it was determined how much time they spent in each room. Patient paths between the rooms were followed. By comparing the similarities in these ways, individual behavior maps of the people were drawn.

With the development of IoT technologies, it has become easy to collect data from people with wearable technologies. The analysis of human behavior is one of the important areas where this data collected through sensors can be used. Three people were given wristbands to wear as part of the study, and their daily activities were tracked and recorded in order to examine the patients' at-home behaviors. The development of general behavioral models is complicated by numerous elements that affect human behavior. Process mining techniques have been selected to investigate unique behavioral patterns for these reasons. Sensors integrated into IoT devices ensure that the process proceeds quickly, reliably and smoothly with automatic monitoring and configuration features. The data is recorded by the devices so that it can be analyzed in real time.

As a result of the analyzes made within the scope of the study, many analysis were made about the behavior of the patients. The first of these findings is that the participants spend the majority of their time in their bedrooms and enter and leave the living room most frequently. It is seen that most of the time spent at home is spent in sleep. Considering the kitchen usage times, it is seen that it is 5.7 min on weekends and 4.1 min on weekdays. Despite the increase in the use of the kitchen on weekends, these times seem to be quite low. Based on this, it can be deduced that people are mostly fed from outside. It has been found that over the weekend, bathroom usage has practically doubled. There aren't any noticeable differences when examining the use of hall.

For future studies, researchers can plan to analyze a more comprehensive map of individual behaviors by using advanced wristbands that can provide measurements of patients' whole body movements, blood pressure, and heart rate. It would be more insightful to frame the study based on various demographic groups but the number of patients included in the research has limited his aim. Further studies can consider to extend the sample size. One of the limitations of the study is data loss due to possible network problems. In addition, the study was carried out with a limited number of people. Reliability of the results should be verified. For example, the time spent in the kitchen of 4–5 minutes is rather low compared to the amount of time someone needs to prepare a meal. It was concluded that the participants were fed from outside. However, it was not verified that the conclusion is correct.

References

1. van der Aalst, W., et al.: Process mining manifesto. In: Daniel, F., Barkaoui, K., Dustdar, S. (eds.) BPM 2011. LNBIP, vol. 99, pp. 169–194. Springer, Heidelberg (2012). https://doi.org/10.1007/978-3-642-28108-2_19
2. Alland, A.: Evolution and Human Behaviour: An Introduction to Darwinian Anthropology. Routledge, London (2012)
3. Álvarez-García, J.A., Barsocchi, P., Chessa, S., Salvi, D.: Evaluation of localization and activity recognition systems for ambient assisted living: The experience of the 2012 EvAAL competition. J. Ambient Intell. Smart Environ. 5(1), 119–132 (2013)
4. Basu, K., Debusschere, V., Bacha, S.: Appliance usage prediction using a time series based classification approach. In: IECON 2012–38th Annual Conference on IEEE Industrial Electronics Society, pp. 1217–1222. IEEE (2012)
5. Basu, K., Hawarah, L., Arghira, N., Joumaa, H., Ploix, S.: A prediction system for home appliance usage. Ener. Build. **67**, 668–679 (2013)
6. Bose, R.J.C., Van der Aalst, W.M.: Context aware trace clustering: towards improving process mining results. In: Proceedings of the 2009 SIAM International Conference on Data Mining, pp. 401–412. SIAM (2009)

7. Byrne, C.A., Collier, R., O'Hare, G.M.: A review and classification of assisted living systems. Information 9(7), 182 (2018)
8. Clement, J., Ploennigs, J., Kabitzsch, K.: Detecting activities of daily living with smart meters. In: Wichert, R., Klausing, H. (eds.) Ambient Assisted Living. ATSC, pp. 143–160. Springer, Heidelberg (2014). https://doi.org/10.1007/978-3-642-37988-8_10
9. Conca, T., et al.: Multidisciplinary collaboration in the treatment of patients with type 2 diabetes in primary care: analysis using process mining. J. Med. Int. Res. 20(4), e8884 (2018)
10. DANIŞ, A.G.M.Z.: Davranış bilimlerinde ekolojik sistem yaklaşımı. Sosyal Politika Çalışmaları Dergisi 9(9), 45–54 (2006)
11. Dogan, O.: Process mining for check-up process analysis. IIOABJ 9(6), 56–61 (2018)
12. Dogan, O.: Discovering customer paths from location data with process mining. Euro. J. Eng. Sci. Technol. 3(1), 139–145 (2020)
13. Dogan, O.: Process mining based on patient waiting time: an application in health processes. Int. J. Web Inf. Syst. (ahead-of-print) (2022)
14. Dogan, O., Bayo-Monton, J.L., Fernandez-Llatas, C., Oztaysi, B.: Analyzing of gender behaviors from paths using process mining: a shopping mall application. Sensors 19(3), 557 (2019)
15. Duda, R.O., Hart, P.E., et al.: Pattern Classification. John Wiley & Sons, Inc. (2006)
16. Fang, S.H., Wang, C.H., Huang, T.Y., Yang, C.H., Chen, Y.S.: An enhanced ZigBee indoor positioning system with an ensemble approach. IEEE Commun. Lett. 16(4), 564–567 (2012)
17. Fernández-Llatas, C., Benedi, J.M., García-Gómez, J.M., Traver, V.: Process mining for individualized behavior modeling using wireless tracking in nursing homes. Sensors 13(11), 15434–15451 (2013)
18. Fernandez-Llatas, C., Lizondo, A., Monton, E., Benedi, J.M., Traver, V.: Process mining methodology for health process tracking using real-time indoor location systems. Sensors 15(12), 29821–29840 (2015)
19. Gajowniczek, K., Zabkowski, T.: Data mining techniques for detecting household characteristics based on smart meter data. Energies 8(7), 7407–7427 (2015)
20. Günther, C.W., Rozinat, A.: Disco: discover your processes. BPM (Demos) 940(1), 40–44 (2012)
21. Hawarah, L., Ploix, S., Jacomino, M.: User behavior prediction in energy consumption in housing using bayesian networks. In: Rutkowski, L., Scherer, R., Tadeusiewicz, R., Zadeh, L.A., Zurada, J.M. (eds.) ICAISC 2010. LNCS (LNAI), vol. 6113, pp. 372–379. Springer, Heidelberg (2010). https://doi.org/10.1007/978-3-642-13208-7_47
22. Hiremath, S., Yang, G., Mankodiya, K.: Wearable internet of things: concept, architectural components and promises for person-centered healthcare. In: 2014 4th International Conference on Wireless Mobile Communication and Healthcare-Transforming Healthcare Through Innovations in Mobile and Wireless Technologies (MOBIHEALTH), pp. 304–307. IEEE (2014)
23. Holmström, J., Holweg, M., Lawson, B., Pil, F.K., Wagner, S.M.: The digitalization of operations and supply chain management: theoretical and methodological implications (2019)

24. Jalal, A., Quaid, M.A.K., Hasan, A.S.: Wearable sensor-based human behavior understanding and recognition in daily life for smart environments. In: 2018 International Conference on Frontiers of Information Technology (FIT), pp. 105–110. IEEE (2018)
25. Li, N., Becerik-Gerber, B.: Performance-based evaluation of RFID-based indoor location sensing solutions for the built environment. Adv. Eng. Informat. **25**(3), 535–546 (2011)
26. Li, S., Xu, L.D., Zhao, S.: The internet of things: a survey. Inf. Syst. Front. **17**(2), 243–259 (2015)
27. Li, Z.: Research on human behavior modeling of sports culture communication in industrial 4.0 intelligent management. Comput. Intell. Neurosci. 2022 (2022)
28. Ma'arif, M.R.: Revealing daily human activity pattern using process mining approach. In: 2017 4th International Conference on Electrical Engineering, Computer Science and Informatics (EECSI), pp. 1–5. IEEE (2017)
29. Manzoor, A., et al.: Analyzing the impact of different action primitives in designing high-level human activity recognition systems. J. Ambient Intell. Smart Environ. **5**(5), 443–461 (2013)
30. Marsch, L.A.: Digital health data-driven approaches to understand human behavior. Neuropsychopharmacology **46**(1), 191–196 (2021)
31. Maruster, L., Faber, N.R., Jorna, R.J., van Haren, R.J.: A process mining approach to analyse user behaviour. In: WEBIST (2), pp. 208–214 (2008)
32. Nakatumba, J., van der Aalst, W.M.P.: Analyzing resource behavior using process mining. In: Rinderle-Ma, S., Sadiq, S., Leymann, F. (eds.) BPM 2009. LNBIP, vol. 43, pp. 69–80. Springer, Heidelberg (2010). https://doi.org/10.1007/978-3-642-12186-9_8
33. Ni, Q., Garcia Hernando, A.B., De la Cruz, I.P.: The elderly's independent living in smart homes: a characterization of activities and sensing infrastructure survey to facilitate services development. Sensors **15**(5), 11312–11362 (2015)
34. Rida, M.E., Liu, F., Jadi, Y., Algawhari, A.A.A., Askourih, A.: Indoor location position based on bluetooth signal strength. In: 2015 2nd International Conference on Information Science and Control Engineering, pp. 769–773. IEEE (2015)
35. Riley, W.T., Nilsen, W.J., Manolio, T.A., Masys, D.R., Lauer, M.: News from the NIH: potential contributions of the behavioral and social sciences to the precision medicine initiative. Transl. Behav. Med. **5**(3), 243–246 (2015)
36. Sanchez-Calzon, A.B., Meneu, T., Traver, V.: Semantic technologies for the modelling of human behaviour from a psychosocial view. Semantic Interoperability: Issues, Solutions, and Challenges, p. 49. River Publishers, Roma, Italy (2012)
37. Sztyler, T., Carmona, J., Völker, J., Stuckenschmidt, H.: Self-tracking reloaded: applying process mining to personalized health care from labeled sensor data. In: Koutny, M., Desel, J., Kleijn, J. (eds.) Transactions on Petri Nets and Other Models of Concurrency XI. LNCS, vol. 9930, pp. 160–180. Springer, Heidelberg (2016). https://doi.org/10.1007/978-3-662-53401-4_8
38. van der Aalst, W.: Data Science in Action. In: Process Mining, pp. 3–23. Springer, Heidelberg (2016). https://doi.org/10.1007/978-3-662-49851-4_1
39. van Dongen, B.F., de Medeiros, A.K.A., Verbeek, H.M.W., Weijters, A.J.M.M., van der Aalst, W.M.P.: The ProM framework: a new era in process mining tool support. In: Ciardo, G., Darondeau, P. (eds.) ICATPN 2005. LNCS, vol. 3536, pp. 444–454. Springer, Heidelberg (2005). https://doi.org/10.1007/11494744_25

40. Veiga, G.M., Ferreira, D.R.: Understanding spaghetti models with sequence clustering for ProM. In: Rinderle-Ma, S., Sadiq, S., Leymann, F. (eds.) BPM 2009. LNBIP, vol. 43, pp. 92–103. Springer, Heidelberg (2010). https://doi.org/10.1007/978-3-642-12186-9_10

41. De Weerdt, J., Caron, F., Vanthienen, J., Baesens, B.: Getting a grasp on clinical pathway data: an approach based on process mining. In: Washio, T., Luo, J. (eds.) PAKDD 2012. LNCS (LNAI), vol. 7769, pp. 22–35. Springer, Heidelberg (2013). https://doi.org/10.1007/978-3-642-36778-6_3

String Matching Based Framework for Online Hindi Question Answering System

Shikha Mehta[(⊠)], Sakshi Gupta, Raashi Agarwal, Shrashti Trivedi,
and Prajjwal Dubey

Computer Science Engineering and Information Technology, Jaypee Institute of
Information Technology, Noida, India
shikha.mehta@mail.jiit.ac.in

Abstract. With the expansion of digital communication on the web,
the necessity to process specific details has grown enormously in the last
few years. However, communication is not restricted to one language and
the WWW is loaded with multilingual data. Nevertheless, the national
Indian languages, i.e., Hindi is under-represented and overlooked on the
web. Therefore, in this paper a hindi language specific question answering framework is developed. This work is mainly focused on questions
with single line or one word answers. The work employs an information
retrieval based approach to process the questions and answers. The performance of the proposed model is evaluated with respect to the three
types of queries and the results show 93.33% of accuracy.

Keywords: Hindi text processing · Question answering system ·
String matching · Levenshtein algorithm · Multilingual system keyword

1 Introduction

The mushrooming growth of interactions over online platforms has paved the
way for new opportunities and challenges. There has been a paradigm shift from
scarcity or unavailability of data to abundance of data. In the last few decades,
the problem of getting data has been transformed to getting precise or relevant
data and information. Another challenge is that WWW is full of multilingual
information as people across the world prefer to interact in their own native languages. Extracting, understanding and processing natural languages to answer
the queries of billions of users escalates the complexity of information retrieval
systems further. Approximately 6,500 languages are spoken across the world and
around 22 languages are spoken in India itself. India is the second most populated country in the world but national Indian languages like Hindi are still
under-represented on the WWW. Research community is more inclined towards

developing the natural language processing models for the English language. The reason being scarcity of tools and resources for Indian Languages [7,10,11]. Authors also emphasized that despite Hindi being the fourth most spoken language, development in the field of Hindi is not proportional. Prevailing Natural Language Understanding (NLU) models do not perform well on Indian languages which further discourage developers from building web applications in Hindi or other native languages. On Dec 2, 2021, Linkedin [10], one of the largest online professional networking platforms launched in Hindi to cater the 600 million Hindi speaking community [8] It was launched in 2003 in English language and after 18 years it is launched in Hindi. This clearly indicates the research gaps in natural language processing models for native languages.

The main objective of every question answering system is to answer "WHO", "WHAT", "WHOM", "WHERE", "WHEN", "HOW" and "WHY" [1]. With the emergence of various voice based tools like Siri, Alexa etc, natural language processing is becoming more prominent in all fields. In the literature, various English/Hindi question answering systems have been proposed. In [1,2] QA was developed which mainly focused on extracting the passages by computing lexical similarity. They ignored semantic relationships of sentences in the text which could probably help in identifying the sentences with answers. This set of curated sentences is also known as a snippet that was taken care of in [4]. Authors developed algorithms for automatic generation of snippets and used deep neural network algorithms to develop the model. In [5] authors employed a multilingual variant of BERT (m-BERT) to develop a question-answering (QA) system which extracts the answer from the snippet available in any language. But this application does not serve the purpose as the user may not be aware of the language of answer. In [6] authors employed a similarity function based approach to develop hindi question answer based system. In [7,9] authors presented a review of linguistic resources, tools and techniques which are required to develop a Hindi QA system. In [8] authors evaluated the performance of similarity techniques used for classification in hindi language question answering systems.

This work presents a pattern matching based method for processing the Hindi questions and answers. The performance of the model is evaluated for 3 types of queries. Rest of the paper is organized as follows: Sect. 2 presents question answering model in detail. Section 3 presents experiments and analysis of results followed by limitations and conclusion in Sect. 4 and Sect. 5 respectively.

2   Question Answering System Based on Pattern Matching

This work utilizes the strength of information retrieval(IR) to develop a Hindi question answering system. IR is the art and science of searching for information in documents, searching for documents themselves, searching for metadata which describes documents, or searching within databases, whether relational standalone databases or hypertext networked databases such as the Internet or intranets, for text, sound, images or data. Text similarity is one of the efficient ways to retrieve the relevant text by computing the relationship between two snippets. The similarity can be computed between sentences, words, paragraphs and documents for proper categorization to retrieve the most relevant answers to users' queries. Finding similar texts is not a simple task as slight change in text leads to dissimilarity. Therefore, approximate text matching algorithms are more suitable as compared to the exact string matching algorithms. Approximate algorithms help in handling grammatical mistakes and minor differences in the text. This work employs the Levenshtein distance method to compute the similarity. The overall architecture of Hindi language based question answering system shown in Fig. 1:

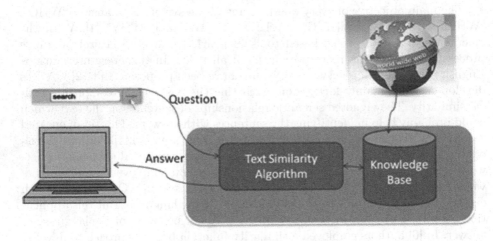

Fig. 1. Architecture of question answering system

The detailed working of question answering system is as follows: Proposed model uses the power of information retrieval to solve this problem efficiently. The main idea is to build a Hindi question answering web application for multiple domains where a user can post a query and get an answer to the desired question from a particular domain. For the same, the prime task is to prepare a repository of Question-Answer pairs beforehand which is a multi-domain knowledge base for the application and then whenever a query is passed, all that is required is to look for similar kinds of questions in the knowledge base for a particular domain in order to return the corresponding appropriate answer. The whole working of the system is divided into two phases- Online phase and Offline phase. Knowledge base creation is the prerequisite and is done offline. Rest of the task is performed online when users post a query to the application.

Knowledge Base Creation(Offline Phase): Despite the fact that petabytes of data is available over the WWW, availability of desired data is a challenging task. Due to unavailability of domain specific question answer databases, it was scrapped from WWW. Parsers were created to extract the question and answers from the various web applications. As each and every website has a different design, a separate parser had to be created for each website. The database was created for about 5000 question answers covering various major domains like science, history, sports, and other miscellaneous questions related to general knowledge.

"किस गवर्नर जनरल कोभारतीय प्रेस का मुक्तिदाता कहा जाता है?"

Fig. 2. Question 1

"जब नींबू के रस को खाने के सोडे पर डाला जाता है तो क्या होता है?"

Fig. 3. Question 2

Step 1: Similar Question Extraction(Online Phase): The Online phase begins when the users posts a query to the web application. The users are liberal to frame their queries in any format without any strict rules to be followed and from any domain. For example, a user who has to know about history may ask as mentioned in Fig. 2. Or if user wants to know about some science facts may ask as mentioned in Fig. 3. The user completes the question processing module by pressing the enter key so that the system will consequently alert the server of an incoming request.

Step 2: Answer Processing at the Server(Online Phase): In the answer processing phase, the primary task is to find the similar question from the knowledge base; the input question here is further passed as input to the Levenshtein distance algorithm. Levenshtein Distance algorithm(LDA) is an algorithm employed for approximate string matching that is to calculate the distance between two strings. LDA is used to measure the difference between two sequences of sentences/words that is Levenshtein distance between two strings a, b (of length |a| and |b| respectively) is given by lev(a,b) where

$$
\mathrm{lev}(a,b) = \begin{cases} |a| & \text{if } |b| = 0, \\ |b| & \text{if } |a| = 0, \\ \mathrm{lev}(\mathrm{tail}(a), \mathrm{tail}(b)) & \text{if } a[0] = b[0] \\ 1 + \min \begin{cases} \mathrm{lev}(\mathrm{tail}(a), b) \\ \mathrm{lev}(a, \mathrm{tail}(b)) \\ \mathrm{lev}(\mathrm{tail}(a), \mathrm{tail}(b)) \end{cases} & \text{otherwise.} \end{cases}
$$

where the tail of some string x is a string of all but the first character of x, and x[n] is the nth character of the string x, starting with character 0. The first element in the minimum corresponds to deletion (from a to b), the second to insertion and the third to replacement.

In approximate string matching, the objective is to find matches for short strings in many longer texts, in situations where a small number of differences are to be expected. This algorithm is widely used in NLP for spell checking and other tasks. However the high computation cost of this algorithm prohibits it from being applied in large-scale applications. The Levenshtein distance algorithm does a character-character comparison of strings. The result (edit distance) of each comparison is then stored temporarily (e.g. in an array) alongside the extracted question-subjects. The question subjects are stored temporarily in order to serve as key for each edit distance calculated. This enhances seamless retrieval of the best QA pair from the database later. Once all question subjects have been extracted, compared and temporarily stored, the minimum of these edit distances is calculated. This is done to identify the question-subject that is the shortest and most similar to the query. Then this minimum score obtained is compared to the threshold value set in the code and only if the obtained score is below the threshold value then the QA pair matching this minimum edit distance is then selected and returned to the user using the temporarily stored question-subjects as the key for retrieval.

After the system has successfully looked for a similar question in the knowledge base, the system returns the corresponding answer stored for that question. And if no such question was found, an appropriate message is displayed.

3   Experiments and Results

For any system to be successful, a strong knowledge database is the key. Developing a Question Answering system for Hindi language is a challenging task as it requires a comprehensive knowledge base. Due to limitations in the availability of the questions for diverse domains, the primary task was to create the data set or knowledge base for questions and answers. To create this knowledge base, up to five thousand questions and answers pairs were scrapped from various resources on WWW. The database had questions and answers under various major domains like science, history, sports, and other miscellaneous questions related to general knowledge. Figure 4 shows the snapshot of the knowledge base.

3.1   Analysis and Discussion of Results

Finding similar texts is not an easy task since a slight change in text can result in dissimilarity and thus to avoid this, approximate string matching LDA is used. This enables the system to accept grammatical mistakes and minor differences in the text. As every user has its own unique way to frame the query, the performance of the system is evaluated by formulating the similar but not same queries as shown in Fig. 5. In the figure, first column represents query framed by user,

Question	Answer
पौधे क्या उत्सर्जित करते हैं	रात में कार्बन
ISRO में S का अर्थ क्या है	Space
हड्डियों एवं दाँतों में मुख्य रूप से कौन-सा रसायन होता है	कैल्शियम
रसोई गैस का मिश्रण क्या है	ब्यूटेन एवं प्रोपेन का
'हीमोफीलिया' एक आनुवंशिक रोग है; जिसका क्या परिणाम है	रक्त का नहीं जमना
मनुष्य का सामान्य रक्तचाप कितना होता है	120/80
ऊष्मा का सबसे कम ऊष्मारोधी धातु कौन सी है	एल्युमीनियम
वाट को किसमें प्रकट कर सकते हैं	जूल प्रति सेकण्ड में
एल्कोहॉल उद्योग में किस कवक का प्रयोग होता है	यीस्ट
कपड़े से स्याही और जंग के धब्बे छुड़ाने के लिए किसका प्रयोग होता है	ईधर
मलेरिया किसके द्वारा होता है	मादा ऐनोफिलीज द्वारा
डॉक्टरों के द्वारा प्रयुक्त शब्द 'CAT' स्कैन का क्या अर्थ है	कम्प्यूटराइज्ड एक्सियल
पौधे किस विधि से भोजन का निर्माण करते हैं	प्रकाश संश्लेषण
ट्रांसफॉर्मर का प्रयोग किसके नियन्त्रित करने में होता है	धारा
पसीना निकलने से शरीर का सबसे उपयोग कार्य क्या होता है	शरीर का ताप नियन्त्रित होना
ग्रेनाइट; ग्रेफाइट और बैसाल्ट में से कौन-सी रूपान्तरि चट्टान है	ग्रेनाइट
टिटेनस रोग किस जीवाणु से होता है	क्लोस्ट्रीडियम टिटेनी
एन्थ्रोपोलॉजी क्या है	मानव विज्ञान का अध्ययन
मानव शरीर की किस ग्रन्थि को 'मास्टर ग्रन्थि' कहा जाता है	पीयूषिका

Fig. 4. Sample of Hindi question answer knowledge base

the second column is the most similar question stored in our knowledge base and the last columns represent the corresponding answer with the score that basically tells how similar the questions are, 1 being exactly same. The results of all the queries came out to be correct. But there might be cases in which no similar question is found or the score is below 0.6, in such cases an appropriate message is displayed to the user.

In order to perform in-depth analysis, the system is tested for 3 main categories of queries. The tests on all the categories were performed and the results were accurate according to the requirements specified. Figure 6, Fig. 7 and Fig. 8 below depicts the responses of the different types of queries returned to the user. In the figures, the first column represents a query proposed by the user, the second column is the most similar question our model could find and the last columns represent the corresponding answer with the score that basically tells how similar the questions are, 1 being exactly the same.

Queries that Exactly Match from the Knowledge Base. Figure 6 shows that the model could correctly identify the exact similar questions which confirms that the Levenshtein model is working perfectly fine. Also score comes out to be 1 rightly since queries are identical to questions in the knowledge base.

Q	Prediction	A	Score
0	आजाद हिन्द फ़ौज की स्थापना कहाँ की गयी थी?	आजाद हिन्द फ़ौज की स्थापना किस देश में की गयी थी?	[सिंगापुर]	0.853933
1	भारत का राष्ट्र गान क्या है?	भारत के राष्ट्र गान का शीर्षक क्या था?	[भारत विधाता]	0.757576
2	'आनंदमठ' किसने लिखा है?	'आनंदमठ' के लेखक कौन हैं?	[बंकिमचन्द्र चटर्जी]	0.750000
3	भारतीय होम रूल लीग के पहले अध्यक्ष कौन थे?	भारतीय होम रूल लीग के पहले अध्यक्ष कौन थे?	[जोसिफ़ बपिस्ता]	0.976744
4	सुप्रीम कोर्ट के जजों की संख्या को कौन बदलता है?	सुप्रीम कोर्ट के जजों की संख्या को कौन बदल सकत...	[कानून द्वारा संसद]	0.969697
5	भारत के वित्त मंत्रालय के परिसर को क्या कहा जा...	भारत के वित्त मंत्रालय के परिसर को ___ भी कहा ...	[उत्तरी ब्लॉक]	0.905660
6	विदेशियों को कौन से अधिकार नहीं दिये जाते हैं?	७ कौन से अधिकार विदेशियों को नहीं दिये जाते ह...	विचार और अभिव्यक्ति की स्वतंत्रता का अधिकार	0.701031
7	संविधान के भाग IV का अनुच्छेद 43B के बारे में ...	संविधान के भाग IV का अनुच्छेद 43B किससे संबं...	[सहकारी समितियां]	0.745455
8	किस सागर/महासागर में सुंडा ट्रेंच स्थित है?	सुंडा ट्रेंच किस सागर/महासागर में स्थित है?	[हिन्द महासागर]	0.697674
9	अकबर की जीवन-गाथा किसने लिखी थी	अकबर की जीवन कथा किसने लिखी थी?	अबुल फजल	0.875000

Fig. 5. Depiction of answer extraction

Q	Prediction	A	Score
0	अल्टीमीटर से क्या नापते हैं	अल्टीमीटर से क्या नापते हैं	भूतल से ऊँचाई	1.0
1	ट्रैकोमा रोग किस अंग से सम्बन्धित रोग है	ट्रैकोमा रोग किस अंग से सम्बन्धित रोग है	आँख	1.0
2	मनुष्य के मस्तिष्क का सबसे बड़ा भाग क्या होता है	मनुष्य के मस्तिष्क का सबसे बड़ा भाग क्या होता है	प्रमस्तिष्क (Cerebrum)	1.0
3	खट्टे फलों में कौनसा विटामिन पाया जाता है	खट्टे फलों में कौनसा विटामिन पाया जाता है	विटामिन C	1.0
4	भारत में 28 फरवरी को विज्ञान दिवस किस उपलक्ष्य...	भारत में 28 फरवरी को विज्ञान दिवस किस उपलक्ष्य...	सी.वी.रमन द्वारा रमन प्रभाव की खोज करने के दिन...	1.0
5	सन् 1902 में कार्ल लैन्डस्टीनर (Karl Landstein...	सन् 1902 में कार्ल लैन्डस्टीनर (Karl Landstein...	रक्त समूह की (Blood Group)	1.0
6	शरीर का सारा रक्त किसके माध्यम से शुद्ध होता है	शरीर का सारा रक्त किसके माध्यम से शुद्ध होता है	वृक्क (किडनी) के माध्यम से	1.0
7	दूध से दही बनाने वाले जीवाणु का नाम है	दूध से दही बनाने वाले जीवाणु का नाम है	बैक्टेरियम लैक्टिसि एसीडाइ(Bacterium dactici a...	1.0
8	यदि किसी लेंस से अक्षरों का आकार छोटा दिखाई दे...	यदि किसी लेंस से अक्षरों का आकार छोटा दिखाई दे...	अवतल है	1.0
9	जब नींबू के रस को खाने के सोडे पर डाला जाता है...	जब नींबू के रस को खाने के सोडे पर डाला जाता है...	कार्बन डाइऑक्साइड	1.0

Fig. 6. Results on queries that exactly match knowledge base

Queries with Some Dissimilarity. Figure 7 depicts that all the differences in the queries are being caught by the model. It can be observed that for the eighth query, the computed score is 0.6 which couldn't pass the threshold value and was actually pointing to the wrong question. Therefore a sorry message was displayed to user instead of returning the answer in this case.

Random Queries Which May or May Not be in the Database. Figure 8 depicts the results of posting random queries that is it evaluates the performance of model unforeseen scenarios. Results on these queries are not all correct and the limitations of using approximate matching algorithms can be precisely seen here. In queries 2, 4, 8 and 9 results came out to be wrong because the sentences were very similar except one or two words resulting in greater score and thus wrong output.

	Q	Prediction	A	Score
0	अल्टीमीटर से क्या नापा जाता है	अल्टीमीटर से क्या नापते हैं	भूतल से ऊँचाई	0.877193
1	ट्रैकोमा रोग किस अंग में पाया जाता है	ट्रैकोमा रोग किस अंग से सम्बन्धित रोग है	आँख	0.716049
2	मानव जाती में मस्तिष्क का सबसे बड़ा भाग क्या क...	मनुष्य के मस्तिष्क का सबसे बड़ा भाग क्या होता है	प्रमस्तिष्क (Cerebrum)	0.823529
3	कट्टे फलों में कौनसा गुण पाया जाता है	कट्टे फलों में कौनसा विटामिन पाया जाता है	विटामिन C	0.871795
4	28 फरवरी को भारत में विज्ञान दिवस किस उपलक्ष्य...	भारत में 28 फरवरी को विज्ञान दिवस किस उपलक्ष्य...	सी.वी. रमन द्वारा रमन प्रभाव की खोज करने के दिन...	0.859375
5	कार्ल लैण्डस्टीनर (Karl Landsteiner) ने किसकी ...	सन् 1902 में कार्ल लैण्डस्टीनर (Karl Landstein...	रक्त समूह की (Blood Group)	0.894309
6	शरीर का सारा रक्त कौन शुद्ध करता है	शरीर का सारा रक्त किसके माध्यम से शुद्ध होता है	वृक्क (किडनी) के माध्यम से	0.756098
7	वह कौन से जीवाणु हैं जो दूध को दही बनाने का का...	वह कौन सा प्रोटीन है जो दूध में पाया जाता है	Sorry, I didn't get you.	0.600000
8	कौन सा लेंस अक्षरों का आकार छोटा दिखाता है	यदि किसी लेंस से अक्षरों का आकार छोटा दिखाई दे...	अवतल है	0.737864
9	नींबू के रस को खाने के सोडे पर डालने से निकलने...	जब नींबू के रस को खाने के सोडे पर डाला जाता है...	कार्बन डाइऑक्साइड	0.605714

Fig. 7. Results on Queries with some dissimilarity

	Q	Prediction	A	Score
0	हिरा क्यों चमकता है	हीरा किस कारण चमकता है	पूर्ण आंतरिक परावर्तन के कारण	0.682927
1	ISRO का फुल फॉर्म क्या है	MCB का फुल फॉर्म क्या है	Miniature Circuit Breaker	0.857143
2	रक्त लाल रंग का क्यों होता है	रक्त में लाल रंग किसके कारण होता है	7 हीमोग्लोबिन	0.738462
3	भारत की राजधानी क्या है	गुजरात की राजधानी है	गांधीनगर	0.727273
4	मानव शरीर में कितनी हड्डियां पायी जाती हैं	मानव शरीर में कितनी हड्डी होती है?	206	0.805195
5	२६ जनवरी कौन सा दिवस है	जंग लगना कौन सा परिवर्तन है	Sorry, I didn't get you.	0.588235
6	पिंक सिटी	पिंक सिटी	जयपुर	1.000000
7	भारत के प्रधानमंत्री कौन हैं	भारत के प्रथम सिख प्रधानमंत्री कौन है	मनमोहन सिंह	0.818182
8	पंजाब में कुल कितनी नदियां हैं	भारत में कुल कितने उच्च न्यायालय हैं	21	0.656716
9	सूर्य के बाद आने वाला ग्रह	सूर्य से सबसे नजदीक ग्रह है-	बुध	0.603774

Fig. 8. Results on random queries

Comparison with Existing Technique: Table 1 compares the accuracy of string matching technique with the existing methods. Sahu et. al. used Proglog for query processing and computed the performance based on various types of questions like what, when, where etc with 15 questions from each category. Authors had not provided any details about the dataset. So the average accuracy of their system is mentioned here. Nanda et. al. employed Naive bayes algorithm which is a machine learning technique. Accuracy was evaluated on two test sets whose size and other details are not provided in the published work. Stalin et al. have not computed the accuracy of their web application for Hindi question answering.

Table 1. Comparison with existing techniques

Algorithm	Mean accuracy(%)
String matching (Proposed)	93.33
Sahu et al. [2]	68.2
Nanda et al. [6]	90
Stalin et al. [3]	Not computed

4   Limitations of the Model

Although the results on the specified test cases came out to be satisfactory, still there are some limitations to this approach.

- First being the data itself since to make this system work perfectly fine like a search engine, a huge amount of data is needed which is not possible to accumulate and even if we do, still it won't be enough.
- Here, no alterations are made to the answer while returning it to the user which is a decent thing to do but can also create problem in some cases.
- Sometimes a score below 0.6 might have also got the right question but instead model returns no answer to the user.
- Using Levenshtein distance seems a right approach to follow here but at last it is just a character to character matching so it might not always find similarity or find similar but avoided the main keyword as shown in fourth query of third set of test cases. Using Semantic similarity algorithms can be helpful where comparisons are done on the basis of meaning and not just characters but again it also has its limitations. No algorithm can give one hundred percent accuracy and so these exceptions can be made.
- Semantic query processing [11,12] may also be applied to get semantically correct answers to questions.
- Performance of the proposed approach may be assessed on benchmark [12] data sets for fare comparison.

5   Conclusion

Question Answering is one of the most popular and user-friendly ways for humans to interact with computers. These systems enable easy and natural access of information to users by providing an interface to ask the question and receive response. This paper presented the question answering system in Hindi language. There are a wide range of rules that are employed to extract all possible set of answers from Hindi text for the input question. The model is analyzed for 3 types of queries and according to the results 100% correct answers were given for the queries that exactly match with the questions of our database, for the queries that were quite dissimilar with the questions of the database also gave 90% accurate answers. For the random queries our model was 50–60% efficient giving some wrong answers due to the limitations of the approximate matching algorithm used in the model. The data sets used in this work may be made available on demand. In future various machine learning algorithms, semantic query processing techniques will be used o improve the performance and comparisons will be done on benchmark data sets.

References

1. Moreda, P., Llorens, H., Saquete, E., Palomar, M.: Combining semantic information in question answering systems. Inf. Process. Manage. **47**(6), 870–885 (2011)
2. Sahu, S., Vasnik, N., Roy, D.: Prashnottar: a Hindi question answering system. Int. J. Comput. Sci. Inf. Technol. **4**(2), 149 (2012)
3. Stalin, S., Pandey, R., Barskar, R.: Web based application for Hindi question answering system. Int. J. Electron. Comput. Sci. Eng. **2**(1), 72–78 (2012)
4. Gupta, D., Ekbal, A., Bhattacharyya, P.: A deep neural network framework for English Hindi question answering. ACM Trans. Asian Low-Resour. Lang. Inf. Process **19**(2), 1–22 (2019)
5. Gupta, S., Khade, N.: Bert based multilingual machine comprehension in English and Hindi. ACM Trans. Asian Low-Resour. Lang. Inf. Process. (TALLIP) **1**, 1–13 (2020)
6. Nanda, G., Dua, M., Singla, K.: A Hindi question answering system using machine learning approach. In: International Conference on Computational Techniques in Information and Communication Technologies (ICCTICT), New Delhi, India, pp. 311–314 (2016)
7. Ray, S.K., Ahmad, A., Shaalan, K.: A review of the state of the art in Hindi question answering systems. In: Shaalan, K., Hassanien, A.E., Tolba, F. (eds.) Intelligent Natural Language Processing: Trends and Applications. SCI, vol. 740, pp. 265–292. Springer, Cham (2018). https://doi.org/10.1007/978-3-319-67056-0_14
8. Kaur, A., Nishi, Kaur, B.: A review on Hindi question answering system. Int. J. Mod. Trends Sci. Technol. **5**(11), 181–185 (2019)
9. https://news.linkedin.com/2021/december/linkedin-launches-in-hindi
10. Tiwari, S., Siarry, P., Mehta, S., Jabbar, M. A.: Tools, Languages, Methodologies for Representing Semantics on the Web of Things. Wiley, Hoboken (2022)
11. Gaurav, D., Tiwari, S.M., Goyal, A., Gandhi, N., Abraham, A.: Machine intelligence-based algorithms for spam filtering on document labeling. Soft Comput. **24**(13), 9625–9638 (2020)
12. Höffner, K., et al.: Survey on challenges of question answering in the semantic web. Seman. Web **8**(6), 895–920 (2017)

Towards an Ontological Approach to Business Continuity Assessment

Oussema Ben Amara[1][✉], Antonio de Nicola[2], Daouda Kamissoko[1],
and Frederick Benaben[1]

[1] Centre Génie Industriel, IMT Mines Albi, University of Toulouse, Campus Jarlard,
Route De Teillet, Albi, France
{oussema.ben_amara,daouda.kamissoko,
frederick.benaben}@mines-albi.fr
[2] ENEA - CR Casaccia, Via Anguillarese, 301, 00123 Rome, Italy
antonio.denicola@enea.it

Abstract. Business Continuity (BC) methods use threat identification, continuous improvement, and recommendations to ensure running the organization's main activities in case of disruptive events. Information Systems, on the other hand, are increasingly based on service-based structures and are seen as fundamental instruments to guarantee business continuity. The paper presents the ontological foundations for representing business continuity semantics, which are based on a widely adopted information systems research framework. The overall aim of this work is to provide BC with formal semantics and business people with an emerging informal BC modeling method.

Keywords: Business continuity · Ontology design · Framework design · Risk management

1 Introduction

The repercussions of damaging and vicious events that occur at the source of crisis situations (pandemics, fire, flood, terrorist attack, etc.) are constantly increasing. Matter of fact, our society's increasing complexity and densification (i) generates more natural events, (ii) disrupts the environment, resulting in more natural events, and (iii) increases its sensitivity to events [1].

Preparation, prevention, response, and recovery are the four phases of crisis management [2]. The preparation phase is the one that is primarily considered in the Business Continuity Management (BCM). In fact, anticipatory management has become a top priority for many public and private stakeholders, including the government, public services, private businesses, law enforcement organizations, operators, and so on [3, 4]. To deal with these uncertainties, organizations tend to build strategies for their Business Continuity (BC) and to execute them. BC is the planning and management of contingency measures aimed at keeping critical operational business processes running in case of disruptive events. Business continuity planning is a subset of organizational mitigation strategies, which is often referred to as risk management. Contingency planning and management in information security usually involve incident response, disaster recovery, and continuity management. Overall, the term BCM refers to both planning and management in an organizational setting [5].

The Business Continuity Plan (BCP) is one of the strategic steering instruments for senior executives [6], but it is frequently neglected [7]. Decisions on strategic IT- and Information Systems (IS), including business continuity planning, should not be delegated to a single member of senior management, the IT-department, or the like [6]. Because IT- and information security is becoming increasingly important in most organizations, the strategic components must be integrated into the senior management agenda on a continuous basis in order to be maintained and improved [8]. Senior management should own and devote time to strategic aspects of the business, as strategic decisions affect operational decisions at a lower level in an organization when working top-down [9].

For modelling BC some approaches were discussed within a handful of papers. The framework designed by Y. Asnar and P. Giorgini in [10] supports modeling and analysis of BCP from an organizational standpoint. In fact, the approach is based on the Tropos Goal-Risk Framework, which emphasizes the interdependence of assets in achieving business objectives. However, this approach is somehow formal, which makes it difficult to deploy in practical situations. Another work done by Griffith University in [11] defines a framework for BCM that, although being detailed and comprehensive, is still quite domain-specific because it only deals with the case of the university with its particular specificities. Another work done by A. Bialas in [12] exhibits an ontological approach to the BCM that is innovative and promising but still quite preliminary. Matter of fact, this paper's proposal tends to meet a proper level of formality but still usable by business professionals.

This research paper aims thus at answering the following main research question: **"How to better understand and assess the Business Continuity using an ontological approach?"** In other terms, this means designing an ontology in order to better comprehend the BC and be then able to assess it.

As a result, this paper is structured as the following: (i) the methodology that was followed to design the BC ontology, (ii) the findings of this paper and (iii) the next steps for this research work.

2 Methodology

To model business continuity, we refer to the Information Systems (IS) research framework defined by Hevner et al. in [10]. According to it, organizational strategies, structure, culture, and existing business processes are used to assess and evaluate business needs. They are situated in relation to current technological infrastructure, applications, communication architectures, and development capabilities. Together, these define the researcher's perception of the business need or "issue". Research relevance is ensured by framing research activities to address business needs [13].

Given such a coherent and consistent business need, IS research is carried out in two complementary stages. Behavioral science approaches research by developing and justifying theories that explain or predict phenomena associated with the identified business need as illustrated by Fig. 1 below. Design science approaches research by creating and evaluating artifacts that are intended to meet a specific business need. The goal of behavioral science research is to discover the truth. Theory is informed by truth, and design is informed by utility. An artifact may be useful because of a previously unknown truth. A theory may not have progressed sufficiently to the point where its truth can be incorporated into design [13].

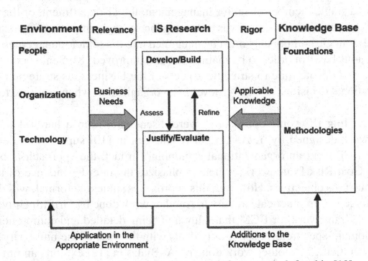

Fig. 1. A sketchy view of the IS research framework defined in [13]

Information technologies (IT) have recently been identified as enablers of business continuity. Almost since the beginning of the e-business era, both researchers and IT-professionals have considered the application of various information technologies in improving levels of business continuity. From an economic standpoint, IT can be used to reduce downtime (increase uptime) and, thus, contribute to better financial results, as each minute of downtime has a cost. In this regard, continuous computing solutions are the most important requirement for business continuity [14]. For these reasons IS and BC are very complementary and useful one to another in many ways such as: (a) documenting and continuously editing the BCP requirements and validations, (b) implementing the BCP and integrating its features within the firm's SI and (c) testing the BCP and assessing its efficiency and robustness by troubleshooting the implemented BC features within the IS.

Although some ontology development methodologies already exist, such as NeOn [15] and Ontology 101 development [16] which are speeding-up and knowledge resources reusing ontology development processes, this work doesn't integrate any of them. The reasons behind that are basically: (a) the ontology within this work is built from scratch and forms a knowledge base only from the considered research query results, (b) it's a single ontological format and doesn't merge two or more ontologies, (c) it's formed manually referring to the IS research framework but exploited automatically and

(d) the concepts of the ontology are the papers' keywords and thus these are beforehand retrieved and classified manually in order to satisfy the notion of business continuity and its various requirements.

Therefore, a BC ontology was designed by this work which was essentially based on the IS research framework (Fig. 1) explained previously. In fact, no previous ontology that contributes to this work, by especially characterizing the BC environment and foundations was found. Thus, to create a Business Continuity ontology, some scientific papers referring to "Business Continuity" were pruned through a research query on SCOPUS. Other research engines such as Web of Science (WoS) were used to get a better overview of information such as country, year and WoS category of the publications. The purpose of this analysis is to deepen the understanding of the domain, the information was gathered from 2547 articles. For reasons of coverage, all papers have been retained but will be later, depending on the exploitation of the ontology, eliminated or kept according to their relevance. Next, the research papers were analyzed and all their keywords were extracted serving as concepts for the designed ontology. Thus, a preliminary BC knowledge graph was created. That includes the concepts or "topics", authors, article types (journals, meetings, books, etc.), and publication locations related to the article. Topics are associated with other topics based on the general occurrence of the topic. We also know the number of occurrences of each topic (that is, the number of papers). The graphic also includes the title, summary, DOI (Digital Object Identifier) and contributors.

The class hierarchy of the implemented ontology contains mostly the classes and subclasses of IS research framework and has the retrieved paper topic as concepts. Some hierarchal properties were implemented as well such as: papers *"has_Author/has_Type"* or *"concerns"* some topics, topic *"is_Related_To; provides_business_needs_to* or *provides_rigor_to"* other topics, and other relationships. The concepts were afterward classified within the ontology to form a better understanding of the BC environment and methodologies. Since the concepts or "topics" are directly linked to the papers, accessing the rest of the information regarding a considered topic is relatively feasible. The ontology gets up to 4 levels (a 5th one is being created, but not finished yet) and is available for review here: https://tinyurl.com/4pc2ejkf.

The OWL visualization of the first two layers of the implemented ontological framework for the BC is given by the Fig. 2 below. It shows the classes and the relations between them within the first two layers. In fact, OWLViz is designed and integrated to work with the protégé owl-editor; it allows class hierarchies in the OWL (The W3C Web Ontology Language) ontology to be viewed and incrementally navigated, allowing comparison of the asserted and inferred class hierarchies. In order to develop the proposed ontology, a comprehensive examination for source of componential analysis was conducted [17].

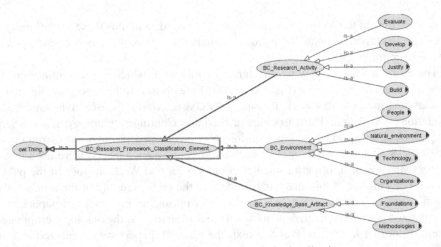

Fig. 2. The BC ontology class graph visualization

3 Findings

One fundamental issue in sustaining the BC ontology development is understanding how the BC is currently conceived, what technologies have contributed to shaping its concept, and what the current challenges for further development are. When used for crisis responsiveness, integration, and disambiguation of ground data and information, ontologies for BC processes are undoubtedly a viable means of providing such answers because they reflect current knowledge while also contributing to the creation of new knowledge. Such as the case in for smart cities in [18], designing an ontological framework is a useful thing to do. An ontology also allows for knowledge sharing, handling, reusing, collecting, interaction, and reasoning [19].

As a result, the following finding were possible:

1. Apprehending the BC environment, research activities and knowledge base. In fact, by classifying the concepts within the BC ontological framework such aspects could be identified and more information could be extracted using the designed framework. As the concepts point directly or indirectly on the relevant research papers, it's completely possible to peruse the methods and environment components for the BC in general or within a particular sector as well.
2. Building a knowledge base for the BC. In fact, all the classified topics as a whole and the filled ontological framework, especially its knowledge base artifacts part form an interesting and complete knowledge base for the BC. The research papers linked to the concepts could be thus forming the different parts of the knowledge base subclasses such as methods, constructs, theories, study cases and so on.
3. Identifying threats for the business continuity: The threats for the BC could be identified easily using the accident subclasses of the BC environment components. An example of the present threats within the ontology could be pandemics, terrorism, floods, cybersecurity attacks and so on. These threats are essential to help building a BCP since the BC process identifies mostly the potential threats to an organization

and provides a framework for building resilience and the capability for an effective response that protects the interests of its key stakeholders, reputation, brand, and value creating activities [20].

4   Further Research and Use of the Methodology

Based on the findings in the previous section, and on the horizons of this research work, the following future results are expected:

1. Use the ontology to define the semantics of BC rules; procedures/instructions that carry out the BCPs such as basic rules of communication in the face of an emergency, rules of reaction to typical threats, scenarios of expected extensive disruptions and how to react to them, rules for including current disruption experiences in future versions of emergency plans as well [21].
2. Query the data through ontology using a query language such as SPARQL which is a query language and protocol for searching, adding, modifying or deleting RDF data. In fact, the Resource Description Framework (RDF) is a graph device intended to formally describe Web resources and their metadata so that such descriptions can be processed automatically. RDF is the Semantic Web's basic language [22].
3. Combining (1.) and (2.) to implement a systematic literature review process and build a survey article on the BC. This process given by Fig. 3 below is mainly inspired by the research work done in [18]. It mainly consists in selecting the papers and extracting their keywords so that they can be later used in the ontology and queried using SPARQL queries. This will allow: (a) obtaining the relevant paper for the selected concepts in order to answer the research questions of the survey and (b) evaluating the relevance, consistency and integrability of the created ontology.

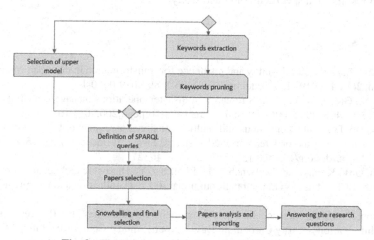

Fig. 3. The BC systematic literature review process

4. Apply the metamodel defined by this ontology and the extra classes such as BC threats to form BC indicators and use them to assess the BC within a case study following these steps: (a) define the socio-technical company profile, (b) identify the BC threats, (c) define socio-technical questions for the firm and design BC indicators, (d) instantiate the indicators into calculated metrics, and (e) use these indicators to assess the BC within the considered case study. This assessment does not need to be scientifically precise, but might be based on assumed likelihood [23].

5 Conclusion

This research work proposes an ontological framework for assessing and creating a knowledge base for the BC. Whereas the proposal was launched by designing and implementing the ontological framework. Further steps such as formalizing and executing the semantic literature review process and testing the metamodel's findings on a study case are still to be made.

Another limitation of this work, and thus a potential improvement for future work is that the framework consists only of the knowledge base, environment and research activities for the BC. Matter of fact, it could be complemented by creating extra classes such as BC threats, indicators and questions such as explained previously in the 4th point of Sect. 4. This could help to assess the BC using the ontological framework but also test and implement the results on a real case study.

At this stage, this research work allows to conceptualize the BC aspects in an ontological framework but proposes on the other hand, a possibility to implement important features to assess the BC such as threats and indicators as an extension to the ontology by following the same design methodology. This gives the possibility to three future works: (a) using the ontological framework to build a BC survey, (b) implement BC assessing features (threats, questions and indicators), and (c) evaluate the BC ontology's theoretical results on a specified field case study.

References

1. Elmqvist, T., et al.: Sustainability and resilience for transformation in the urban century. Nat. Sustain. **2**(4), 4 (2019). https://doi.org/10.1038/s41893-019-0250-1
2. Altay, N., Green, W.G.: OR/MS research in disaster operations management. Eur. J. Oper. Res. **175**(1), 475–493 (2006). https://doi.org/10.1016/j.ejor.2005.05.016
3. Kamissoko, D., et al.: Continuous and multidimensional assessment of resilience based on functionality analysis for interconnected systems. Struct. Infrastruct. Eng. **15**(4), 427–442 (2019). https://doi.org/10.1080/15732479.2018.1546327
4. Amara, O.B., Kamissoko, D., Benaben, F., Fijalkow, Y.: Hardware architecture for the evaluation of BCP robustness indicators through massive data collection and interpretation, p. 8 (2021)
5. Peterson, C.A.: Business continuity management & guidelines. In: Proceedings of the 2009 Information Security Curriculum Development Annual Conference, InfoSecCD 2009, Kennesaw, GA, pp. 114–120 (2009). https://doi.org/10.1145/1940976.1940999
6. Lindström, J., Hägerfors, A.: A model for explaining strategic IT-and information security to senior management, vol. 2009, p. 13 (2009)

7. Kajava, J., Anttila, J., Varonen, R., Savola, R., Roning, J.: Senior executives commitment to information security - from motivation to responsibility. In: 2006 International Conference on Computational Intelligence and Security, November 2006, vol. 2, pp. 1519–1522. https://doi.org/10.1109/ICCIAS.2006.295314

8. Anttila, J., Kajava, J., Varonen, R.: Balanced integration of information security into business management. In: Proceedings 30th Euromicro Conference, September 2004, pp. 558–564 (2004). https://doi.org/10.1109/EURMIC.2004.1333422

9. Lindström, J., Samuelsson, S., Hägerfors, A.: Business continuity planning methodology. Disaster Prevent. Manage. Int. J. 19(2), 243–255 (2010). https://doi.org/10.1108/096535610 11038039

10. Asnar, Y., Giorgini, P.: Analyzing business continuity through a multi-layers model. In: Dumas, M., Reichert, M., Shan, M.-C. (eds.) BPM 2008. LNCS, vol. 5240, pp. 212–227. Springer, Heidelberg (2008). https://doi.org/10.1007/978-3-540-85758-7_17

11. Business Continuity Management and Resilience Framework, p. 14

12. Bialas, A.: Ontological approach to the business continuity management system development, pp. 386–392 (2010)

13. Hevner, A.R., March, S.T., Park, J., Ram, S.: Design science in information systems research. MIS Q. 28(1), 75–105 (2004). https://doi.org/10.2307/25148625

14. Bajgoric, N.: Information technologies for business continuity: an implementation framework. Inf. Manag. Comput. Secur. 14(5), 450–466 (2006). https://doi.org/10.1108/096852206107 17754

15. Gómez-Pérez, A., Suárez-Figueroa, M.C.: NeOn methodology for building ontology networks: a scenario-based methodology, p. 8

16. Noy, N.F., McGuinness, D.L.: Ontology development 101: a guide to creating your first ontology, p. 25

17. Alshehab, A., Alazemi, N., Alhakem, H.A.: Semantic integration sharing for egovernment domains ontology: Design and implementation using owl. J. Theor. Appl. Inf. Technol. 97, 1820–1831 (2019)

18. De Nicola, A., Villani, M.L.: Smart city ontologies and their applications: a systematic literature review. Sustainability 13(10), 10 (2021). https://doi.org/10.3390/su13105578

19. Edgington, T., Choi, B., Henson, K., Raghu, T.S., Vinze, A.: Adopting ontology to facilitate knowledge sharing. Commun. ACM 47(11), 85–90 (2004). https://doi.org/10.1145/1029496. 1029499

20. Reuvid, J.: The Secure Online Business Handbook: A Practical Guide to Risk Management and Business Continuity. Kogan Page Publishers (2006)

21. Zawiła-Niedźwiecki, J.: Business Continuity. Found. Manage. 2(2), 2010. https://doi.org/10. 2478/v10238-012-0031-x

22. De Nicola, A., Missikoff, M.: A lightweight methodology for rapid ontology engineering. Commun. ACM 59, 79–86 (2016). https://doi.org/10.1145/2818359

23. Niemimaa, M., Järveläinen, J., Heikkilä, M., Heikkilä, J.: Business continuity of business models: Evaluating the resilience of business models for contingencies. Int. J. Inf. Manage. 49, 208–216 (2019). https://doi.org/10.1016/j.ijinfomgt.2019.04.010

From Ontology to Knowledge Graph Trend: Ontology as Foundation Layer for Knowledge Graph

Fatima N. AL-Aswadi[1,2], Huah Yong Chan[1(✉)], and Keng Hoon Gan[1]

[1] School of Computer Sciences, Universiti Sains Malaysia, 11800 Gelugor, Pulau Pinang, Malaysia
`fnsa15_com016@student.usm.my`, {`hychan,khgan`}`@usm.my`
[2] Faculty of Computer Science and Engineering, Hodeidah University, P.O. Box 3114, Hodeidah, Yemen

Abstract. Ontology and Knowledge Graph (KG) are two hotspot topics in Semantic Web and Artificial Intelligence (AI) fields. They have gained their importance due to the thriving development of the Internet in this century coupled with the explosive growth of data, which leads to increased demand for ontologies or KG to promote the Semantic Web and AI applications. Many researchers conflate ontology with KG, especially when even Wikipedia refers to both terms as synonymous; however, the difference is obvious. In this paper, we will highlight the differences between ontology and KG. Also, we have provided a redefinition for the ontology tuple and explained how the ontology is a foundation layer for KG.

Keywords: Knowledge graph · Ontology · RDF Triple · Real-world relation · Semantic relation

1 Introduction

An ontology is an explicit specification of a conceptualization [14]. Ontologies now fundamentally represent knowledge more meaningfully on the Semantic Web [20]. Ontology construction is a process that entails analyzing the collected data of a specific domain; extracting and determining the relevant terms and concepts; discovering and mapping the relationships among relevant concepts; representing the ontology by representation language (e.g. OWL[1], RDF[2], or RDFS[3]); and eventually evaluating the created ontology [5]. There are three possible ways of

---

[1] Ontology Web Language.
[2] Resource Description Framework.
[3] Resource Description Framework Schema.

© The Author(s), under exclusive license to Springer Nature Switzerland AG 2022
B. Villazón-Terrazas et al. (Eds.): KGSWC 2022, CCIS 1686, pp. 330–340, 2022.
https://doi.org/10.1007/978-3-031-21422-6_25

constructing an ontology; manual, cooperative, or automatic. Manual or cooperative construction mainly creates the concept and relationship of ontology (fully or mostly) through domain experts; in these ways, the constructed ontology is more professional and highly reliable, but it is time-consuming and laborious process [1,10]. The automatic construction of ontology called Ontology Learning (OL). OL has gradually become a research hotspot due to the explosive growth of data on Web coupled with the increase on demand for effective methods to get and organize knowledge from these big noisy data [6,10]. OL is a process that seeks to automatically extract, organize and represent the knowledge in machine-understandable form.

Knowledge Graph (KG) is considered synonymous with Knowledge-Base (KB) with the difference that data can be structured and visualized graphically [17,30]. It is a more comprehensive ontology with reasoned inferencing to extract new knowledge, it is the ontology that is rich in instances and extensions with real-world connections [21]. This paper discusses and points out the differences between ontology and KG concepts. And to present how the ontologies are the foundation layer for KG. The rest of this paper has been organized as follows; Sect. 2 presents the ontology structure, relationships, and triples. Section 3 explains the KG term, phases, KB and RDF in KG. Section 4 presents with example how ontology is a foundation layer for KG. In the end, we concluded in Sect. 5.

2   Ontology Structure and Relationships

2.1   Ontology Scheme

There are two types of Ontology: (i) formal ontologies, which involve taxonomies, concepts with detailed relations between them, and the rules [15]; (ii) informal ontologies that are created by user communities such as the internet encyclopedias [4]. This study focuses on formal ontology, so any indication of ontology in this research refers to formal ontology.

According to R Subhashini and J Akilandeswari [27], the ontology consists of four main components to define and represent a domain. They are: (i) Relevant Concepts which represent a set of classes, objects and entities within a domain. (ii) Relations (taxonomic and non-taxonomic) which specify the classification and interaction among concepts. (iii) Instances which indicate concrete examples and named individuals or entities within a domain. (iv) Axioms which denote a statement that is always true. Amal Zouaq [35] had defined the ontology components through the following tuple:

$$O =< C, H, Rr, A >$$

where O stands for ontology, C stands for classes (concepts), H stands for hierarchical links between concepts (taxonomic relations), Rr stands for conceptual links (non-taxonomic relations), and A stands for axioms.

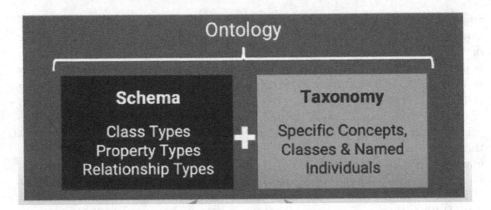

Fig. 1. Ontology components [19].

In other words, the ontology comprises a semantic schema of classes, properties, and relationship types, plus the taxonomy of specific concepts, classes and named individuals. Figure 1 shows the ontology scheme components as defined in [19].

2.2 Ontology Relationships and Triples

The relations extraction task, also called relations discovery, is aimed to discover and extract the taxonomic and non-taxonomic relationships among relevant concepts. Non-taxonomic relations extraction seeks to extract the semantic relations among selected concepts, whereas taxonomic relations extraction aims to construct the hierarchical taxonomy of specific concepts. Discovering the relationships among the relevant concepts within a domain is considered the backbone of an OL system [7, 24].

Most of the related work tried to extract the relations between pairs of concepts or named entities. Many of these studies used OWL, RDF or RDFS models to describe the metadata and semantics of information, such as [8,18,22,28]. This model is efficient in organizing a massive amount of knowledge in triple form as defined by Semantic Web Standards (SWS). A triple is a form of language following a "Subject-Verb-Object" (SVO) word order that mirrors "Subject-Predicate-Object" (SPO). [19,29]. Some of these existing studies had built the graph of concepts and relations lattice along with RDF or OWL such as [8,21,31].

Ontology uses the basic SPO structure to model the conceptual domain semantically. The SPO triple is used for extracting non-taxonomic relations. Also, it can be used along with Lexico-syntactic patterns or syntactic structure analysis for extracting taxonomic relations [5]. Most of the studies related to ontology construction focus primarily on extracting taxonomic relations and attributes, plus some semantic relations. In contrast, KG focuses on extracting

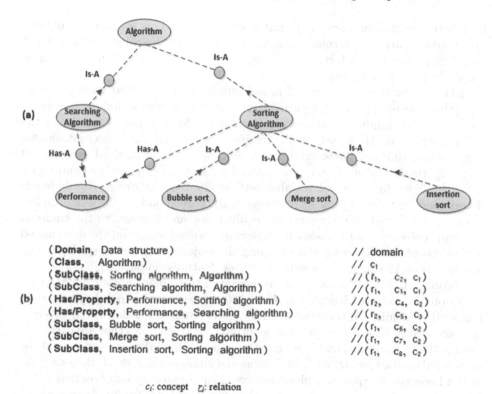

	(Domain, Data structure)	// domain
(b)	(Class, Algorithm)	// c_1
	(SubClass, Sorting algorithm, Algorithm)	// (r_1, c_2, c_1)
	(SubClass, Searching algorithm, Algorithm)	// (r_1, c_3, c_1)
	(Has/Property, Performance, Sorting algorithm)	// (r_2, c_4, c_2)
	(Has/Property, Performance, Searching algorithm)	// (r_2, c_5, c_3)
	(SubClass, Bubble sort, Sorting algorithm)	// (r_1, c_6, c_2)
	(SubClass, Merge sort, Sorting algorithm)	// (r_1, c_7, c_2)
	(SubClass, Insertion sort, Sorting algorithm)	// (r_1, c_8, c_2)

c_i: concept r_i: relation

Fig. 2. An example of Ontology.

relationships in all forms with the same priority; more details are in Sect. 4. Figure 2(a) shows a graph example of ontology relations in "Data Structures" domain, while Fig. 2(b) shows the annotation of this example that can be encoded by ontology representation languages (e.g., OWL or RDF).

3   Knowledge Graph Construction

3.1   Knowledge Graph Term and Phases

Lisa Ehrlinger and Wolfram Wöß [11] have presented a new definition of KG: *"A knowledge graph acquires and integrates information into ontology and applies a reasoner to derive new knowledge."* And Sören Auer, et al. [26] have defined the KG as follows: *"a knowledge graph for science acquires and integrates scientific information in a knowledge base, and may apply a reasoner or other computational methods to derive new information."* Ibrahim A. Ahmed, et al. [1], Shaoxiong Ji, et al. [17] and Xiaohan Zou [34] have defined KG as the KB that is represented in a graph. To put it simply, the KG is a collection of entities, relations, attributes, rules, facts, and other forms of knowledge, that present and define some sort of fact, relation, or connection as a paradigm rather than a specific category of things.

KG construction is a group of linked mining strategies that explicitly consider the objects' links when creating descriptive or predictive linked data models. These link mining strategies are responsible for a variety of tasks in the KG construction. M. Nickel, et al. [25] have introduced three major tasks for constructing KG as follows: *(i) Link Prediction* is identified the feature-based to group them, and it clusters the entities with relations-based depending on their similarity in relational learning setting [25,33]. *(ii) Entity Resolution* is determined whether objects in relational data correspond to the same underlying entities based on their semantic equivalents, and it refers to the common real-world objects. *(iii) Link-based Clustering* is aimed to extend the determined feature-based for clustering and grouping the objects with relations depending on entities' links similarity as well as entities' features similarity [25,33].

Ibrahim A. Ahmed, et al. [1] have presented three main phases for KG construction: *Knowledge Acquisition, Knowledge Fusion* and *Knowledge Storage.* Knowledge Acquisition is extracted entity, attribute and relation for discovering new knowledge. Knowledge Fusion is identified the true triples that are extracted from multiple resources, and it is aimed to enhance the inferring and extend the discovered entities and triples with linking the similarities. Indeed, the sub-tasks of the knowledge acquisition phase are considered ontology construction tasks. While the ontology construction sub-task of the knowledge fusion phase is aimed at assigning and mapping the entities with their attributes and relationships. In contrast, the entities linking sub-task is desired to cluster the entities depending on their similarity. The inferring and reasoning sub-task is aimed at refining the constructed ontology by reducing the conflict and inferring new knowledge.

The last phase of KG construction is Knowledge Storage, it is aimed at creating and storing KG graph schema, e.g., RDF, with linking and visualizing the knowledge in real-world links. The GraphDB[4], FlockDB[5], and Neo4j[6] are some examples of tools that can be used to visualise KG. In addition, this phase is aimed at building the KG queries for visualizing and retrieving the data. The expressivity level of the querying method can be affected by the required usability and efficiency, also by the interaction model [9]. SPARQL is the common query language used for querying KG. However, due to the SPARQL complexity, many studies tried to develop queries framework to reduce the need for SPARQL awareness by users. For example, SPARKLIS [13] built keyword search interfaces to hide the SPARQL complexity. Another example is Query Data of Interest (QueDI) [9], which tried to allow users to build queries without requiring the explicit usage of SPARQL by two-step: querying mechanism (for SPARQL query) and guided workflow (for users interaction). Recently, there are

---

[4] http://graphdb.net/.
[5] https://webscripts.softpedia.com/script/Database-Tools/FlockDB-66248.html.
[6] https://neo4j.com/.

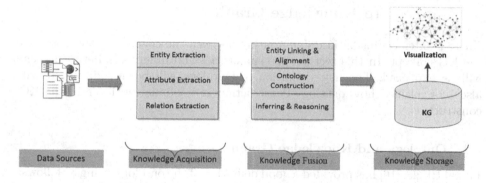

Fig. 3. Phases of knowledge graph construction [1].

new flexible query languages such as Cypher language that is used in Neo4j. Figure 3 shows these three phases and their related tasks in KG construction as introduced in [1].

3.2 Knowledge-Base and RDF in Knowledge Graph

Michael Färber, et al. [12] have defined KG as an RDF graph, as follows: *"The term (knowledge graph) was reintroduced by Google in 2012 and is intended for any graph-based knowledge repository. Since in the Semantic Web RDF graphs are used we use the term knowledge graph for any RDF graph."*

As mentioned above, KG is defined as the KB that is represented in a graph. A KB is a set of rules, facts, and assumptions used to store knowledge in a machine-readable form [23, 27]. These KB facts and rules need to organize in a data model, e.g., RDF graph, to be easy-to-use, easy-to-understand, easy-to-maintain, and easy-to-retrieval [29]. A graph refers to a set of entities, objects, or nodes and their connections (relations) [34].

RDF is typically used to conceptualise knowledge in the SPO triple form. For instance, "Mark Zuckerberg was born in New York" can be represented as (S: Mark Zuckerberg, P: birth-place, O: New York). In the RDF data model, triple is considered an atomic element. It delivers the knowledge utilising web resources with a unique URI. For instance, the corresponding URI for "Mark Zuckerberg" can be shown as https://entityexample.com/MarkZuckerberg, where https://entityexample.com/ is the URI for each entity in the KB. Subjects and Predicates are given through URI, while objects can be URI or literal values. RDF data can be stored in a variety of formats, including RDFa, JSON-LD, NTriples, N-Quads, or XML [29].

The most common application of KG is to refer to more reliable text interpretation for text analysis. Through the semantic annotating of text to show the structured data, a particular notion has a link in the KG. To comprehend the meaning of data and use logic to create new facts, KG has a specialised data format that is more effective in adding new facts regularly and updating dynamically.

4   Ontology to Knowledge Graph

There is often confusion in differentiating between the concept of ontology and the KG concept. In this section, the similarities and differences between them will be explained. The new definition of Ontology will be introduced. It will also be explained how ontology construction is a foundation layer in the KG construction.

4.1   Ontology and Knowledge Graph

Pascal Hitzler [16] has provided a good definition of the ontology using as follows: *"In a Semantic Web context, ontologies are a main vehicle for data integration, sharing, and discovery, and a driving idea is that ontologies themselves should be reusable by others."* On other side, Bob Kasenchak and Ahren E. Lehnert [19] defined KG as follows: *"A knowledge graph is made up of interlinked real-word entities described in a formal structure so they can be used by both computers and people."* In addition, [11,26] studies defined KG as ontologies with applying reasoner computational methods to derive new knowledge. Based on previous studies and our knowledge, we can derive a definition for KG as follows:

> *KG is a combination of one or more ontologies with applying and enriching them to a set of individual real-world data points.*

In addition, based on previous studies regarding ontology and KG, and based on our knowledge, we can redefine the ontology as follows:

> *Ontology is a semantic schema of classes (concepts), attributes, and relationship types with building a taxonomy of specific concepts, plus named individuals.*

As observed from these definitions, that ontology is considered as a foundation layer for KG. More details and the example are in Sect. 4.3.

4.2   Ontology Redefinition

From the KG perspective and other applications on the Semantic Web, most of the existing methods for constructing relations within the ontology are unsatisfactory [2,3,5,7,32]. Most existing approaches do not consider the hierarchical links between the relations in the definition of ontology. Although these hierarchical links between the relations play an important role in improving enriching and linking concepts and relations. Most of these studies extract non-taxonomic relations by extracting any verb according to the predefined patterns and assigning them as non-taxonomic relationships. Still, this assumption is unreasonable and needs pruning (mostly fails to extract semantic relations) [3,24,32], so there is a need to build major classes of relations for pruning them. Therefore, this research redefines the tuple of ontology by introducing the relation hierarchy parameter in the ontology definition, as shown in the following tuple:

$$O =< C, CH, Rr, RH, A >$$

where O stands for ontology, C stands for concepts set, CH stands for a set of hierarchical links between concepts (taxonomic relations), Rr stands for conceptual links set (non-taxonomic relations), RH stands for hierarchical links set (classes) between the relations, and A stands for axioms set. The terms in this tuple can be defined as follows:

1. $C = \{c_1, c_2, c_3, ...c_n\}$, and $CH \subseteq C \times C$ represent a set of concepts, and a concept hierarchy, respectively. $CH(c_1, c_2)$ means that c_1 is sub-concept of concept c_2. Concept c can be defined as follows: $< I, L, E >$ where I refers to the concept intension sign, and L refers to the concept lexical intention (linguistic realization), E refers to concept extension (examples and synonyms).
2. $Rr = \{r_1, r_2, r_3, ...r_m\}$ and $RH \subseteq Rr \times Rr$ represent a set of relations. $F : Rr \rightarrow C \times C$ is a relation function that maps the sets C and Rr. $Rr(c_1, c_2)$ refers to the relation Rr between c_1 and c_2 concepts, while $RH(rx, r_1)$ means that the extracted rx relation is a r_1 relation class.
3. $A = \{a_1, a_2, a_3, ...a_d\}$ represents a set of axioms. The generalization of axioms or rules depends on taxonomic and non-taxonomic relations. Also the new axioms can be generated from other axioms. In the most of the OL systems and approaches, there is no clear definition of axiom learning function. In this research, it can be defined as the next, $FPi : L_1, ...L_m \rightarrow Pi(x_1, ...x_n)$, where FPi is the predication (learning) function, Pi is a predicated target, all x_i are variables, and all L_i are literals (an atomic formula (atom) or its negation). An axiom a_i is consistent if $\{\forall x : x \in c_1 \rightarrow x \in c_2, c_1 \subseteq c_2\} \cup \{\forall x : x \in r_1 \rightarrow x \in r_2, r_1 \subseteq r_2\}$, where $c_1, c_2 \in C$ and $r_1, r_2 \in Rr$.

As observed, this new definition tuple for ontology meets the introduced definition of ontology in Sect. 4.1. Hence, we can draw that the components of the KG consist of three components, as shown in the following tuple:

$$KG =< O, T, Rc >$$

where O stands for ontologies, T stands for a set of real-world individuals (or real-world modelled data) according to the provided ontologies, and Rc stands for a set of connections between entities or concepts in provided ontology with the real-world individuals.

4.3   Ontology as Foundation Layer for Knowledge Graph

Ontology is a better representation schema of, and between, domain knowledge. It is the foundational layer for KG. KG uses an ontology knowledge scheme for connecting concepts to content by utilizing SPO semantic relationships. In other words, KG is the combination of an ontology and content or data linked to real-world individual points or entities. Notably, ontology provides a conceptual schema embedded in a KG. Hence, it allows the queries with logical inference for organising and retrieving explicit and implicit knowledge.

Figure 4 shows an example of ontology and KG for the Publication domain. This example depicts the difference between ontology and KG. Also, it shows how ontology is a foundation layer that serves as a schema layer for KG and how it is enriched with a set of individual real-world data points.

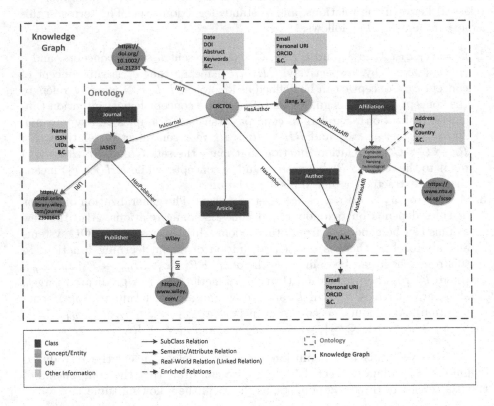

Fig. 4. An example of Ontology as a foundation layer for knowledge graph.

5   Conclusion

Ontology and KG are two important concepts in Semantic Web and AI applications. These two concepts are sometimes confused by scholars. Yet, the distinction is clear. This paper has evidenced the difference between both concepts. Ontology is a semantic schema of classes, properties, and relationship types; plus building a taxonomy of specific concepts, and sometimes named individuals. In contrast, KG is a combination of one or more ontologies with applying and enriching them to a set of real-world individual data points.

References

1. Ahmed, I.A., et al.: Arabic knowledge graph construction: a close look in the present and into the future. J. King Saud Univ. Comput. Inf. Sci. (2022). iSSN: 1319–1578. https://doi.org/10.1016/j.jksuci.2022.04.007
2. Albukhitan, S., Helmy, T., Alnazer, A.: Arabic ontology learning using deep learning. In: Proceedings of the International Conference on Web Intelligence, 3109052, pp. 1138–1142. ACM (2007). https://doi.org/10.1145/3106426.3109052
3. Arefyev, N., et al.: Neural GRANNy at SemEval-2019 task 2: a combined approach for better modeling of semantic relationships in semantic frame induction. In: Proceedings of the 13th International Workshop on Semantic Evaluation, pp. 31–38 (2019)
4. Astrakhantsev, N A., Yu Turdakov, D.: Automatic construction and enrichment of informal ontologies: a survey. Program. Comput. Softw. **39**(1), 34–42 (2013). issn: 1608–3261. https://doi.org/10.1134/S0361768813010039
5. Al-Aswadi, F.N., Chan, H.Y., Gan, K.H.: Automatic ontology construction from text: a review from shallow to deep learning trend. Artif. Intell. Rev. **53**(6), 3901–3928 (2019). https://doi.org/10.1007/s10462-019-09782-9
6. AL-Aswadi, F.N., Chan, H.Y., Gan, K.H.: Extracting semantic concepts and relations from scientific publications by using deep learning. In: Saeed, F., Mohammed, F., Al-Nahari, A. (eds.) IRICT 2020. LNDECT, vol. 72, pp. 374–383. Springer, Cham (2021). https://doi.org/10.1007/978-3-030-70713-2_35
7. Buitelaar, P., Cimiano, P., Magnini, B.: Ontology learning from text: an overview. Ontology Learn. Methods Eval. Appl. **123**, 3–12 (2005)
8. Cimiano, P., Völker, J.: Text2Onto. In: Montoyo, A., Muñoz, R., Métais, E. (eds.) NLDB 2005. LNCS, vol. 3513, pp. 227–238. Springer, Heidelberg (2005). https://doi.org/10.1007/11428817_21
9. De Donato, R., et al.: QueDI: from knowledge graph querying to data visualization. In: SEMANTiCS, pp. 70–86 (2020)
10. Du, Y., et al.: Knowledge extract and ontology construction method of assembly process text. In: MATEC Web of Conferences, vol. 355. EDP Sciences (2022). ISBN: 2274-7214
11. Ehrlinger, L., Wöß, W.: Towards a definition of knowledge graphs. SEMANTiCS (Posters, Demos, SuCCESS) **48**(1-4), 2 (2016)
12. Färber, M., et al.: Linked data quality of DBpedia, Freebase, OpenCyc, Wikidata, and YAGO. Seman. Web **9**(1), 77–129 (2018). issn: 1570–0844
13. Ferré, S.,. Sparklis: an expressive query builder for SPARQL endpoints with guidance in natural language. Seman. Web **8**(3), 405–418 (2017). issn: 1570–0844
14. Gruber, T.R.: A translation approach to portable ontology specifications. Knowl. Acquisition **5**(2), 199–220 (1993). issn: 1042–8143
15. Guarino, N., Oberle, D., Staab, S.: What is an ontology? In: Staab, S., Studer, R. (eds.) Handbook on Ontologies. IHIS, pp. 1–17. Springer, Heidelberg (2009). https://doi.org/10.1007/978-3-540-92673-3_0
16. Hitzler, P.: A review of the semantic web field. Commun. ACM **64**(2), 76–83 (2021). ISSN: 0001–0782. https://doi.org/10.1145/3397512
17. Ji, S., et al.: A survey on knowledge graphs: representation, acquisition and applications. IEEE Trans. Neural Netw. Learn. Syst. **33**(2) 494–514 (2022). https://doi.org/10.1109/TNNLS.2021.3070843
18. Jiang, X., Tan, A.H.: CRCTOL: a semantic-based domain ontology learning system. J. Am. Soc. Inf. Sci. Technol. **61**(1), 150–168 (2010). ISSN: 1532–2890

19. Kasenchak, B., Lehnert, A.E.: ontology for knowledge graphs. Online Webinar (2021). https://www.youtube.com/watch?v=7qIBex7a0kE

20. Kondylakis, H., et al. Delta: a modular ontology evaluation system. Information 12(8), 301 (2021). ISSN: 2078–2489

21. Lan, N., et al.: Research on knowledge graphs with concept lattice constraints. Symmetry **13**(12), 2363 (2021). ISSN: 2073–8994

22. Lehmann, J., et al.: DBpedia-a large-scale, multilingual knowledge base extracted from Wikipedia. Seman. Web **6**(2), 167–195 (2015). ISSN 1570–0844

23. Liu, Z., Han, X.: Deep learning in knowledge graph. In: Deng, L., Liu, Y. (eds.) Deep Learning in Natural Language Processing, pp. 117–145. Springer, Singapore (2018). https://doi.org/10.1007/978-981-10-5209-5_5

24. Browarnik, A., Maimon, O.: Ontology learning from text: why the ontology learning layer cake is not viable. Int. J. Signs Semiot. Syst. **4**(2) 1–14 (2015). ISSN: 2155–5028. https://doi.org/10.4018/ijsss.2015070101

25. Nickel, M., et al.: A review of relational machine learning for knowledge graphs. Proc. IEEE **104**(1), 11–33 (2016). ISSN: 1558–2256. https://doi.org/10.1109/JPROC.2015.2483592

26. Sören, A., et al.: Towards a knowledge graph for science. In: Proceedings of the 8th International Conference on Web Intelligence, Mining and Semantics. Association for Computing Machinery, p. 3227689. https://doi.org/10.1145/3227609

27. Subhashini, R., Akilandeswari, J.: A survey on ontology construction methodologies. Int. J. Enterp. Comput. Bus. Syst. **11**, 60–72 (2011)

28. Suchanek, F.M., Kasneci, G., Weikum, G.: Yago: a core of semantic knowledge. In: Proceedings of the 16th international conference on World Wide Web, 1242667, pp. 697–706. ACM (2007). https://doi.org/10.1145/1242572.1242667

29. Tiwari, S., Al-Aswadi, F.N., Gaurav, D.: Recent trends in knowledge graphs: theory and practice. Soft Comput. **25**(13), 8337–8355 (2021). https://doi.org/10.1007/s00500-021-05756-8

30. Tiwari, S., Gaurav, D., Srivastava, A., Rai, C., Abhishek, K.: A preliminary study of knowledge graphs and their construction. In: Tavares, J.M.R.S., Chakrabarti, S., Bhattacharya, A., Ghatak, S. (eds.) Emerging Technologies in Data Mining and Information Security. LNNS, vol. 164, pp. 11–20. Springer, Singapore (2021). https://doi.org/10.1007/978-981-15-9774-9_2

31. Völker, J., Fernandez Langa, S., Sure, Y.: Supporting the construction of Spanish legal ontologies with Text2Onto. In: Casanovas, P., Sartor, G., Casellas, N., Rubino, R. (eds.) Computable Models of the Law. LNCS (LNAI), vol. 4884, pp. 105–112. Springer, Heidelberg (2008). https://doi.org/10.1007/978-3-540-85569-9_7

32. Wong, W., Liu, W., Bennamoun, M.: Ontology learning from text: a look back and into the future. ACM Comput. Surv. (CSUR) **44**(4), 20 (2012). ISSN: 0360–0300

33. Alavijeh, Z.Z.: The application of link mining in social network analysis. In: 2015. p. 6 (2015), ISSN: 2322–5157

34. Zou, X.: A survey on application of knowledge graph. J. Phys. Conf. Ser. **1487**, 012016 (2020). ISSN: 1742–6588 1742–6596. https://doi.org/10.1088/1742-6596/1487/1/012016

35. Zouaq, A.: An overview of shallow and deep natural language processing for ontology learning. In: Wong, W., Liu, W., Bennamoun, M. (eds.) Ontology Learning and Knowledge Discovery using the Web: Challenges and Recent Advances, vol. 2. USA, Information Science Reference (IGI Global), Chap. 2, pp. 16–37 (2011)

Author Index

Printed in the United States
by Baker & Taylor Publisher Services